Construction Project Management

Construction Project Management

A Practical Guide to Field Construction Management

Sixth Edition

S. Keoki Sears
Glenn A. Sears
Richard H. Clough
Jerald L. Rounds
Robert O. Segner, Jr.

Cover Design: Wiley
Cover Images: Capsher Technology building photograph by Jennifer Olson, design by The Arkitex Studio, Inc.; Mesh Background © iStock.com/Zhenikeyev

This book is printed on acid-free paper. ∞

Copyright © 2015 by John Wiley & Sons, Inc. All rights reserved.

Published by John Wiley & Sons, Inc., Hoboken, New Jersey.
Published simultaneously in Canada.

No part of this publication may be reproduced, stored in a retrieval system, or transmitted in any form or by any means, electronic, mechanical, photocopying, recording, scanning, or otherwise, except as permitted under Section 107 or 108 of the 1976 United States Copyright Act, without either the prior written permission of the Publisher, or authorization through payment of the appropriate per-copy fee to the Copyright Clearance Center, 222 Rosewood Drive, Danvers, MA 01923, (978) 750-8400, fax (978) 646-8600, or on the web at www.copyright.com. Requests to the Publisher for permission should be addressed to the Permissions Department, John Wiley & Sons, Inc., 111 River Street, Hoboken, NJ 07030, (201) 748-6011, fax (201) 748-6008, or online at www.wiley.com/go/permissions.

Limit of Liability/Disclaimer of Warranty: While the publisher and author have used their best efforts in preparing this book, they make no representations or warranties with the respect to the accuracy or completeness of the contents of this book and specifically disclaim any implied warranties of merchantability or fitness for a particular purpose. No warranty may be created or extended by sales representatives or written sales materials. The advice and strategies contained herein may not be suitable for your situation. You should consult with a professional where appropriate. Neither the publisher nor the author shall be liable for damages arising herefrom.

For general information about our other products and services, please contact our Customer Care Department within the United States at (800) 762-2974, outside the United States at (317) 572-3993 or fax (317) 572-4002.

Wiley publishes in a variety of print and electronic formats and by print-on-demand. Some material included with standard print versions of this book may not be included in e-books or in print-on-demand. If this book refers to media such as a CD or DVD that is not included in the version you purchased, you may download this material at http://booksupport.wiley.com. For more information about Wiley products, visit www.wiley.com.

Library of Congress Cataloging-in-Publication Data:

Sears, S. Keoki.
    Construction project management : a practical guide to field construction management.
—Sixth edition/S. Keoki Sears [and 4 others].
      pages cm
   Includes index.
     ISBN 978-1-118-74505-2 (hardback : acid-free paper); ISBN 978-1-118-74525-0 (ebk); ISBN 978-1-118-74526-7 (ebk)
  1. Construction industry—Management.   2. Project management.   I. Title.
  TH438.S43 2015
    690.068'4—dc23
                                        2014041076

Printed in the United States of America

10 9 8 7 6 5 4 3 2 1

Contents

Preface — vii
1 Construction Practices — 1
2 Systematic Project Management — 19
3 Project Estimating — 35
4 Project Planning — 71
5 Project Scheduling Concepts — 97
6 Production Planning — 143
7 Managing Time — 161
8 Resource Management — 191
9 Project Scheduling Applications — 213
10 Project Coordination — 241
11 Project Cost System — 275
12 Project Financial Management — 309
Index — 335

Additional resources for students and instructors are available on the book's companion website at www.wiley.com/go/cpm6e.
The following icon is used throughout the text to indicate content for which a related resource is available on the site:

Preface

This sixth edition of the well-respected text on construction project management represents a significant revision. The intent is to retain the flavor and quality of the classic book while eliminating some of the detail and updating the content. The content is enhanced by the addition of new material introducing some rapidly evolving topics in construction project management. New instructional materials have also been added to each chapter to increase its value in the classroom. A new example project, selected from the building sector, has been incorporated. Finally, this book takes advantage of advances in technology by employing a companion website that contains material that was appended at the end of the book or inserted as oversized figures in previous editions.

Some detail (such as the detailed cost estimate) and some older content, such as that describing the Arrow Diagramming Method for Critical Path representation have been greatly reduced or eliminated. The coverage of other topics has been increased, and new topics have been added. Reference to line-of-balance scheduling, which provides a valuable representation of repetitive projects, has been expanded. An introduction to Building Information Modeling (BIM) and its associated contract basis founded on Integrated Project Delivery (IPD) has been added. Also a section and various references throughout the book have been added to introduce the basics of lean construction.

To support the use of the book in a learning environment, an introductory section for each chapter has been added that includes a bulleted list of learning objectives. At the end of each chapter, a list of key points from the chapter has been provided, together with review questions and problems. To support live delivery of classes, a set of PowerPoint slides has been prepared for each chapter. It is anticipated that these instructional enhancements will enrich the learning experience provided by the book.

To broaden the scope of the sixth edition, a new example project in the form of a commercial building has been added. The bridge example is retained and in many instances is the single example of reference where topics are addressed that represent the building sector as well as the heavy civil sector.

The building example is an actual project, in the form of a single-story office building constructed in 2006–2007 for a technology company engaged in computer programming and consulting. The design and construction information for the building project was graciously provided to the authors for inclusion in this book by the project design team, the contractor, and the owner. A picture of the building, shown on the book cover, was also provided.

A building, by nature, is far more complex than a bridge. Whereas the bridge is primarily composed of structure and earthwork, and can be fully described in a few drawings, a building contains structure and earthwork, but it also contains an envelope to keep the elements out and to enable control of the interior environment. All surfaces in the building, including floors, walls, and ceilings, require finishes. Buildings have openings requiring doors and windows and the associated hardware. They typically have architectural features such as millwork and signage. Moreover, a building contains a number of systems, such as mechanical, plumbing, fire safety, electrical, and low-voltage systems for security and communication. Each of these building elements must be fully designed and represented in graphical models (the drawings). As a result, though the building is relatively small and uncomplicated, the design of our commercial building example project encompasses some 40 pages of drawings. The inclusion of this complete set of drawings is made possible only by linking a companion website to the book. The companion website is accessed through the following URL: www.wiley.com/go/cpm6e.

In addition to the commercial building drawings, a number of other useful items will be found on the companion website. On the website will be found the figures that were large enough in previous editions to require cumbersome fold-out pages. These are referred to as charts in this edition to distinguish them from figures embedded in the text. Providing them in digital format on the companion website is consistent with the industry trend away from paper documentation. Relevant appendix material from previous editions has also been retained on the companion website. Finally, an instructor-only, password accessible location on the website holds an answer key for the end-of-chapter review questions and problems, as well as the PowerPoint presentations.

A new team of authors has taken on the responsibility to develop this sixth edition of *Construction Project Management*. Jerald L. Rounds and Robert O. Segner, Jr. are honored to have been chosen to continue the excellence of *Construction Project Management* initiated by Richard H. Clough and continued through many revisions by Glenn A. Sears and more recently by S. Keoki Sears.

Jerald L. Rounds and Robert O. Segner, Jr.

1 Construction Practices

1.1 Introduction

The objective of this book is to present and discuss the management of field construction projects. These projects involve a great deal of time and expense, so close management control is required if they are to be completed within the established time and cost constraints. The text also develops and discusses management techniques directed toward the control of cost, time, resources, and project finance during the construction process. Emphasis is placed on practical and applied procedures of proven efficacy. Examples relate to field construction practices.

Effective management of a project also requires a considerable background of general knowledge about the construction industry. The purpose of this chapter is to familiarize the reader with certain fundamentals of construction practice that will be useful for a complete understanding of the discussions presented in later chapters. Learning objectives for this chapter include:

- Recognize the size and impact of the construction industry.
- Understand how a construction project moves from concept to reality.
- Identify entities involved in a construction project and various project delivery systems that link the participating entities.
- Introduce the role of the project manager.

1.2 The Construction Industry

In terms of the dollar value of output produced, the construction industry is one of the largest single production activities of the US economy. As of December of 2013, the seasonally adjusted value of construction put in place for 2013 was estimated to be $0.930 trillion. This figure is updated monthly by the US Census Bureau. The current value can be found on the US Census Bureau website under Construction Spending, www.census.gov/construction/c30/c30index.html. The current dollar gross domestic product (GDP) in the fourth quarter of 2013 was $17.103 trillion. This figure is regularly updated by the US Department of Commerce, Bureau of Economic Analysis, and can be found at www.bea.gov/newsreleases/national/gdp/gdpnewsrelease.htm.

Though construction was severely impacted by the economic downturn at the end of the first decade of this century, the annual expenditure for construction still accounts for almost 5.44 percent of the GDP. More than 1 of every 20 dollars spent for goods and services in the United States is spent on construction. The construction industry is also one of the nation's largest employers, so its impact in terms of both dollars and jobs remains considerable.

Not only does the construction industry touch the lives of virtually every human being on a daily basis; it also occupies a fundamental position in many national economies. This large and pervasive industry is regarded as the bellwether of economic growth in the United States. Periods of national prosperity usually are associated with high levels of construction activity. One is the natural result of the other.

The construction industry is heterogeneous and enormously complex. There are several major classifications of construction that differ markedly from one another: housing, nonresidential building, heavy civil, utility, and industrial. In addition, these construction types are further divided into many specialties, such as electrical, concrete, excavation, piping, and roofing.

Construction work is accomplished by contractors who vary widely in terms of size and specialty. Some contractors choose to concentrate on a particular task or aspect of the construction project and are therefore referred to as specialty contractors. Others assume broader responsibility for a comprehensive work package and are referred to as general contractors. Commonly, general contractors will subcontract specific aspects of a project to specialty contractors, forming a contractual web of general contractors and specialty contractors. Within the industry, very large contractors handle annual volumes in excess of $15 billion; their annual budgets rival the gross national products of many small countries. However, the construction industry is typified by small businesses.

1.3 The Construction Project

Construction projects are intricate, time-consuming undertakings. The total development of a project normally consists of several phases requiring a diverse range of specialized services. In progressing from initial planning to project completion, the typical job passes through successive and distinct stages that demand input from such disparate areas as financial organizations, governmental agencies, engineers, architects, lawyers, insurance and surety companies, contractors, material and equipment manufacturers and suppliers, and construction craft workers.

During the construction process itself, even a project of modest proportions involves many skills, materials, and literally hundreds of different operations. The assembly process must follow a natural order of events that constitutes a complicated pattern of individual time requirements and restrictive sequential relationships among the project's many segments.

To a great extent, each construction project is unique—no two jobs are ever exactly the same. In its specifics, each structure is tailored to suit its environment, arranged to perform its own particular function, and designed to reflect personal tastes and preferences. The vagaries of the construction site and the possibilities for creative and utilitarian variation of even the most standardized building product combine to make each construction project a new and different experience. The contractor sets up its "factory" on the site and, to a large extent, custom builds each structure.

Construction is subject to the influence of highly variable and sometimes unpredictable factors. The construction team, which includes architects, engineers, craft workers, specialty contractors, material suppliers, and others, changes from one job to the next. All the complexities inherent in different construction sites—such as subsoil conditions, surface topography, weather, transportation, material supply, utilities and services, local specialty contractors, labor conditions, and available technologies—are an innate part of construction.

However, though construction projects are subject to infinite variety, construction processes tend to be consistent from job to job. Each job goes through mobilization and closeout processes. All materials and installed equipment are subject to a procurement process that includes submittals, approvals, purchase, and shipment to the job site. Contracts are negotiated. Costs are estimated and billed out when a component is completed. Changes happen regularly, but even changes are handled through a consistent change order process. Hence, much of the current focus in construction management is on understanding and managing construction processes more effectively.

The character of construction projects, typified by their complexity and diversity and by the nonstandardized nature of their production, is a result of variable inputs operated on by standard processes yielding a unique product. The use of prefabricated modular units is somewhat limiting this variability, but it is unlikely that field construction will ever be able to adapt completely to the standardized methods and product uniformity of assembly-line production. To the contrary, many manufacturing processes are moving toward mass customization, or "one-off" production and adopting many of the project management tools originating in the construction industry.

1.4 Project Stages

A construction project proceeds in a rather definite order; the stages of development that follow are typical.

A. Planning and Definition

Once an owner has identified the need for a new facility, he or she must define the requirements and delineate the budgetary constraints. Project definition involves establishing broad project characteristics, such as location, performance criteria, size, configuration, layout, equipment, services, and other owner requirements needed to establish the general aspects of the project. Conceptual planning stops short of detailed design, although a considerable amount of preliminary architectural or engineering work may be required. The definition of the work is basically the responsibility of the owner, although a design professional may be called in to provide technical assistance and advice.

B. Design

The design phase involves the architectural and engineering design of the entire project. It culminates in the preparation of final working drawings and specifications for the total construction program. In practice, design, procurement, and construction often overlap, with procurement and construction beginning on certain segments as soon as the design is completed and drawings and specifications become available for those segments.

With the advent of high-speed computing and massive data management capabilities, the various design components can now be incorporated into a single database producing integrated design in a digital format. Production information is being added to the database, resulting in a Building Information Modeling (BIM) model of the project that contains all information on the project in a unique digital format. The use of BIM is still in the early stages, but its use is evolving rapidly throughout the construction industry.

C. Procurement and Construction

Procurement refers to the ordering, expediting, and delivering of key project equipment and materials, especially those that may involve long delivery periods. This function may or may not be handled separately from the construction process itself. *Construction* is, of course, the process of physically erecting the project and putting the materials and equipment into place, and this involves providing the manpower, construction equipment, materials, supplies, supervision, and management necessary to accomplish the work.

This stage moves toward conclusion with substantial completion of the project when the owner gains beneficial use of the facility. The conclusion of the project occurs when the terms of all contracts are fulfilled and the contracts are closed out. This closeout cycle is often part of a commissioning process that accomplishes many things, including bringing the facility on line, facilitating owner occupancy and turnover of facility operations to the owner, and closing out of all construction contracts. Many contractors follow the final closeout of the project with an internal postproject review from which the contractor gleans a great deal of information that helps to improve company processes and hence to mold the evolution of the company.

1.5 Owner

The owner, whether public or private, is the instigating party that gets the project financed, designed, and built. Public owners are public bodies of some kind and range from the federal government down through state, county, and municipal entities to a multiplicity of local boards, commissions, and authorities. Public projects are paid for by appropriations, bonds, or other forms of financing and are built to perform a defined public function. Public owners must proceed in accordance with applicable statutes and administrative directives pertaining to advertising for bids, bidding procedure, construction contracts, contract administration, and other matters relating to administration of the design and construction process.

Private owners may be individuals, partnerships, corporations, or various combinations thereof. Most private owners have the project built for their own use: business, habitation, or otherwise. However, some private owners do not intend to be the end users of the constructed facility; rather, they plan to sell, lease, or rent the completed structure to others. These end users may or may not be known to the owners at the time of construction.

A third classification of ownership in the form of a public-private partnership (PPP) has become popular. This tends to be subject to the rules and regulations governing public ownership since the partnership is typically used as a means to procure private financing for a public project.

1.6 Architect-Engineer

The architect-engineer, also known as the design professional, is the party or firm that designs the project. Because such design is architectural or engineering in nature, or often a combination of the two, the term *architect-engineer* is used in this book to refer to the design professional, regardless of the applicable specialty or the relationship between the architect-engineer and the owner.

The architect-engineer can occupy a variety of positions with respect to the owner for whom the design is undertaken. Many public agencies and large corporate owners maintain their own in-house design capability. In such instances, the architect-engineer is the design arm of the owner. In the traditional and most common arrangement, the architect-engineer is a private and independent design firm that accomplishes the design under contract with the owner. Where the *design-construct* mode of construction is used, the owner contracts with a single party for both design and construction. In such cases, the architect-engineer is a branch of, or is affiliated in some way with, the construction contractor.

1.7 Prime Contractor

A prime contractor, who is typically a general contractor, is a firm that contracts directly with the owner for the construction of a project, either in its entirety or for some designated portion thereof. In this regard, the owner may choose to use a single prime contract or several separate prime contracts.

Under the single-contract system, the owner awards construction of the entire project to one prime contractor. In this situation, the contractor brings together all the diverse elements and inputs of the construction process into a single, coordinated effort and assumes full, centralized responsibility for the delivery of the finished job, constructed in accordance with the contract documents. The prime contractor is fully responsible to the owner for the performance of the subcontractors and other third parties to the construction contract. The prime contractor may choose to self-perform certain parts of the work or may choose to subcontract all of the work to specialty contractors.

When multiple prime contracts are used, the project is not constructed under the centralized control of a single prime contractor. Rather, several independent contractors work on the project simultaneously, and each is responsible for a designated portion of the work. Each of the contractors is under contract with the owner, and each functions independently of the others. Responsibility for coordination of these contractors may be undertaken by the owner, the architect-engineer, a construction manager, or one of the prime contractors who is paid extra to perform certain overall job management duties.

1.8 Competitive Bidding

The owner selects a prime contractor on the basis of competitive bidding, negotiation, or some combination of the two. A large proportion of construction in the United States is done by contractors that obtain their work in bidding competition with other contractors. The competitive bidding of public projects is often required by law and is standard procedure for public agencies. Traditionally, all public construction work has been done by this method, though recently some alternative approaches have been approved. When bidding a project, the contractor estimates how much the project will cost using the architect-engineer's drawings and specifications as a basis for the calculations. The contractor then adds a reasonable profit to this cost and guarantees to do the entire job for the stated price.

Bid prices quoted by the bidding contractors most often constitute the principal basis for selection of the successful contractor, with the low bidder usually receiving the contract award. Most bidding documents stipulate that the work shall be awarded to the "lowest responsible bidder." This gives the owner the right to reject the proposal of a bidding contractor if the contractor is judged to be unqualified for some reason. If its bid is selected, the contractor is obligated to complete the work in exchange for the contract amount.

Competitive bidding can also be used where the successful contractor is determined on a basis other than the estimated total cost of the construction. For example, where the contract involves payment of a prescribed fee to the contractor, the amount of the fee is sometimes used as a basis of competition among contractors. Construction management services are sometimes obtained by an owner using the fees proposed by the different bidders as the basis for contract award. This is often referred to as a fee-based bid.

1.9 Negotiated Contracts

At times it can be advantageous for an owner to negotiate a contract for its project with a preselected contractor or small group of contractors. It is common practice for an owner to forgo the competitive bidding process and to handpick a contractor on the basis of its reputation and overall qualifications to do the job. A contract is negotiated between the owner and the chosen contractor. Clearly, such contracts can include any terms and provisions that are mutually agreeable to the parties. Most negotiated contracts are of the cost-plus-fee type, a subject that will be developed more fully later.

1.10 Combined Bidding and Negotiation

An owner sometimes will combine elements of both competitive bidding and negotiation into an approach known as *best value*. In this approach, a portion of the decision is based on price and the rest on qualifications. In the best-value approach, the competing contractors are required to submit their qualifications along with their bids and are encouraged to tender suggestions as to how the cost of the project could be reduced. Competing contractors are often interviewed, in addition to submitting the bid and the qualifications statement. The owner then scores the bid and the qualifications, awarding the project to the contractor with the best score. The best-value approach is now acceptable in many public-sector environments.

1.11 Subcontracting

The extent to which a general contractor will subcontract work depends greatly on the nature of the project and the contractor's own organization. There are instances where the job is entirely subcontracted, so the general contractor provides only supervision, job coordination, project billing, and perhaps general site services. At the other end of the spectrum are those projects where the general contractor does no subcontracting, choosing to do the work entirely with its own forces. Customarily, however, the prime contractor will perform the basic project operations and will subcontract the remainder to various specialty contractors. Types of work with which the prime contractor is inexperienced or for which it is not properly equipped are usually subcontracted, since qualified specialty contractors generally are able to perform their specialty faster and less expensively than the general contractor. In addition, many construction specialties have specific licensing, bonding, and insurance requirements that would be costly for the general contractor to secure for intermittent use.

When the prime contractor engages a specialty firm to accomplish a particular portion of the project, the two parties enter into a contract called a subcontract. No contractual relationship is thereby established between the owner and the subcontractor. When a general contractor sublets a portion of its work to a subcontractor, the prime contractor remains responsible under its contract with the owner for any negligent or faulty performance by the subcontractor. The prime contractor assumes complete responsibility to the owner for the direction and accomplishment of the total work. An important part of this responsibility is the coordination and supervision of the various subcontractors.

1.12 Design-Bid-Construct

Traditionally, field construction is not begun until the architect-engineer has completed and finalized the design. This sequence is still predominant in the industry and is referred to as the *design-bid-construct* procedure. While completing one step before initiating the next may be acceptable to owners on some projects, it will be unacceptably slow to other owners. A number of financial considerations dictate the earliest possible completion date for many construction projects. It is possible to reduce the total design-construction time required for some projects by starting the construction before complete design of the entire project has been accomplished.

1.13 Fast Tracking

Fast tracking refers to the overlapping accomplishment of project design and construction. As the design of progressive phases of the work is finalized, these work packages are put under contract, a process also commonly referred to as *phased construction*. Early phases of the project are under construction while later stages are still in the design process. This procedure of overlapping the design and construction can appreciably reduce the total time required to achieve project completion. For obvious reasons, fast tracking and phased construction sometimes can offer attractive advantages to the owner and also can be the source of severe coordination problems.

1.14 Construction Contract Services

A myriad of contract forms and types are available to the owner for accomplishing its construction needs, and all of them call for defined services to be provided under contract to the owner. The scope and nature of such services can be made to include almost anything the owner wishes. The selection of the proper contract form appropriate to the situation is an important decision for the owner and is one deserving of careful consideration and consultation.

The construction contract can be made to include construction, design-construct, or construction management services, each of which is discussed in the next three sections.

1.15 Construction Services

A large proportion of construction contracts provide that the general contractor have responsibility to the owner only for accomplishment of the field construction. Under such an arrangement, the contractor is completely removed from the design process and has no input into it. Its obligation to the owner is limited to constructing the project in full accordance with the contract terms.

Where the contractor provides construction services only, the usual arrangement is for a private architect-engineer firm to perform the design in contract with the owner. Under this arrangement, the design professional acts essentially as an independent design contractor during the design phase and as an agent of the owner during construction operations. The architect-engineer acts as a professional intermediary between the owner and contractor and often represents the owner in matters of construction contract administration. Under such contractual arrangements, the owner, architect-engineer, and contractor play narrowly defined roles, and the contractor is basically in an adversarial relationship with the other two.

1.16 Design-Construct

When the owner contracts with a single firm for both design and construction and possibly procurement services, this is referred to as a *design-construct* project. This form of contract is usually negotiated, although occasionally it is competitively bid. Usually, the contractor has its own design section with architects and engineers as company employees. In other cases, however, the architect-engineer can be a contractor's corporate affiliate or subsidiary, or the contractor can enter into a joint venture arrangement with an independent architect-engineer firm for a given project or contract.

The team concept is basic to design-construct. The owner, designer, and builder work cooperatively in the total development of the project. The contractor provides substantial input into the design process on matters pertaining to materials, construction methods, cost estimates, and construction time schedules. In recent years, owners have shown increasing acceptance and usage of this concept, largely due to the economies of cost and time that can be realized by melding the two functions of design and construction. Injecting contractor experience and expertise into the design process offers the possibility of achieving cost savings for the owner. Because fast tracking is possible under a design-construct contract, the owner may well have the beneficial use of the structure considerably before it would have under the more traditional design-bid-construct arrangement.

A *turnkey* contract is similar to a design-construct contract. The difference lies in the greater range of responsibilities that the contractor undertakes on behalf of the owner under a turnkey arrangement. For example, a turnkey contract often includes such services as land selection and acquisition, project financing, project equipage procurement, and leasing of the completed facility.

1.17 Construction Management

The term *construction management* is applied to the provision of professional management services to the owner of a construction project with the objective of achieving high quality at minimum cost. Such services may encompass only a defined portion of the construction program, such as field construction, or they may include total project responsibility. The objective of this approach is to treat project planning, design, and construction as integrated tasks within a construction system. Where construction management is used, a nonadversarial team is created consisting of the owner, construction manager, architect-engineer, and contractor. The project participants, by working together from project inception to project completion, attempt to serve the owner's best interests in optimum fashion. By striking a balance between construction cost, project quality, and completion schedule, the management team strives to produce for the owner a project of maximum value within the most economical time frame. Construction management does not include design or construction services per se but involves management direction and control over defined design and construction activities.

Construction management services can be performed for the owner for a stipulated fee by a range of firms, including design firms, contractors, and professional construction managers. Such services range from merely coordinating contractors during the construction phase to broad-scale responsibilities over project planning and design, project organization, design document review, construction scheduling, value engineering, field cost monitoring, and other management services. Selection of the construction manager by the owner is sometimes accomplished by a best-value approach, including both fee and qualifications as bases for contract awards. Usually, however, the construction management arrangement is considered to be a professional services contract and is negotiated. These contracts normally provide for a fixed fee plus reimbursement of management costs.

The Construction Management approach is practiced in two distinct variations. With Agency Construction Management, the construction manager operates throughout the project as the agent of the owner to manage the entire project in the best interest of the owner and does not perform any of the construction work. With Construction Management At Risk, the construction manager will, at some relatively early stage of design, provide cost and schedule commitments to the owner and take on the responsibility of completing the job within the time frame and cost stated. At that point, the project takes on the nature of a negotiated, fixed-sum project.

1.18 Fixed-Sum Contract

A fixed-sum contract requires the contractor to complete a defined package of work in exchange for a sum of money fixed by the contract. Should the actual cost of the work exceed this figure, the contractor absorbs the loss. The owner is obligated to make only such total payment as is stipulated in the contractual agreement. A fixed-sum contract may be either lump sum or unit price.

With a lump-sum contract, the contractor agrees to complete a stipulated package of work in exchange for a single lump sum of money. Use of this form of contract is obviously limited to those construction projects where both the nature and quantity of each work type can be accurately and completely determined before the contract sum is set.

A unit-price contract requires the contractor to perform certain well-defined items of work in accordance with a schedule of fixed prices for each unit of work put into place. The total sum of money paid to the contractor for each work item is determined by multiplying the contract unit price by the number of units actually done on the job. The contractor is obligated to perform the quantities of work required in the field at the quoted unit prices, whether the final quantities are greater or less than those initially estimated by the architect-engineer. This is subject to any contract provision for redetermination of unit prices when substantial quantity variations occur. Unit-price contracts are especially useful on projects where the nature of the work is well defined but the quantities of work cannot be accurately forecast in advance of construction.

1.19 Cost-Plus-Fee Contracts

Cost-plus-fee contracts provide that the owner reimburse the contractor for all construction costs and pay a fee for its services. How the contractor's fee is determined is stipulated in the contract, and a number of different procedures are used in this regard. Commonly used are provisions that the fee shall be a stipulated percentage of the total direct cost of construction or that the fee shall be a fixed sum. Incentive clauses are sometimes included that give the contractor an inducement to complete the job as efficiently and expeditiously as possible through the application of bonus and penalty variations to the contractor's basic fee. A guaranteed maximum cost is frequently included in cost-plus contracts. Under this form, the contractor agrees that it will construct the total project in full accordance with the contract documents and that the cost to the owner will not exceed some total price.

1.20 Work-by-Force Account

The owner may elect to act as its own constructor rather than have the work done by a professional contractor. If the project is being built for the owner's own use, this method of construction is called the force-account system. In such a situation, the owner may accomplish the work with its own forces and provide the supervision, materials, and equipment itself. Or the owner may choose to subcontract the entire project, assuming the responsibility of coordinating and supervising the work of the subcontractors. Because public projects generally must be contracted out on a competitive-bid basis, force-account work by a public agency usually is limited to maintenance, repair, or cases of emergency. Force-account work can also be coupled with other contracting methods discussed earlier in this chapter to handle specific aspects of the project that cannot be clearly defined or have undergone significant change. In such cases, the contractor performs the associated work at the direction of the owner and bills for these services on a time and materials basis.

Over the years, many studies have revealed that most owners cannot perform field construction work nearly as well or as inexpensively as professional contractors. The reason for these findings is obvious: The contractor is intimately familiar with materials, equipment, construction labor, and methods. It maintains a force of competent supervisors and workers and is equipped to do the job. Only when the owner conducts a steady and appreciable volume of construction and applies the latest field management techniques is it economically feasible for it to carry out its own construction operations.

1.21 Turnkey and BOT Contracts

Fixed-sum, cost-plus-fee, and work-by-force account contracting methods all require owners to coordinate initial planning, design, construction, and facilities start-up. These tasks distract the owners from their core business responsibilities. For this reason, some owners also contract these responsibilities to the contractor. Turnkey and build-operate-transfer (BOT) contracts provide a vehicle for complete project delivery by the contractor.

In a turnkey arrangement, the owner provides the facility design requirements to the contractor, which designs and constructs the facility under a single contract. The single contract eliminates the need for owner coordination and reduces project duration. Upon completion, the key to the project is turned over to the owner and the contract is closed out.

BOT contracts are an extension of the turnkey method. The contractor designs, constructs, operates, and maintains the facility for a predetermined concessionary period. In most cases, the contractor receives no payment from the owner for these services but retains all or a portion of the revenues earned by the project during the concession. This contracting method generally is used for bridges, highways, power plants, and similar projects that generate a long-term revenue stream. At the end of the concession period, ownership transfers from the contractor to the owner.

1.22 Integrated Project Delivery

A new type of project delivery system is emerging with the development of BIM. As previously indicated, in a BIM environment, all data for the project go into a massive database that is shared by all stakeholders in the project. Because the data are integrated into a single entity, all stakeholders who contribute and use the data become active participants in the entire project. The result is shared risk and shared reward across all stakeholders, which, in turn, requires a new type of contract that integrates all stakeholders into a single operational structure. Individual contracts between various stakeholders are no longer efficient in a BIM environment.

Integrated Project Delivery (IPD) ties all stakeholders together based on a single contract that is signed by each stakeholder. This includes the owner, designer, contractors, and subcontractors. Such an arrangement of shared data, shared risk, and shared reward imposes the requirement upon the project of an efficient team approach. All succeed together, or all fail together. Though BIM and IPD are just emerging, as they evolve, they portend dramatic changes to the management of construction projects.

1.23 Speculative Construction

When owners build structures for sale or lease to other parties, they engage in what is commonly referred to as *speculative construction*. Housing and commercial properties like shopping centers and warehousing facilities are common examples of such construction. In tract housing, for instance, "merchant" builders develop land and build housing for sale to the general public. This is a form of speculative construction through which developers act as their own prime contractors. They build dwelling units on their own account and employ sales forces to market their products. In much speculative housing, contractors build for unknown buyers. In commercial applications, however, construction does not normally proceed until suitable sale or lease arrangements have been made. Leases are usually necessary so that the developers can arrange their financing. Leases also enable them to build to the lessee's individual specifications. Most speculative builders function more as land or commercial developers than as contractors, choosing to subcontract all or most of the actual construction work.

1.24 Management during the Design Phase

Project cost and time control actually begin very early in the project with the project owner's developing basic cost and time requirements necessary to justify the project. The design team respects the owner's limits, constraining the design, as it evolves, to meet the owner's cost and time boundaries. In the initial design stages, estimates such as annual cost to the owner and total life-cycle costs of the facility are made. Technical job standards are weighed against cost, function, maintenance, and appearance with the objective of minimizing the full cost of constructing, operating, and maintaining the new facilities over their useful life. As the design develops, construction methods and material alternatives are subjected to value analysis as a rational means of optimizing the entire construction process in terms of cost and time. Cost budgets—ranging from preliminary to final—are prepared as the design approaches completion.

Time control during the design stage is directed toward minimizing construction time consistent with project quality, safety, and cost. The delivery times of materials and project equipment are checked. Where long delivery periods are involved, the design is changed or procurement is initiated as soon as the design has progressed sufficiently to allow detailed purchasing specifications to be drawn. Construction methods are chosen whose cost characteristics are favorable and for which adequate labor and construction equipment will be available as needed. A preliminary project time schedule usually is prepared as the design progresses.

1.25 Management of Field Construction

Discussions up to this point have demonstrated that owners have the option of using many different project delivery systems to get their projects built. Regardless of the variability of these systems, however, one party assumes management responsibility for the field construction process. Depending on the methods used by the owner, this party may be the owner, the architect-engineer, a construction manager, or a general contractor.

The management of field construction customarily is done on an individual project basis, with a project manager being made responsible for all aspects of the construction. For large projects, a field office usually is established directly on the job site for the use of the project manager and staff. A good working relationship with a variety of outside persons and organizations, such as architects, engineers, owners, subcontractors, material and equipment suppliers, labor unions, and regulatory agencies, is an important part of guiding a job through to its conclusion. Field project management is directed toward pulling together all the diverse elements necessary to complete the project satisfactorily. Management procedures presented later will, in general, be discussed only as they apply to field construction, although they are equally applicable to the entire project, from concept to commissioning.

1.26 Project Manager

The project manager organizes, plans, schedules, and controls all the work of the project and is responsible for getting the project completed within the time and cost limitations. The project manager acts as the focal point for all facets of the project and brings together the efforts of all organizations contributing to the construction process. He coordinates matters relevant to the project and expedites project operations by dealing directly with the individuals and organizations involved. In any such situation where events progress rapidly and decisions must be consistent and informed, the specific leadership of one person is needed. Because the project manager has the overall responsibility, this person must have broad authority over all elements of the project. The nature of construction is such that the manager often must take action quickly on his own initiative, and it is necessary that he be empowered to do so. To be effective, he must have full control of the job and be the one voice that speaks for the project. Project management is a function of executive leadership and provides the cohesive force that binds together the several diverse elements into a team effort for project completion.

Though a single project manager is recognized for overall responsibility for a project, there are, in reality, numerous project managers involved in most projects. Generally, each entity will have its own project manager, designated to be the top person within the context of that entity that is responsible for the successful completion of its portion of the project. Thus, in addition to the general contractor, each specialty contractor will have a designated project manager for its portion of the project. The architect-engineer will have a designated person responsible for successful completion of the design of the project, and the owner will also have a designated owner's project manager. Though the focus of this text is the general contractor's project manager, the techniques and procedures presented are equally applicable to each of the project managers working throughout the project.

Large projects normally will have a full-time project manager who is a member of the firm's top management or who reports to a senior executive of the company. The manager may have a project team to assist it or may be supported by a central-office functional group. When smaller contracts are involved, a single individual may act as project manager for several jobs simultaneously. An important aspect of a project manager's position is that the duties normally are separate from those of field supervision. The day-to-day direction of field operations is handled by a site supervisor or field superintendent whose duties involve working with the foremen, coordinating the subcontractors, directing construction operations, and keeping the work progressing safely, smoothly, and on schedule. The fact is that construction project authority is a partnership effort between the project manager and the field superintendent, who work very closely together. Nevertheless, centralized authority is necessary for the proper conduct of a construction project, and the project manager retains that central authority.

1.27 Project Manager Qualifications

The effective project manager must possess four essential attributes. The first is a considerable background of practical construction experience to provide thorough familiarity with the workings and intricacies of the industry. In the case of project managers for specialty contractors, they must be thoroughly familiar with the details of construction in that specialty area, in addition to having an overall understanding of how construction operates at the project level. Without such a basic grounding in construction fundamentals, the project manager would be completely unprepared to carry out the responsibilities of the job.

Second, the project manager must have, or have available, persons with expertise and experience in the application of specialized management techniques to the planning, scheduling, and control of construction operations. These procedures have been developed specifically for application to construction projects and are those discussed in this book. Because much of the management system is computer based, the project manager must have access to adequate computer support services.

Third, the project manager must have the capacity to step back from the complex details of daily construction operations and look into the future—planning for upcoming activities, checking material deliveries, determining manpower and training requirements, identifying possible changes to the work, and recognizing other potential problem areas that may develop.

Fourth, the project manager must have the personality and insight to work harmoniously with other people, often under very strained and trying circumstances. The manager, after all, cannot accomplish everything through its efforts alone. The project manager must work with and through people in the performance of its duties. Doing this requires an appreciation and understanding of the human factor. Without this, the other attributes, however commendable, will be of limited effectiveness.

Key Points and Questions

Key Points

- The construction industry forms a large part of a national economy and affects all people, organizations, and institutions.
- A construction project is complex, requiring extensive resources and masterful execution by a variety of skilled professionals and craft workers under the leadership of the project manager.
- The structure of a specific construction project is based on one of many project delivery systems that best matches the characteristics of the project.
- Resources a project manager employs to effectively manage a construction project include experience, both technical and human skills, vision, and the support of other experts.

Review Questions and Problems

1. What is the approximate annual value of construction, and how does this compare to the gross domestic product?
2. If each construction project is unique, how can it be governed by processes that are consistent from job to job?
3. What are the three stages of development of a construction project?
4. What unique part does each of the following play in a construction project?
   a. Owner
   b. Designer

c. General contractor

d. Specialty contractor

5. What are the distinguishing characteristics of the following project delivery systems?

   a. Design-bid-construct

   b. Design-construct

   c. Construction Management

   d. Work-by-force account

   e. Turnkey

   f. Integrated Project Delivery

6. How can an aspiring project manager develop, or a practicing project manager improve, the four essential attributes required for an effective project manager?

2 Systematic Project Management

2.1 Introduction

The focus of this chapter is on management of the construction project as a whole. A construction project is made up of many small components that are integrated to form a single complex project. To have a successful project, various characteristics of that project, including time, quality, cost, and safety, need to be managed. This chapter introduces the various components that need to be managed for a successful project, which will be treaded individually in detail in the chapters that follow.

Learning objectives for this chapter include:

- Recognize the various components of a construction project that need to be managed if the project as a whole is to be successful.
- Understand the nature of a construction project that consists of various construction processes that can be tracked, measured, and managed.
- Introduce valuable tools employed to manage the project and its various processes.

2.2 Need for Project Management

On most construction projects, the contractor is given only one opportunity to establish cost and schedule objectives. The price is set and schedule milestones are agreed to when a bid is submitted or the project contract is negotiated. From that point on, profits are determined by the project manager's ability to save money through wise procurement, better planning of daily operations, and the ability to make sound decisions. If a project is to be constructed within its established budget and schedule, close management control of field operations is essential. Project conditions such as technical complexity, timely completion, resource limitations, and substantial costs put great emphasis on the planning, scheduling, and control of construction operations. Unfortunately, the construction process, once set into motion, is not a self-regulating mechanism and requires expert guidance if events are to conform to plans.

It must be remembered that projects are one-time and largely unique efforts of limited time duration that involve work of a nonstandardized and variable nature. Field construction work can be affected profoundly by events that are difficult, if not impossible, to anticipate. Under such uncertain and shifting conditions, field construction costs and time requirements are changing constantly and can seriously deteriorate with little or no advance warning. The presence of uncertainty in construction does not suggest that planning is impossible but rather that it will assume a monumental role in the success or failure of the project. The greater the level of uncertainty in the project, the greater is the need for exhaustive project planning and skilled and unremitting management.

Under most competitively bid, fixed-sum contracts calling for construction services only, the general contractor exercises management control over construction operations. Self-interest is the essential motivation in such cases, the contractor being obligated by contract to meet a prescribed completion date and to finish the project for a stipulated sum. The surest way for the contractor to achieve its own objectives, and those of the owner in the bargain, is by applying systematic project management.

Serving the best interests of the owner is the primary emphasis of project control under other forms of contract. Field management under design-construct, construction management, and many cost-plus contracts is directed principally toward providing the owner with professional advisory and management services to best achieve the owner's objectives.

2.3 Project Management Characteristics

In its most common context, the term *management* relates to the planning, organizing, directing, controlling, and staffing of a business enterprise. Business management is essentially a continuing and internal activity involving that company's own personnel, finances, property, and other resources. Construction project management, however, applies to a specific project, the various phases of which usually are accomplished by different organizations. Therefore, the management of a construction project is not so much a process of managing the internal affairs of a single company as it is one of coordinating and regulating all of the elements needed to accomplish the job at hand. Thus, the typical project manager must work extensively with organizations other than its own. In such circumstances, much of the authority is conferred by contractual terms or power of agency and is therefore less direct than that of the usual business manager. In the case of a subcontractor, authority of the project manager is further limited, and effective project management is accomplished much more through influence and persuasion rather than direct exercise of authority. In any case, project management is accomplished largely through the personnel of different employers working closely together to achieve common goals.

2.4 Discussion Viewpoint

As mentioned previously, the responsibility for field construction management rests with different parties, depending on owner preference and the choice of a project delivery system. Whether the owner, architect-engineer, general contractor, construction manager, or specialty contractor performs such duties is very much a matter of context. The basics of project management are essentially the same, however, regardless of the implementing party. Nevertheless, to show detailed workings and examples of such management methods, it is necessary to present the material from the specific viewpoint of one of these parties. Thus, where the nature of the discussion requires such designation, the treatment of management methods herein will be from the particular viewpoint of the general contractor.

2.5 Management Procedures

Day-to-day construction in the field has little in common with the assembly-line production of standardized products. Standard costs, time-and-motion studies, process flowcharts, and line-of-balance techniques—all traditional management devices used by the manufacturing industries—have limited applicability to general construction. Historically, construction project management has been a rudimentary and largely intuitive process, aided by useful but inadequate adaptations from manufacturing, the bar chart being a prime example (see Section 5.29).

Over the years, however, new scientific management concepts have been developed and applied. Application of these concepts to construction has resulted in the development of techniques for the control of construction cost, time, resources, and project finance, which recognize construction as a series of repetitive processes that can be managed and improved. The result is treatment of the entire construction process as a unified system. Comprehensive management control is applied from inception to completion of construction operations.

Construction project management starts at the point at which the contractor is brought into the project. Initial activities include development of a comprehensive construction budget and a detailed schedule of operations. These cost- and time-based models of the project establish the accepted cost and time goals used as a blueprint for the tangible construction operations. After the project has begun, monitoring systems are established that measure the actual costs and progress of the work at periodic intervals. The reporting system provides progress information that is measured against the planned targets. Comparison of field investments and progress with the established plan quickly reveals exceptions that must receive prompt management attention. In addition to using data from the system to design corrective action and exert control over project operations, the data can also be used to make corrected forecasts of costs and time to complete the work.

The process just described is often called *management-by-exception*. When applied to a given project, it emphasizes the prompt and explicit identification of deviations from an established plan or norm. Reports that highlight exceptions from the standard enable the manager to recognize quickly those project areas requiring attention. As long as an item of work is progressing in accordance with the plan, no action is needed, but there are always plenty of problem areas that do require attention. Management-by-exception devices are useful, and this book emphasizes their application.

In addition to cost and time, the project management system is necessarily concerned with the management of job resources, with safe execution of operations, and with project financial control. Resources in this context refer to materials, labor, construction equipment, and subcontractors. Resource management is primarily a process of advanced recognition of project needs, scheduling and expediting of the resources required to meet those needs, and adjusting the demands where necessary. Safe execution of construction operations involves the use of labor, tools, and construction equipment to properly install the materials and equipment brought to the workface in an incident-free environment. Project financial control relates to the responsibility of the project manager for the total cash flow generated by the construction work and the terms of the contract.

As indicated by the preceding discussion, there are several different elements of a project control system. Though each of these major management topics is treated separately in the chapters that follow, it must be recognized that these elements are highly interrelated segments of a total project management process.

2.6 Time and Cost Management

Project time and cost management are based on time and cost models developed for the project and an information system that will provide data for comparing expected with actual performance. The information system measures, evaluates, and reports job progress, comparing it with the planned performance expressed in the models. This keeps the project manager apprised of the nature and extent of any deviation. When deviations do occur, the manager identifies the cause or causes of the deviation, develops corrective action that can be taken to correct the deviation, and then executes whatever action is considered feasible and effective to correct the situation. The manager then verifies the effectiveness of the corrections in future reports.

Costs and time can get out of hand quickly on construction projects where production conditions are volatile. Job monitoring must detect such aberrations quickly. Cost and time control information must be timely, with little delay between fieldwork and management review of performance. This timely information gives the project manager a chance to evaluate alternatives and take corrective action while an opportunity still exists to rectify problem areas.

In a sense, all management efforts are directed toward cost control because expedient completion of safe and high-quality projects represents both construction savings for the contractor and beneficial usage for the owner. In practice, however, time and cost management are spoken of and applied as separate, although interrelated, procedures. One aspect of this separation is the difference in job breakdown structure used for time- and cost-control purposes. The distinctive character of the two procedures requires that the project be divided into two different sets of elements: project components for time control and work classifications for cost control.

The realities of a field project make the strict control of every detail unattainable in a practical sense. Consequently, it must be recognized that the time and cost management methods discussed in this book are imperfect procedures, affording results of reasonable accuracy to managers whose powers to control are far from absolute. Project management procedures offer no panacea for construction problems. They provide no magic answers, and the management information generated is no better than the quality of the input data. Nevertheless, a reasonably good basis is established for informed decision making.

2.7 Planning and Scheduling

Planning, the first step in the process of construction time control, is discussed at length in Chapter 4, "Project Planning." On the basis of a detailed study of job requirements, planning establishes what is to be done, how it is to be done, and the order in which it will proceed. The planning function is accomplished by dividing the project into many components or time-consuming steps, called activities, and establishing the sequence in which they will be performed. An example of an activity might be "Install boiler" or "Set bar joists." The results of project planning are shown graphically in the form of a network diagram. This diagram can be drawn using either of two different graphical notation systems, "precedence" or "arrow." Since precedence notation has become the predominant way to represent schedule networks, it is emphasized herein and is used throughout for discussion purposes. However, since older schedules in arrow diagram formatting are still encountered, examples of arrow networks will be found on the companion website.

A detailed time study of the planning network is then conducted, with adjustments to the plan being made as necessary to meet the project milestone and completion dates. Some selective shortening of key construction activities may be in order at this point. Manpower and construction equipment requirements are evaluated for individual job activities, with adjustments made to minimize unbalanced or conflicting demands. On the basis of these studies, the contractor establishes a calendar-date schedule of the anticipated start and finish times for each activity. The resulting time schedule, subject to periodic revision and correction during construction, is the essential basis for the day-to-day time control of the project. Such a schedule serves as an exceptionally effective early-warning device for detecting when and where the project is falling behind and the impact that the delays will have on the project as a whole. The several facets of project scheduling are the focus of Chapters 5, "Project Scheduling Concepts"; 7, "Managing Time"; and 9, "Project Scheduling Applications."

2.8 CPM Procedure

The planning and scheduling of construction projects normally uses a network-based management procedure referred to as the *Critical Path Method (CPM)*. CPM was developed especially to provide an effective and workable procedure for planning and scheduling construction operations. Widely used by the construction industry, and frequently a contract requirement, CPM involves a definite body of management procedures and is the basis for the planning and scheduling methods discussed in this book.

The heart of CPM is a graphical job plan that shows all the construction activities necessary for job completion and the order in which they will be done. This graphical network portrays, in simple and direct form, the complex time relationships and constraints among the various segments of a project. It has the tremendous advantage of easily accommodating modifications, refinements, and corrections. It provides the project manager with 12 invaluable elements of time control information and devices for control:

1. *Concise information* regarding the planned sequence of construction operations.

2. A *means to predict* with reasonable accuracy the time required for overall project completion and the times to reach intermediate construction goals (commonly called milestones).
3. *Proposed start and finish calendar dates* for project activities consistent with the construction plan.
4. *Identification* of those "critical" activities whose expedient execution is crucial to timely project completion.
5. A *guide* for reducing project time.
6. A *basis* for scheduling subcontractors and material deliveries to the job site.
7. A *basis* for balanced scheduling of manpower and construction equipment on the project.
8. The *rapid evaluation* of time requirements for alternative construction methods.
9. An *effective model* for numerically computing project status.
10. An *essential vehicle* for progress reporting, recording, and analysis.
11. A *basis* for evaluating the time effects of construction changes and delays.
12. A *language* for the communication of plans, processes, and goals for the entire project team.

2.9 Time Monitoring and Control

When field operations begin, the order in which the project proceeds is in accordance with an approved job plan. During the construction period, advancement of the work is monitored by measuring and reporting field progress at regular intervals. These data are analyzed and time-control measures are taken as appropriate to keep the work progressing on schedule.

Progress measurement for time-control purposes is an approximate process and is based on determining the time status of each individual job activity. Progress normally is measured by noting those activities that have been completed and estimating the remaining time required to complete those in process. When compared with the latest planned schedule, these data give the manager an immediate indication of the time status of each job activity. Because activities seldom start or finish exactly as scheduled, the field information also serves as the basis for occasional updates that yield revised project completion dates and corrected time schedules for the construction yet to be done. The workings of project time control are discussed in Chapter 7, "Managing Time."

2.10 Project Cost System

The project cost system is concerned with the control of expenses on current projects and the gathering of production information for use in estimating the cost of future work. The application of cost controls to a construction project actually begins when the costs are estimated initially. It is then that the project budget is established. This is the budget used by the project manager for cost-control purposes during field construction.

If there is to be an opportunity for genuine cost control, it must be possible to detect cost overruns promptly by making frequent comparisons between actual and budgeted expenses of production during the construction process. In addition, the actual costs must be determined in sufficient detail to enable project management to locate the source of cost overruns. During construction, cost accounting methods are applied to obtain the actual production rates and costs as they occur. Specifically designed summary reports are prepared periodically to pinpoint work areas where costs are exceeding the budget. This management-by-exception cost system immediately identifies for the project manager where production costs are unsatisfactory and management action is needed. If the project manager takes timely and suitable corrective measures, cost overruns often can be minimized and future expenditures brought into line with budget estimates. In addition to maintaining a continuous check on production costs for cost-control purposes, the project cost system yields valuable information needed for estimating future construction work. Average production rates and unit costs are obtained from completed projects and maintained in permanent and easily accessible databases. These records of past cost experience are a valuable resource to the estimator when new projects are being estimated.

For both cost-control and estimating purposes, a construction project must be broken down into standardized and categorized building blocks, often called cost codes, work types, or work breakdown structure (WBS) elements. Hence, cost information gathered during the construction phase must be tracked using the same cost codes that will be used in producing future estimates. This allows the historical cost information to be recalled and assembled in a variety of different ways to produce reasonable cost estimates of future projects. Some examples of work types might be "footing concrete, place (cy)" or "structural steel, erect (ton)." These classifications are used throughout a company's cost system. Each work type is assigned a unique and permanent cost code number that is used consistently by all company personnel and that does not change from project to project. Chapter 11, "Project Cost System," presents a detailed treatment of a project cost system.

2.11 Estimating the Project

When the project design has been finalized, a complete and detailed cost estimate is prepared. The contractor uses this estimate for bidding and subsequent cost-control purposes. With cost-plus and construction management contracts, a similar estimate is compiled, essentially for the owner's cost-control purposes during construction. The final estimate is based on a detailed quantity takeoff that is a compilation of the total amounts of elementary work classifications required. The costs of labor, construction equipment, and materials are computed on the basis of the work quantities involved. Subcontract amounts are obtained from bids submitted by subcontractors to the general contractor. Taxes, overhead, and surety bonds are added as required.

Of all the costs involved in the construction process, those of labor and construction equipment are the most difficult to estimate and control. Fundamentally, the estimating of such costs is based on production rates. A production rate can be expressed as hours of labor or equipment time required to accomplish a unit amount of a given work type. An example of this is the number of labor hours required to erect a ton of structural steel. Production rates also can be expressed as the number of units of a work type that can be done per unit of time, such as per hour. An example of such a production rate is the number of cubic yards of excavation that a power shovel can perform in one hour. In addition to estimating, these production rates are also used to determine durations of construction activities for scheduling. For quick and convenient application, production rates frequently are converted to costs per unit of work. The source of production rates and unit costs is the company's cost accounting system. When the cost estimate has been completed, the project control budget is prepared. This schedule of costs is the standard to which the actual costs of production are compared during field operations. Chapter 3, "Project Estimating," discusses estimating project costs and preparing the project budget.

2.12 Project Cost Accounting

Project cost accounting is the process of obtaining actual production rates and unit costs from ongoing projects. This system provides the basic information for project cost control and for estimating new work, as well as for scheduling future work. Because of the uncertain nature of labor and equipment costs, these two items of expense are subjected to detailed and frequent analysis during the construction period. They are the main emphasis of a construction cost accounting system. Basic elements of this system are labor and equipment work hours, hourly expense rates for labor and equipment, and the quantity of work accomplished during the specified period. These data are analyzed and periodic cost reports are produced. These cost reports compare budgeted with actual costs of production for each work type and are used for cost-control purposes. These reports not only enable comparisons to be made between budgeted and actual expenses as the work proceeds, but also provide a basis for making forecasts of the final project cost. They also record production rates and unit costs for future use in estimating and scheduling.

Cost accounting, unlike financial accounting, is not conducted entirely in terms of cost. To establish production rates and unit costs, work quantities and hours of labor and equipment usage must also be determined. Consequently, accurately measuring and reporting work quantities completed, and the associated hours of labor and equipment expended in the field, are integral parts of a cost accounting system.

2.13 Resource Management

The term *resource* refers to manpower, construction equipment, materials, and subcontractors. These resources totally control job progress and must be managed carefully during the construction process. Schedules of future resource needs are prepared and positive steps taken to assure that adequate resources will be available as required. Favorable material deliveries require skilled attention to procurement, shop drawing review and approval, expediting, and quality control. Labor crews and construction equipment must be scheduled and arranged. Subcontractors must be kept informed of the overall job schedule and given advance notice when their services are required, and their work must be coordinated with the total project effort.

Resource management involves other aspects as well. Job schedules occasionally must be adjusted to reduce the daily demand for certain resources to more practical levels. Impracticable bunching of job resources must be leveled to a smoother and more attainable demand profile. Resource management and its procedures are discussed in Chapter 8, "Resource Management."

2.14 Project Financial Control

For management purposes, a construction project is treated essentially as a separate and autonomous effort, requiring resources and input from a variety of sources. For income and expense, profit or loss, and general financial accounting purposes, each project is handled separately and individually. The significance of this method is that the project manager generally has responsibility for the control of project financial matters. Of concern here are considerations ranging from total project cash flow to everyday matters of contract administration. Monthly pay requests, an estimated schedule of payments by owner, project cash forecasts, changes to the contract, and disbursements to material dealers and subcontractors are all examples of project financial matters subject to management control procedures. Methods and procedures applicable to project financial management are discussed in Chapter 12, "Project Financial Management."

2.15 Automating Project Management Tasks

The management and control of project time, cost, resources, and finance by the contractor during the field construction process require that the project manager originate, manipulate, summarize, and interpret large volumes of numerical data. In order to generate such information and apply it in optimum fashion, the project manager customarily relies on computer software to provide a wide range of data processing and to automate routine tasks. Project managers must react quickly to changing conditions, and their decisions must be made with the secure knowledge that they are acting on the basis of adequate, accurate, and current intelligence. Modern management software can greatly assist in making this information available and in providing information for the evaluation of alternate courses of action. Because most of the time-consuming data manipulation is automated, the project manager can devote more time to problem solving and developing more profitable approaches.

However, because of the continuous emergence of new technology in terms of both hardware and software, this book makes no effort to discuss specific hardware configurations or the workings of software in current usage for project management purposes. Rather, emphasis is placed on sources of information, the management significance of the data generated by various computational devices, and how these data are applied to the control of a construction project.

2.16 Manual Methods

The preceding discussion of computer applications to job management is not meant to imply that manual methods have no place in the system. The project manager may rely on hand methods for limited portions of a project and to carry out computations for making quick checks, to determine the effect of changes, or to study a specialized portion of the work.

Even when the calculations are automated, the project manager must understand the computational procedures that are an innate part of the techniques applied. The manager's intimate familiarity with the workings of the procedures will provide an intuitive feel and grasp of a project that cannot be obtained in any other way. Because manual methods are useful in their own right and a thorough understanding of the computational methods involved with the computer generation of management data is crucial to the proper application of project control methods, this book discusses in step-by-step fashion the manual calculation of the several kinds of construction management tools.

2.17 Discussion Format

Several different management procedures are presented in the chapters that follow. In an attempt to provide a sense of continuity while going from one topic to another, two example projects are used as a continuing basis for the succeeding series of discussions. The primary example project is chosen from the heavy civil sector. For those with a primary interest in building construction rather than heavy civil construction, a second example project is provided in the form of a commercial building.

In order to acquaint the reader both with the detailed workings of certain procedures and the broad applicability of others, examples of construction work ranging from modest to comprehensive in extent are needed. To provide examples of both macro- and micro-work packages, the large-scale heavy civil project consisting of several separate segments or subprojects serves as the primary example project. Two segments of the heavy civil example project are used for illustrative purposes where a considerable scope of construction activity is needed to present a given management application. However, where the level of detail is such that the procedure is best explained by using an example of limited proportions, an individual subproject of the heavy civil project is used. Several project management actions are presented subsequently using the two limited examples as the basis for discussion. Each major management responsibility is the subject of a different chapter. The changes, modifications, revisions, and corrections that are discussed in any one chapter are limited to that chapter and do not carry forward to the next. For purposes of clarity, the methods presented in each chapter are discussed independently of one another and are applied, in turn, to the original, unchanged example project.

In a similar manner, the example commercial building project is used to illustrate project level management principles applicable to the building sector of the construction industry. This project is a single-story office building built in 2006–2007 for a technology company engaged in computer programming and consulting. The design and construction information for the building project were graciously provided to the authors for inclusion in this book by the project design team, the contractor, and the owner.

2.18 Example Projects

As indicated, two example projects are provided. The first focuses on heavy civil construction and involves the construction of an earth dam and some appurtenant structures. The completed project will serve a number of purposes, including flood control, irrigation, and recreation. The dam will be constructed across an existing river, the flow of which is highly variable with the seasons. A permanent reservoir will be formed, the extent of which will vary considerably during the year. The job site covers a considerable geographical area that is undeveloped and unpopulated.

The second example project is a 12,200-square-foot commercial building designed specifically to consolidate various operations of a technology company focused on consulting and software development. It was constructed on a previously developed site on the outskirts of a moderate-sized western town and designed with the opportunity to move on to a Phase II expansion when the company grew. Phase II was started soon after Phase I was completed.

As we consider the heavy civil project, the design has been completed, and working drawings and specifications are now available. The owner is a public agency, and the project will be competitively bid by prequalified contractors. Accordingly, project field management will be carried out by the successful prime contractor. Public notice of the receipt of proposals has been given as required by law, bidding documents are available, and the bidding process is about to begin. The entire heavy civil project will be awarded by a single contract to the low-bidding prime contractor. Because of the nature of the work involved, unit-price bidding will be used. The contract will contain a time requirement for completion of the construction, and liquidated damages will apply in the event of late contract completion.

For the purposes of this book, we will consider that the design of the building project has been completed. The contractor has also been selected. The owner is a private owner. The basis of the construction contract is a negotiated general contract.

The flowcharts in Figure 2.1a and 2.1b depict the overall example projects and the principal operations that will be covered by the construction contracts. These figures show the major aspects of the work for each project. The general sequence of each project will be described.

For the heavy civil project, one of the first operations must be the diversion of the river away from the dam site area. The borrow areas from which the earth dam material is to be obtained are located some distance from the dam site and must be stripped of surface soil and vegetation. Haul roads between the dam and the borrow areas must be developed, keeping grades to a minimum and providing hard, smooth, rolling surfaces. After the river diversion, borrow development, and haul roads have been accomplished, construction of the dam itself can proceed.

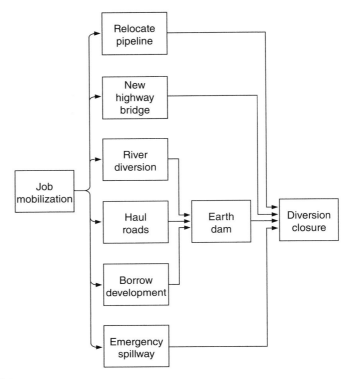

Figure 2.1a
Example heavy civil construction project

While this preparatory work and the dam are under way, other segments of the heavy civil project can progress simultaneously. A concrete emergency spillway is to be built at a location removed from the main dam itself. The new reservoir area necessitates the relocation of five miles of existing natural gas pipeline, and a new bridge must be built where an adjoining highway crosses a reservoir inlet. The closure and removal of the river diversion will be the final major construction operation. The highway bridge and the pipeline relocation will be used to illustrate several construction management applications.

The commercial building is located within a commercial development area so that roads and utilities are available to the site. After the job is mobilized by securing the site and bringing in temporary utilities and facilities to support the construction site, work specific to this building must be done to configure the site to meet the specific needs of this project. Structural work will then commence by drilling pier foundations, pouring

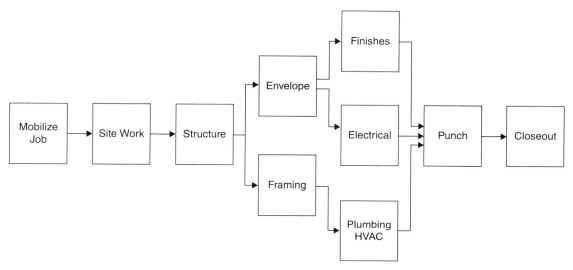

Figure 2.1b
Example commercial building project

the grade beams and structural slab, and erecting the steel framework. When the structure is in place, the building envelope (roof and exterior walls) will be constructed and interior framing can begin. After the building is dried in, electrical work and finishes can begin, and after framing is in place, mechanical work can proceed. When the finishes, electrical work, and mechanical work are completed, the job can be punched out, after which the owner can obtain beneficial occupancy and the project can be closed out.

It is important to note that this is a very simple representation of a complex project with many overlapping activities throughout the project. For example, though much site work is accomplished prior to commencing structural work, a good deal of site work is completed toward the end of the project, such as sidewalks, landscaping, and parking areas. Another example is electrical work, which begins early with rough-in below the slab, then further rough-in in partitions after they are framed, then trimming out electrical devices, and finishing with plates on the walls after the finishes are completed. The purpose of this simple model is to introduce the project at a very high level. As we proceed through this book, we will learn many techniques to help represent the complexities of the project much more accurately.

Key Points and Questions

Key Points

- Construction projects are dynamic and complex, requiring skilled and competent managers for successful execution.
- Construction project managers employ a variety of models to understand projects, including cost models, time-based models, graphical models, and production models.
- Standard procedures can increase efficiency or production.
- Useful management tools are available to the construction project manager, such as project financial systems and critical path schedules.
- The critical path schedule is one of the most useful construction management tools for controlling time and cost, as well as for managing resources and production.

Review Questions and Problems

1. At what point in the project are project cost and time objectives defined for the contractor?
2. The construction process consists of numerous standardized and repetitive procedures. Why is expert guidance by a project manager necessary for a successful project?
3. What differences are there between constructing a large project and manufacturing a large batch of a product?
4. Describe the basis of management-by-exception in construction.
5. What is the first step in the process of time control?
6. Why is it critical to identify cost and time anomalies early?
7. What resources are the primary target of "resource management"?

3 Project Estimating

3.1 Introduction

The topic of this chapter is estimating, with the primary focus on estimating applications that the project manager will be involved with. Many smaller companies have estimator/project managers that have the responsibility to estimate the cost of potential projects, and when a new project is brought in, the estimator becomes the project manager for that project, as well as estimating new potential projects. As companies become larger and more sophisticated, estimating becomes a specialty area within the company, and when new projects are acquired, the estimator hands the project off to a project manager and continues to focus on estimating new projects. In this case, the project manager will still be involved in estimating such things as change orders and claims, but the primary focus is to effectively manage the project at hand. The focus of this book is the project manager, rather than the estimator/project manager of smaller companies.

When a project manager is assigned to a recently acquired project, one of the first tasks is to become thoroughly familiar with the project, including the estimate. There will be a hand-off meeting between the estimator and the project manager, and from then on the project manager takes responsibility for the project, including the estimate. The project manager sets up the project cost system (see Chapter 11, "Project Cost System") to document and track costs throughout the project. The project manager will also make estimates of work accomplished as a basis for monthly pay requests and, as previously indicated, when changes come up, will estimate the cost of those changes as the basis to negotiate change orders.

This chapter begins with an overview of cost-estimating procedures and how the final project budget is obtained. Then it considers developing monthly progress estimates and change order estimates. It should be noted that estimating is a high-level, complex task. Many excellent texts focus expressly on construction cost estimating. The purpose of dealing with estimating in this book is to acquaint the project manager with the fundamentals of estimating and to provide some detail regarding specific estimates that the project manager typically prepares.

Learning objectives for this chapter include:

- Understand how the project estimate is developed.
- Recognize ways in which the project manager uses estimates.
- Learn about several types of specialty estimates that are developed during execution of the project.

3.2 Project Cost System

During the design phase of a construction project, the project costs are continuously approximated and reviewed following each design change to ensure that they will not exceed the owner's budget. This working budget is generally referred to as the engineer's or architect's estimate. Upon design completion, the field cost-control system is initiated by making a final, detailed cost estimate of the entire work. The construction contractor or another party who will be directly involved in the field operations normally prepares this "contractor's estimate." The contractor's estimate is then reduced to a working construction budget and forms the basis of the construction cost-control system.

During the construction process, cost accounting methods (discussed in Chapter 11, "Project Cost System") are used to retrieve actual construction expenses from ongoing construction operations. This information is then used for cost-control purposes on the current project and for estimating the cost of future projects. Additionally, the cost system provides considerable information pertinent to project financial control (discussed in Chapter 12, "Project Financial Management").

3.3 Preliminary Cost Estimates

Often, project managers are approached by a client to provide some preliminary cost information on a potential project that is under consideration. Preliminary estimates of future construction expenditures, made during the project planning and design phases, are necessarily approximate because they are compiled before the project is completely defined.

Making such conceptual estimates is an art quite different from determining the final detailed estimate of construction costs.

Fundamentally, all conceptual price estimates are based on some system of gross unit costs obtained from previous construction work. These unit costs are extrapolated forward in time to reflect current market conditions, project location, and the particular character of the job presently under consideration. Some of the methods commonly used to prepare preliminary estimates include:

Cost per Function Estimate

> This analysis is based on the estimated expenditure per unit of use, such as cost per patient, student, seat, or car space. Construction expense may also be approximated as the average outlay per unit of a plant's manufacturing or production capacity. These parameters are generally used as a method of quickly defining facilities costs at the inception of a project when only raw marketing information is known, such as the number of patients that a planned hospital will hold. This broad method of developing costs can also provide a powerful check on more detailed estimates once they have been generated.

Index Number Estimate

> This method involves estimating the price of a proposed project by updating the construction cost of a similar existing facility. It is done by multiplying the original construction cost of the existing project by a national price index that has been adjusted to local conditions, such as weather, labor expense, materials costs, transportation, and site location. A price index is the ratio of present construction cost to the original construction outlay for the type of project involved. Many forms of price indexes are available in various trade publications.

Unit Area Cost Estimate

> This method of estimating facilities costs is an approximate cost obtained by using an estimated price for each unit of gross floor area. The method is used frequently in building and residential home construction. It provides an accurate approximation of costs for buildings that are standardized or have a large sampling of historical cost information from similar buildings. This type of estimate is used often in the industry to compare the relative worth of various facilities.

Unit Volume Cost Estimate

> This estimate is based on an approximated expenditure for each unit of the total volume enclosed. This estimating method works well in defining the costs of warehouses and industrial facilities. It can also be quite valuable in developing specialty estimates for heating, ventilating, and air conditioning (HVAC) systems.

Panel Unit Cost Estimate

> This analysis is based on unit costs per square unit area of floors, unit length of perimeter walls, partition walls, and unit roof area. Generally,

this form of estimating is used to improve the preceding estimates once additional detailed information about the facility is known.

Parameter Cost Estimate

This estimate involves unit costs, called parameter costs, for each of several different building components or systems. The prices of site work, foundations, floors, exterior walls, interior walls, structure, roof, doors, glazed openings, plumbing, heating and ventilating, electrical, and other items are determined separately by the use of estimated parameter costs. These unit expenses can be based on dimensions or quantities of the components themselves or on the common measure of building square footage.

Partial Takeoff Estimate

This analysis uses quantities of major work items taken from partially completed design documents. These are priced using estimated unit prices for each work item taken off. During the design stage, this type of estimate is considered to provide the most accurate preliminary costs. Yet the feasibility of this method is highly dependent on the availability of completed design documents. Generally, this estimate cannot be made until the project is well into the design process. Hence, it is often used to refine the previously discussed estimating methods.

3.4 Final Cost Estimate

The final cost estimate of a project is prepared when finalized working drawings and specifications are available. This detailed estimate of construction expense is based on a complete and detailed survey of work quantities required to accomplish the work. The process involves the identification, compilation, and analysis of the many items of cost that will enter into the construction process. Such estimating, which is done before the work is actually performed, requires careful and detailed study of the design documents, together with an intimate knowledge of the prices, availability, and characteristics of materials, construction equipment, and labor. A thorough knowledge of construction field operations required to build the specific type of project under consideration is also essential.

It must be recognized that even the final construction estimate is of limited accuracy and that it bears little resemblance to the advance determination of the production costs of mass-produced goods. By virtue of standardized conditions and close plant control, a manufacturer can arrive at the future expense of a unit of production with considerable precision. Construction estimating, by comparison, is a relatively crude process. The absence of any appreciable standardization together with a myriad of unique site and project conditions make the advance computation of exact construction expenditures a matter more of accident than design.

Nevertheless, a skilled and experienced estimator, using cost accounting information gleaned from previous construction work of a similar nature, can do a credible job of predicting construction disbursements despite the project imponderables normally involved. The character or location of a construction project sometimes presents unique problems, but some basic principles for which there are precedents almost always apply.

When pricing a job of some size, there will undoubtedly be more than one person involved in the quantity takeoff and pricing phases. The term *estimator* is used herein to refer to whoever may be involved with the estimating procedure being discussed.

There are probably as many different estimating procedures as there are estimators. The use of computerized estimating systems has imposed some uniformity on the estimating process, but even standardized systems are tailored for unique characteristics of specific contractors. In any process involving such a large number of intricate manipulations, variations naturally result. The form of the data generated, the sequential order followed, the nature of the elementary work classifications used, the mode of applying costs—all are subject to considerable diversity. Individual estimators develop and mold procedures to fit their own context and to suit their own preferences. In this book, rather than attempt any detailed discussion of estimating methods, only the general aspects of construction estimating are presented.

3.5 Example Projects

To illustrate the workings of several major aspects of project management, including the cost system, it will be useful to have a construction job large enough to be meaningful, but not so large that the sheer size and complexity will obscure the basic objectives. The highway bridge, a segment of the heavy civil example project, has been selected to serve as the basis for discussion of an estimating and cost system. Although the highway bridge would be estimated and bid as a part of the total heavy civil example project, it is easy to isolate and has been isolated here for demonstration purposes.

The commercial building project is a relatively small commercial building. However, due to the nature of building construction, even a small project is very complex, including the building shell; structural, mechanical, and electrical systems; and interior finishes. The components of the building are also highly integrated and do not lend themselves easily to the isolation of individual elements that can be the subject of detailed analysis. Because of the relative complexity of the building as compared to the bridge, detailed analysis will focus on the bridge with reference to the building only for more general principles.

The highway bridge is a single-span vehicular bridge that will cross a small ravine. The bridge is of a deck-girder type and is of composite steel-concrete construction. Figures 3.1a and b show the bridge profile and a

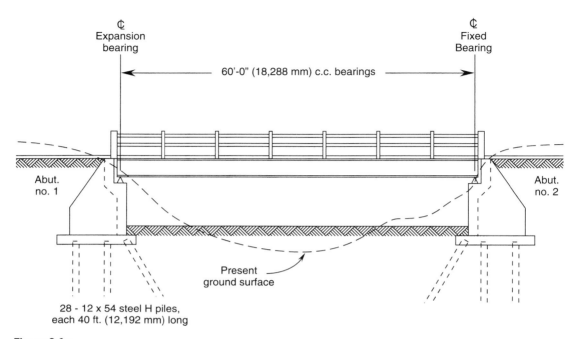

Figure 3.1a
Highway bridge, profile

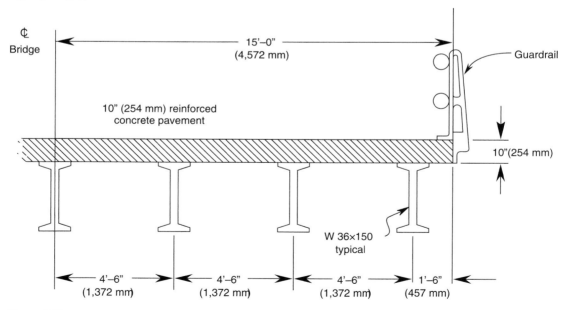

Figure 3.1b
Highway bridge, transverse section

transverse section. The two abutments are of reinforced concrete, each consisting of a breast wall and two wing walls. Each abutment rests on a heavy concrete footing supported by twenty-eight 40-foot-long, H-section steel piles. The 10-inch-thick reinforced concrete paving slab is supported by seven W36×150 steel floor girders. A steel guardrail is required to extend the length of the bridge along each side. All exposed structural concrete is to be given a rubbed finish, and all of the metal surfaces are to be painted.

The commercial building project is a 12,200-square-foot, single-story office building comprised of steel post and beam construction on drilled pier foundations. The exterior is masonry. Figures 3.2a and b show the architectural floor plan and elevations. These are actual construction drawings that, when combined with the other drawings of the complete set, provide all the detail needed by a contractor to construct the building. Because of the detail and the scale required for the book, these drawings are difficult to read, but a quick examination will enable the reader to understand the general scope of the commercial building example project. For further study, the full set of drawings is found on the companion website (www.wiley.com/go/cpm6e).

The owner of the heavy civil example project is a public agency, and the entire project, including the highway bridge, is to be competitively bid on the basis of unit prices. The final estimate and working job budget that are shown in the sections that follow are those prepared by the prime contractor for its bidding and cost-control purposes. The design has been completed, and bidding documents, including finalized drawings and specifications, are in the hands of the bidding contractors.

Figure 3.3, the proposal form to be used for the bidding of the highway bridge, shows 12 bid items and the engineer's quantity estimate of each. The bidding contractor fills in the unit price, column 5, and estimated amount, column 6, after the estimating process has been completed (see Figure 3.9)

It should be noted that because of the scope and complexity of the commercial building example project, any type of detailed estimate for this project would be beyond the scope of this book. For this reason, detailed cost-estimating examples are restricted to the highway bridge example project, and only a summary cost estimate of the commercial building example project is provided later in this chapter as a basis for the project budget.

3.6 Quantity Survey

The first step in preparing the final estimate of a construction project is the preparation of a quantity survey. This survey is simply a detailed compilation of the nature and quantity of each work type required. Taking off quantities is done in substantial detail, with the project being divided into many different work types or classifications. In the case of heavy civil

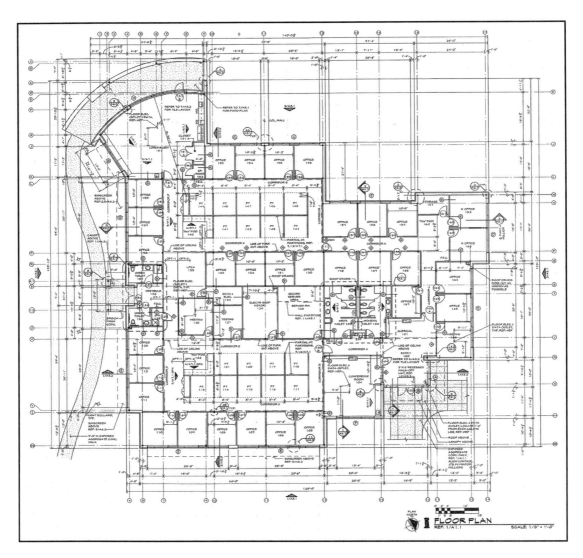

Figure 3.2a
Commercial building architectural floor plan
The Arkitex Studio, Incorporated.

3.6 Quantity Survey

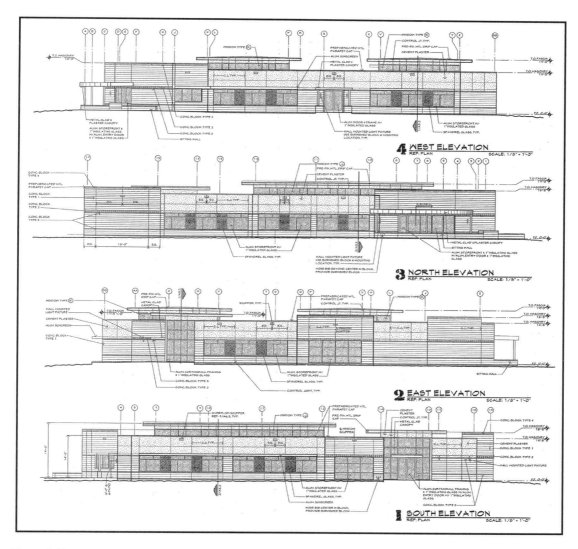

Figure 3.2b
Commercial building elevations
The Arkitex Studio, Incorporated.

Unit Price Schedule					
Item No. [1]	Description [2]	Unit [3]	Estimated Quantity [4]	Unit Price [5]	Estimated Amount [6]
1	Excavation, unclassified	cy	1,667		
2	Excavation, structural	cy	120		
3	Backfill, compacted	cy	340		
4	Piling, steel	lf	2,240		
5	Concrete, footings	cy	120		
6	Concrete, abutments	cy	280		
7	Concrete, deck slab, 10 in.	sy	200		
8	Steel, reinforcing	lb	90,000		
9	Steel, structural	lb	65,500		
10	Bearing plates	lb	3,200		
11	Guardrail	lf	120		
12	Paint	ls	job		
				Total Estimated Amount =	

Figure 3.3
Highway bridge, bid form

projects where unit-price bidding is involved, even though the architect-engineer customarily provides estimated quantities of each bid item with the bidding documents, a detailed quantity survey is still required. A basic reason for making a quantity survey for unit-price bidding is that most bid items cannot be priced without breaking the work down into smaller subdivisions. Another reason is that the architect-engineer's quantity estimates are specifically stated to be approximations only. Finally, for any estimate, developing the quantity survey provides the estimator with the intimate familiarity of job requirements so vital to realistic project pricing.

The estimator takes the dimensions and numbers of units of each work type from the design drawings, enters them on quantity sheets, and extends them into totals. The summarized results of the quantity survey on the highway bridge are shown in Figure 3.4. These figures do not show any quantities for items, such as painting. This is because the prime contractor normally limits its takeoff to work items it might self-perform. The contractors intend to subcontract the painting and thus did not take off detailed quantities for this specialized work category.

A comparison of Figure 3.3 with 3.4 will show that the architect-engineer's estimate and the contractor's takeoff are the same insofar as the quantities of the individual bid items are concerned. Quantities estimated by the architect-engineer and by the bidding contractors do not always check; at times, the differences can be substantial. However, for such a relatively precise work package as the highway bridge, it is reasonable to assume that agreement of the quantity figures will be relatively close in all cases.

Cost Code	Work Quantities		
	WorkType	Unit	Quantity
	Sitework		
32220.10	Excavation, unclassified	cy	1,667
32222.10	Excavation, structural	cy	120
32226.10	Backfill, compacted	cy	340
31350.00	Pile driving rig, mobilization & demobilization	ls	job
31361.10	Piling, steel, driving	lf	2,240
	Concrete		
03150.10	Footing forms, fabricate	sf	360
03150.20	Abutment forms, prefabricate	sf	1,810
03157.10	Footing forms, place	sf	720
03159.10	Footing forms, strip	sf	720
03157.20	Abutment forms, place	sf	3,620
03159.20	Abutment forms, strip	sf	3,620
03157.30	Deck forms, place	sf	1,800
03159.30	Deck forms, strip	sf	1,800
03200.00	Steel, reinforcing, place	lb	90,000
03251.30	Concrete, deck, saw joints	lf	60
03311.10	Concrete, footings, place	cy	120
03311.20	Concrete, abutments, place	cy	280
03311.30	Concrete, deck, place & screed	sy	200
03345.30	Concrete, deck, finish	sf	1,800
03346.20	Concrete, abutments, rub	sf	1,960
03370.20	Concrete, abutments, curing	sf	3,820
03370.30	Concrete, deck, curing	sf	1,800
	Metals		
05120.00	Steel, structural, place	lb	65,500
05520.00	Guardrail	lf	120
05812.00	Bearing plates	lb	3,200

Figure 3.4
Highway bridge, work quantities

3.7 Management Input

Early in the estimating process, certainly before the work is priced, a number of important management decisions must be made concerning the project and how construction operations are to be conducted. When the job is being priced, the estimator must exert every effort to price each work type as realistically as possible. To do this, major decisions must be made concerning project organization, the major construction methods to be used, the sequential order of operations, and what construction equipment will be utilized. These four considerations require management attention and are of consuming importance to the bidding process.

A new project cannot be intelligently priced until some major management determinations have been made concerning the conduct of the work. It is clear, therefore, that there must be some regular and usual procedure for the estimator to precipitate such decision making. Typically, the lead estimator has the background and company knowledge, as well as the authority, to make such decisions. Sometimes the estimator will want to enlist the guidance of a senior project manager and/or field supervisor as internal "consultants" to expand the creative opportunities. Even more effective, where possible, is to have a prebid meeting of company personnel who will actually be involved in the project, including the proposed project manager and the field superintendent, as well as the company purchasing agent. At this meeting, details of the job are discussed, job requirements are reviewed, alternative choices are evaluated, and decisions are made.

3.8 Field Supervision

Before pricing the job, it is always good practice to identify the top supervisors who will be assigned to it. The reason for this is not only to establish specific salary requirements but also to recognize that most job supervisors do better in terms of construction costs on some portions of a project than on others. Some superintendents are known to be very good at getting a job off the ground but do not perform as well during later phases of construction. A given supervisor may be experienced with one type of equipment but not with another. The pricing of a project must take into account the special abilities of key field personnel.

The matching of field supervisory talents with the demands of a particular project is one of the most important management actions. The best management system in the world cannot overcome the handicap of poor supervision. The importance of an experienced, skilled, and energetic field supervisory team cannot be overemphasized.

3.9 Construction Methods

Seldom is there a job operation that can be performed in only one manner. Preferably, the choice of method is made after evaluating the time and cost characteristics of the feasible alternatives. This is not to say that a detailed cost study must be undertaken before making decisions concerning alternative ways of doing every construction operation. Frequently, convention, together with experience and equipment availability, prescribes most of these choices to the estimator. However, there are times when these operational choices are sufficiently important to justify conducting a detailed comparative study.

Many examples of alternative construction methods can be cited, such as procedures to be followed in underpinning an adjacent structure, how best to brace an excavation, what method of scaffolding to use, the type and placement of hoisting equipment, and how to dewater the site. All of these involve judgments of major import to the conduct of the work. The proper evaluation of alternatives can require considerable time and extensive engineering studies. It is obvious, however, that the principal construction procedures to be used must be identified before the job can be intelligently priced.

3.10 General Time Schedule

When a new project is being estimated, it is necessary to devise a general plan and operational time schedule. Estimators customarily do this during the takeoff stage, although often in an informal and almost subconscious way. Small jobs may require little investigation in this regard, but larger projects deserve more than a cursory time study. Time is of prime importance on all projects, in part because most contracts impose a required project completion date on the contractor.

An approximate construction schedule is also important for project pricing purposes. Many of the job overhead expenses are directly related to the duration of the construction period. When a calendar of work operations is prepared, the time periods required for each of the major job parts can be established, as well as the kinds of weather to be anticipated. This calendar provides invaluable information to the estimator concerning equipment and labor productivity, cold weather operations, necessity of multiple shifts or overtime, and other such matters.

Devising a general job plan and time schedule must start with a study of project requirements. This study will enable an approximation to be made of the times necessary to accomplish each of the major job segments. Next, the sequence in which these segments must proceed is established. The result is a bar chart—a series of bars plotted against a horizontal time scale showing the completion date of the overall project and the approximate calendar times during which the various parts of the job will proceed. Each bar represents the beginning, duration, and completion of some designated segment of the total project. Together, the bars make up a time schedule for the entire job. If the bar chart completion date is not consonant with owner requirements, the estimator will have to rework the job plan.

Although the bar chart has had the advantage of some general planning, it cannot be regarded as the equivalent of a detailed network analysis. However, the development of a Critical Path Method (CPM) job schedule requires considerable time and effort. Consequently, on a competitively bid job, a contractor usually will not make a full-scale time study until it has been proclaimed the successful bidder and awarded the contract. This practical fact emphasizes the need for making a reasonably accurate general time schedule of the project during the bidding period.

Figures 3.5a and 3.5b show the general time schedules for the highway bridge and the commercial building. The bridge schedule indicates that

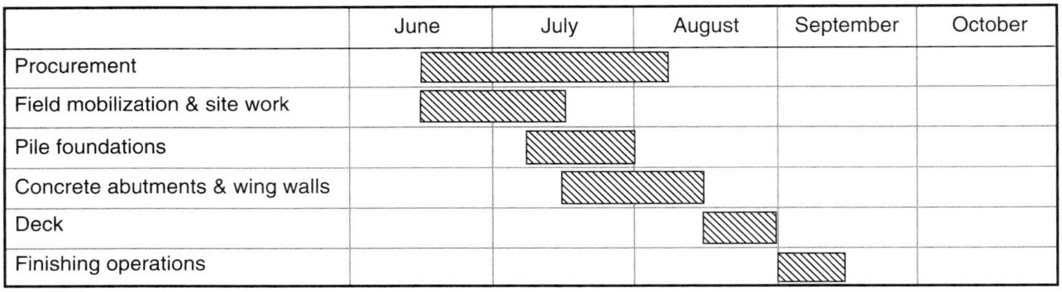

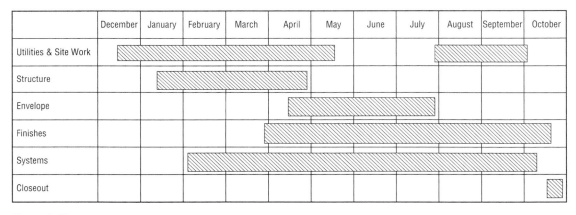

Figure 3.5a
Highway bridge, general time schedule

Figure 3.5b
Commercial building, time schedule

a construction period of about 15 weeks will be necessary. The building project schedule indicates a construction period of 46 weeks. Subsequent refinement and development of the job plan will undoubtedly disclose imperfections. Nevertheless, if the development of these initial schedules has received adequate consideration, they will serve as an acceptably accurate job picture for pricing purposes. Considerable job planning has already occurred to develop the job schedule to this stage.

3.11 Construction Equipment

Projects that are of the highway, heavy, or utility category normally require considerable amounts of construction equipment for their accomplishment. A substantial proportion of the total cost of these projects is associated with such equipment; however, equipment expenses vary greatly with the type and size of the individual unit. Commensurately, the detailed pricing of equipment on the bridge project cannot proceed very far until equipment selection decisions have been made in fairly specific terms. It is essential that the estimator be able to price the job with reasonable assurance that this is being done on the basis of the equipment types that actually will be used during construction operations.

To illustrate the workings of equipment decisions, consider the equipment commitments made for the highway bridge. A decision to use transit-mix concrete obtained from a commercial concern obviates the need for a field concrete plant. For pouring concrete, placing structural steel, and driving steel piles, a 50-ton crane equipped with an 80-foot boom will be used. A 7,200-foot-pound double-acting hammer and a 900-cubic-foot-per-minute portable air compressor with hose and connections will be used for the pile driving. A tractor with a low-boy trailer and a 25-ton crane will be needed for transport and assembly of the pile driving rig. A crawler tractor with bulldozer blade will do the unclassified excavation, and a 1-cubic-yard backhoe will be used for the structural excavation. A flatbed truck, troweling machine, and concrete saw will complete the list of larger equipment needs. Smaller equipment such as concrete vibrators and assorted small tools will be provided as needed and priced elsewhere. These are the kinds of advance equipment decisions that must be made during the estimating stage.

3.12 Summary Sheets

After the quantity survey has been completed and decisions concerning methods and equipment have been made, the total quantities are transferred from the quantity sheets to summary sheets for pricing purposes. On lump-sum projects, like our commercial building example, it is standard practice to transfer all quantities of work pertaining to a single construction classification to the same summary sheet before pricing them. As an example, concrete quantities and prices would appear on a concrete summary sheet. Similar summary sheets are prepared for the other work classifications, such as excavation, concrete forms, masonry, and carpentry.

On a unit-price job like the highway bridge, each summary sheet lists the work types necessary to accomplish the total quantity of a single-bid item and may include several different classifications of work. Figure 3.6 is the summary sheet for Bid Item No. 6, Concrete, abutments, on the highway bridge. In Figure 3.6, an identifying number designates each work type. These are the contractor's standard cost account numbers that are basic to the workings of its cost accounting system. When the takeoff is being compiled for a new project, the work is broken down into the standard elementary classifications established by this system. Each of these standard work

Summary Sheet

Job: Highway Bridge
Estimator: GAS

Bid Item No. 6: Concrete, abutments

Cost Code	Work Type	Quantity	Unit	Calculations	Labor Cost Direct	Labor Cost Indirect	Equipment Cost	Material Cost	Total Cost
03150.20	Abutment forms, prefabricate	1,810	sf	**Labor** labor unit cost = $1.96 per sf 1,810 × $1.96 = $3,548 **Material** plyform: $1.05 per sf 10% waste 50% salvage 1,810 × $1.05 × 1.10 × 50% = $1,045 lumber: $0.85 per bf 2 uses 50% salvage 1.44 bf/sf 1,810 × $0.85 × 1.44 × 50% = $1,108	$3,548	$1,596		$1,045	$5,144
								$1,108	$1,045
									$1,108
				Total this account	$3,548	$1,596	$0	$2,153	$7,297
03157.20	Abutment forms, place	3,620	sf	**Labor** labor unit cost = $2.02 per sf 3,620 × $2.02 = $7,312 **Equipment** 3 days of 50-ton crane time 3 × 8 × $105 = $2,520 **Materials** nails, form ties, coating $0.28 per sf 3,620 × $0.28 = $1,014	$7,312	$2,559			$9,872
							$2,520		$2,520
								$1,014	$1,014
				Total this account	$7,312	$2,559	$2,520	$1,014	$13,405
03159.20	Abutment forms, strip	3,620	sf	**Labor** labor unit cost = $0.97 per sf 3,620 × $0.97 = $3,511 **Equipment** 1 day of 50-ton crane time 1 × 8 × $105 = $840	$3,511	$878			$4,389
							$840		$840
				Total this account	$3,511	$878	$840	$0	$5,229

Figure 3.6
Highway bridge, bid-item summary sheet

03311.20	Concrete, abutment place	280	cy	**Labor**				
				foreman	1 @	$34.00 = $34.00		
				mason	1 @	$32.00 = $32.00		
				laborers	6 @	$22.00 = $132.00		
				operator	1 @	$33.00 = $33.00		
				oiler	1 @	$22.00 = $22.00		
				carpenter	1 @	$31.00 = $31.00		
					Crew hourly rate =	$284.00		
					Production rate	8.75 cy/hr		
				280 ÷ 8.75 × $284.00 = $9,088			$9,088	
				Equipment				
				50-ton crane	1 @	$105.00 = $105.00		
				vibrators	2 @	$3.00 = $6.00		
				buckets	2 @	$3.00 = $6.00		
					Equipment hourly rate =	$117.00		
				280 ÷ 8.75 × $117.00 = $3,744				$3,744
				Material				
				transit mix	@	$92.00 per cy		
				280 × 1.05 × $92.00 = $27,048		5% waste		$27,048	$27,048
						Total this account		$43,061	$12,269
03346.20	Concrete, abutment rub	1,960	sf	**Labor**				
				production rate	=	16.67 sf/hr (1 mason)		
				1,960 ÷ 16.67 × $32.00 = $3,762			$3,762	
				Material				
				material unit cost	=	$0.12 per sf		
				1,960 × $0.12 = $235				$3,181
						Total this account		$3,181	
								$3,744
								$235	$235	$5,456
03370.20	Concrete, abutment curing	3,820	sf	**Labor**				
				production rate	=	500 sf/hr (1 mason)		
				3,820 ÷ 500 × $32.00 = $244			$244	
				Materials				
				curing compound:	coverage =	270 sf/gal		
				3,820 ÷ 270 × $6.50 = $92				$1,693
						Total this account		$1,693	$0
								$86	$86	$330
									$92	$92
							$244		$422
							$27,466	$9,993	$30,542	$5,691
Total Bid Item No. 6							$37,459	$7,104	$75,105

Figure 3.6
(Continued)

types bears a unique cost account number. The coding system is discussed further in Chapter 11, "Project Cost System."

3.13 Material Costs

It is customary for the contractor to solicit and receive specific price quotations for most of the materials required by the job being priced. Exceptions to this generality are stock items such as plyform, nails, and construction lumber, which the contractor purchases in large quantities and of which a running inventory is maintained. Written quotations for special job materials are desirable so that such important considerations as prices, freight charges, taxes, delivery schedules, and guarantees are explicitly understood. Most material suppliers tender their quotations on printed forms that include stipulations pertaining to terms of payment and other considerations. For both example projects, the contractor will receive, during the bidding period, written price quotations from material dealers covering specific job materials, such as transit-mix concrete, structural and reinforcing steel, steel pilings, and guardrails. These must all meet the specifications developed by the architect-engineer and are subject to approval before they can be delivered to the job site. However, if the quantity survey has been prepared with precision, materials usually can be accurately priced based on the supplier quotations.

When material costs are entered on the summary sheets, they must all have a common basis, for example, delivered to the job site and without sales tax. Prices as entered will ordinarily include freight, drayage, storage, and inspection. It is common practice to enter material prices without tax, adding this as a lump-sum amount on the final recap sheet (see Section 3.23).

It is not unusual in some types of projects for the owner to provide certain materials to the contractor for use on the project, although this does not occur in either of our example projects. In such a case, contractors need not add this material price into their estimates. However, all other charges associated with the material, such as handling and installation expense, must be included.

3.14 Labor Costs

The real challenge in pricing construction work is determining labor and equipment expenditures. These are the categories of construction expense that are inherently variable and the most difficult to estimate accurately. To do an acceptable job of establishing these outlays, the estimator must make a complete and thorough job analysis, maintain a comprehensive library of unit costs and production rates from past projects, and obtain advance decisions regarding how construction operations will be conducted.

Contractors differ widely in how they estimate labor costs. Some choose to include all elements of labor expense in a single hourly rate. Others evaluate direct labor cost separately from indirect cost. Some contractors compute regular and overtime labor costs separately, while others combine scheduled overtime with straight time to arrive at an average hourly rate. Some evaluate labor charges using production rates; others use labor unit costs. There are usually good reasons for a contractor to evaluate its labor expense as it does, and there certainly is no single correct method that must be followed. The procedures described in this chapter are commonly used and are reasonably representative of general practice.

Basic to the determination of the labor cost associated with any work category is the production rate. Figure 3.6, in which representative labor expenses of pouring concrete for the bridge project is determined, is a good example of production rates. Figure 3.6 shows that a placement rate of 8.75 cubic yards of concrete per hour is used as the production rate for the prescribed concrete crew. The direct labor cost to pour the entire 280 cubic yards of abutment concrete is obtained by multiplying the time required to pour the concrete (280 ÷ 8.75 = 32 hours) by the direct hourly wage rate of the entire crew ($284), giving a direct labor cost of $9,088. To this must be added indirect labor costs of $3,181, giving a total labor charge of $12,269 to accomplish this particular item of work. The distinction between direct and indirect labor expenses is discussed in Section 3.15.

The most reliable source of labor productivity information is obtained from cost accounting reports compiled from completed projects. Labor cost information is also available from a wide variety of published sources. While information of this type can be very useful at times, it must be emphasized that labor productivity differs from one geographical location to another and varies with the seasons and many other job factors. Properly maintained labor records from recent jobs completed in the locale of the project being estimated reflect, to the maximum extent possible, the effect of local and seasonal conditions.

3.15 Indirect Labor Costs

Direct labor cost is determined from the workers' basic wage rates, that is, the hourly rates used for payroll purposes. Indirect labor costs are those expenses that are additions to the basic hourly rates and that are paid by the employer. Indirect labor expense involves various forms of payroll taxes, insurance, and a wide variety of employee fringe benefits. Employer contribution to Social Security, unemployment insurance, workers' compensation insurance, and contractor's public liability and property damage insurance are all based on payrolls. Though the specifics are changing rapidly, employers in the construction industry typically provide various kinds of fringe benefits, such as pension plans, health and welfare funds, employee insurance, paid vacations, and apprenticeship programs. The charge for these benefits customarily is based on direct payroll costs. Premiums for workers' compensation insurance and most fringe benefits differ considerably from one craft to another.

Indirect labor costs are substantial in amount, often constituting a 35 to 55 percent addition to direct payroll expenses. Exactly when and how indirect labor costs are added into a project estimate is unimportant so long as it is done. For estimating purposes, total labor outlay can be computed in one operation by using hourly labor rates, which include both direct and indirect costs. However, this procedure may not interrelate well with labor cost accounting methods. For this reason, direct and indirect labor charges often are computed separately when job prices are being estimated. One commonly used scheme is to add a percentage allowance for indirect costs to the total direct labor expense, either for the entire project or for each major work category. Because of the appreciable variation in indirect costs from one classification of labor to another, it may be preferable to compute indirect labor costs at the same time that direct labor expense is obtained for a given work type. This is the method followed in Figure 3.6.

3.16 Labor Unit Costs

Direct labor cost was computed in Section 3.14 for a given work type using hourly crew payroll expense together with the work quantity and appropriate production rate. A widely used alternative to this procedure involves the use of "labor unit costs." A labor unit cost is the direct labor expense per unit of production of a work type. To illustrate, reference is again made to concrete placing in Figure 3.6. This figure shows that 32 hours of crew time are required to place 280 cubic yards of abutment concrete at a total direct labor outlay of $9,088. Dividing $9,088 by 280 yields $32.46. This value of $32.46 is a labor unit cost; it is the average direct labor charge of pouring 1 cubic yard of abutment concrete. Thus, the direct labor cost of pouring the abutment concrete could have been computed in Figure 3.6 simply by multiplying the total of 280 cubic yards of abutment concrete by the labor unit expense of $32.46. The same value of $9,088 would have been obtained.

The use of labor unit costs in estimating practice usually is limited to the determination of direct labor expense. Once the direct labor cost is computed for a given work category, the applicable indirect labor cost is determined by multiplying the direct labor expense by the appropriate percentage figure. Labor unit prices are used in Figure 3.6 to compute the direct labor costs of fabricating the abutment forms and for placing and stripping the abutment forms.

When labor unit costs are being used, care must be exercised to see that they are based on the appropriate levels of work productivity and the proper wage rates. Also, estimators must be very circumspect when using labor unit costs that they have not developed themselves. For the same work items, different estimators will include different expenses in their labor unit costs. It is never advisable to use a labor unit cost derived from another source without knowing exactly what categories of expense it does and does not include.

3.17 The Cost of Heavy Equipment

Unfortunately, the term *equipment* does not have a unique connotation in the construction industry. A common usage of the word refers to scaffolding, hoists, power shovels, paving machines, and other such items used by contractors to accomplish the work. However, the term *equipment* also is used with reference to various kinds of mechanical and electrical furnishings that become a part of the finished project, such as boilers, escalators, electric motors, and hospital sterilizers. In this text, *equipment* refers only to the contractor's construction equipage. The term *materials* will be construed to include all items that become a part of the finished structure, including the electrical and mechanical plant.

Equipment costs, like those of labor, are difficult to evaluate and price with precision. Equipment accounts for a substantial proportion of the total construction expense of most engineering projects but is less significant for buildings. When the nature of the work requires major items of equipment, such as earth-moving machines, concrete plants, and truck cranes, detailed studies of the associated costs must be made. Expenses associated with minor equipment items, such as power tools, concrete vibrators, and concrete buggies, are not normally subjected to detailed study. A standard expense allowance for each such item required is included, usually based on the duration it will be required on the job. The cost of small tools, wheelbarrows, water hoses, extension cords, and other such items are covered by a lump-sum allowance sometimes obtained as a small percentage of the total labor cost of the project.

Because estimating the cost of major equipment used on a job is typically done by the estimator and is highly specialized, depending on the industry sector and the way a specific company operates, project managers who would like a detailed analysis of pricing heavy equipment should reference specific texts that focus on heavy equipment costs. However, any construction project manager should keep in mind the fact that efficient use of heavy equipment on a job is paramount to attaining cost and schedule objectives. Early project planning should include analysis of exactly what equipment is best for the job and how this equipment is to be acquired: by using the company's own equipment, purchasing new equipment, or renting or leasing the equipment. Project managers do well to remember that company-owned equipment is not "free equipment." It represents an investment of capital by the company, and its use on the project should show a return on investment to the company for the time it is on the job.

3.18 Bids from Subcontractor

If the prime contractor intends to subcontract portions of the project to specialty contractors, the compilation and analysis of subcontractor bids is an important aspect of making up the final project estimate. Bids from subcontractors sometimes contain qualifications or stipulate that the general contractor is to be responsible for providing the subcontractor with certain job-site services, such as hoisting, electricity and water, storage facilities for materials, and many others. Before estimators can identify the low subcontractor bid (subbid) for any particular item of work, they must analyze each bid received to determine exactly what each such proposal includes and does not include. The checking of subbids can be a considerable chore when substantial portions of the project are to be sublet.

For the highway bridge project, the general contractor has made an advance decision that the painting will be subcontracted. Painting is a specialty area for which the general contractor is ill equipped and has had no past field experience. When such a decision is made, the contractor does not compile the cost of doing the work with its own forces. Rather, the lowest subcontract bid received from a responsible subcontractor will be included with the contractor's other expenses. The lowest acceptable painting subcontract bid received was $8,550 and will cover all such services as required by the project drawings and specifications.

The advance decision to subcontract certain areas of work does not necessarily mean that the general contractor will perform all of the other work with its own forces. Other specialty areas may also be subcontracted, depending on a number of circumstances. In this regard, the contractor may specifically request subbids from selected subcontractors, or it may merely await receipt of such bids that subcontractors voluntarily submit. In any event, the contractor must compile its own cost of doing the work involved and normally will be interested only in those subbids whose amounts are less than its own estimated direct cost.

When the general contractor receives a subbid whose amount is less than its own estimated direct outlay for doing the same work, it cannot accept such a subbid until consideration is given to several factors. Has the contractor had past experience with that subcontractor, and can it be expected to carry out its work properly? Does the subcontractor have a history of reliability and financial stability? Is the subcontractor experienced and equipped to do the type of work involved? Does the company have a good safety record? The general contractor must remember that it is completely responsible by contract with the owner for all subcontracted work as well as that performed by its own forces.

In compiling its bid for the bridge, the prime contractor received a subbid for Bid Item No. 8, Steel, reinforcing. Subcontract bids for reinforcing steel often include only the cost of labor. However, in this case, the subbid includes the prices of all materials and labor. The general contractor estimated its own direct expense of providing and placing reinforcing steel for the bridge to be $81,045. The subbid received for the same items was for $66,240, low enough to merit serious consideration. The general contractor has worked with this subcontractor before and has found the company to be honest, reliable, safe, and of good reputation. It is a complete bid and does not require the general contractor to provide the subcontractor with specific job-site services. The use of this subbid by the general contractor rather than its own estimated direct cost will reduce the contractor's bid by a significant amount. Consequently, the prime contractor decides to use the reinforcing steel subbid in the final compilation of its proposal to the owner. At this point, it should be noted that the bid item summary sheets have now been completed. From the price information contained in these summary sheets, the contractor will prepare its working project budget (see Section 3.25) if it becomes the successful bidder. At the moment, however, the estimating process must continue in order to obtain the necessary 12 bid unit prices.

3.19 Project Overhead

Overhead or indirect expenses are outlays that are incurred in achieving project completion but that do not apply directly to any specific work item. Two kinds of overhead pertain to a contractor's operations: project overhead and office overhead, the latter of which is discussed in the next section. Project overhead, also referred to as job overhead or field overhead, pertains to indirect field expenses that are chargeable directly to the project. Some contractors figure their job overhead outlay as a percentage of the total direct job cost; common values for the job overhead allowance range from 5 to 15 percent or more.

The use of percentages when computing field overhead is not generally considered to be good estimating practice because different projects can and do have widely varying job overhead requirements. The only reliable way to arrive at an accurate estimate of project overhead is to make a detailed analysis of the particular demands of that project. It is standard estimating procedure to list and price each item of indirect expense on a separate overhead sheet. Figure 3.7 is the project overhead sheet for the bridge project cost estimate. The job overhead amount of $69,868.40 was compiled using an estimated project duration of 15 weeks (from Section 3.10).

Project Overhead Estimate

Job: Highway Bridge
Estimator: GAS

Overhead Item	Calculations						Amount	Totals
Project manager	$5,500	x	3.5 mo	x	0.5 time	=	$9,625.00	
Project superintendent	$5,000	x	3.5 mo			=	$17,500.00	
Utilities								
Electricity	$225	/	mo					
Telephone	$350	/	mo					
Fax	$100	/	mo					
	$675	/	mo	x	3.5 mo	=	$2,362.50	
Utility installation charges	(job)						$1,520.00	
Facilities								
Job office	$350	/	mo					
2 ea. tool sheds	$600	/	mo					
Toilet	$125	/	mo					
	$1,075			x	3.5 mo.	=	$3,762.50	
Travel expense	$212	/	wk	x	15 wk	=	$3,180.00	
Water tank & water service	$80	/	wk	x	15 wk	=	$1,200.00	
Soil & concrete testing	$600	/	mo	x	3.5 mo	=	$2,100.00	
Scaffolding	$480	/	mo	x	1 mo	=	$480.00	
Trash removal	$190	/	mo	x	3.5 mo	=	$665.00	
Tire repair	$100	/	mo	x	3.5 mo	=	$350.00	
Photographs	$130	/	mo	x	3.5 mo	=	$455.00	
Computer	$140	/	mo	x	3.5 mo	=	$490.00	
					Subtotal of time-variable overhead expenses =			$43,690.00
Surveys	(job)						$1,600.00	
Project insurance	(job)						$1,164.00	
First aid	(job)						$220.00	
Sign	(job)						$570.00	
Reproductions	(job)						$400.00	
Fence	$2.00	/	ft	x	680 ft	=	$1,360.00	
Move in	See calculations, Appendix A						$14,155.20	
Clean up	See calculations, Appendix A						$6,709.20	
					Subtotal of time-constant overhead expenses =			$26,178.40
					Total project overhead =			**$69,868.40**

Figure 3.7
Highway bridge, overhead estimate

3.20 Home Office Overhead

Home office overhead includes general company expenses such as office rent, office insurance, heat, electricity, office supplies, furniture, telephone, legal costs, donations, advertising, office travel, association dues, and office salaries. The total of this overhead expense usually ranges from 2 to 8 percent of a contractor's annual business volume. An allowance for such indirect expense must be included in the cost estimate of each new project. Home office overhead is made up of charges that are incurred in support of the overall company construction program and that cannot be charged to any specific project. For this reason, home office overhead is normally included in the job estimate as a percentage of the total estimated project expense. The allowance for office overhead can be added as a separate line item in the cost estimate, or a suitable "markup" percentage can be applied, or a fee can be established that includes both home office overhead and profit.

The home office overhead is typically calculated annually. When an accounting cycle is completed and an audit is run for tax purposes, the total cost of overhead items is calculated. Targeted annual volume of work for the next year will be established. To determine the allowance for company overhead, the overhead cost will be divided by the targeted volume to yield the percent that must be added to each job to offset the company overhead cost. Since the cost of home office overhead tends to be stable independent of whether the actual volume is above or under the targeted volume, it should be noted that if volume falls short of the target, insufficient funds will be recovered to offset the overhead cost unless the percent added is increased. However, if actual volume exceeds targeted volume, excess funds will be recovered for home office overhead, and the percentage could actually be diminished or the excess funds otherwise invested.

3.21 Markup

On competitively bid projects, markup or margin is added at the close of the estimating process and is an allowance for profit plus possibly other items, such as office overhead and contingency. Regarding contingency as a separate component of markup is a matter of management philosophy. The profit included in a job bid represents the minimum acceptable return on the contractor's investment. Return on investment is a function of risk, and greater risk calls for a greater profit allowance in the proposal. Whether recognition of risk is in the form of a higher profit percentage or the inclusion of a contingency allowance seems to be a matter of personal preference.

Markup, which may vary from 5 percent to more than 20 percent of the estimated project cost, represents the contractor's considered appraisal of a whole series of imponderables that may influence its chances of being the low bidder and of its making a reasonable profit if it is. Many factors must be considered in deciding a markup figure, and each can have an influence on the figure chosen. The size of the project and its complexity, its location, provisions of the contract documents, the contractor's evaluation of the risks and difficulties inherent in the work, the competition, the contractor's desire for the work, the identity of the owner and/or the architect-engineer, and other intangibles can have a bearing on how a contractor marks up a particular job.

The contractor is required to bid under a form of contract that has been specifically written to protect the owner. In an attempt to afford the owner, and often the architect-engineer, protection against liabilities and claims that may arise from the construction process, the drafters of contract documents often incorporate a great deal of boilerplate, including disclaimers of one sort or another. The writers of contract documents sometimes force the contractor to assume liability for all conceivable contingencies, some of which are not subject to its control and which are not rightfully its responsibility. For example, some contracts may include provisions that absolve the owner from all claims for damages arising from project delay, even those caused through its own fault or negligence. Contractors frequently are made to assume full responsibility for any and all unknown physical conditions, including subterranean, that may be found at the construction site. It suffices to say that the markup figure selected must take into account the risks created by such contract provisions.

By adding the markup to the project cost, the estimator develops the project price or the bid price. This is the price that will be submitted to the owner in an effort to win the contract. When competitively bidding for projects, often it is useful to calculate project cost and project price individually and then determine whether the difference is adequate to cover the markup considerations already discussed. Project price is frequently determined by looking at current market conditions, such as the state of the local economy, the workload and backlog of competing contractors, the value of similar contracts awarded in recent months, and the number of bidders on the bid list. Using this method, the estimator is better able to determine the value of the service being provided, rather than the cost, and make the appropriate adjustments to the bid.

3.22 Contract Bonds

Many construction contracts, especially those involving public owners, require that the prime contractor provide the owner with a specified form of financial protection against contractor default. Two forms of surety bonds, called contract bonds, are used for this purpose. A contract bond is an agreement, the terms of which provide that a surety company will carry out the contractor's obligations to the owner if the contractor itself fails to do so. A surety is a party that assumes the legal liability for the debt, default, or failure in duty of another.

By the terms of its contract with the owner, the contractor accepts two principal responsibilities: to perform the objective of the contract and to pay all expenses associated with the work. Where contract bonds are called for, the general contractor is required to provide the owner with a performance bond and a labor and material payment bond. By the terms of these two bonds, the surety guarantees the owner that the work will be completed in accordance with the contract and that all construction costs will be paid should the contractor not perform as promised.

When the bidding documents provide that the successful contractor shall furnish the owner with performance and payment bonds, as is the case with the heavy civil example project, the contractor must purchase these bonds if it is awarded the contract. The contractor obtains these bonds from the surety company with which it customarily does business. Sureties are large corporate firms that specialize in furnishing many forms of surety bonds, including contract bonds for contractors. The premium charged for these bonds is substantial and varies with the type of work involved and the contract amount. This cost is paid by the contractor and must be included in the price estimate of the project. Because the bond outlay is based on the total contract amount, it is normally the last item of expense to be added into the project estimate, as shown in Figure 3.8.

3.23 Recap Sheet

The last computation for the project estimate is typically the recapitulation or recap sheet. This summarizes all of the estimated hard costs of the project and then provides the opportunity to add markups for the job. The recap sheet for the bridge is shown in Figures 3.10a.

For the highway bridge, in order to calculate the needed bid unit prices, all costs associated with the highway bridge are brought together in summary form on a recap sheet, which is shown in Figure 3.8. The expenses of labor, equipment, material, and subcontracts have been entered on the recap sheet from the summary sheets, an example of which is shown in Figure 3.6.

On the highway bridge recap sheet, the direct cost of the entire quantity of each bid item is obtained as the sum of its labor, equipment, material, and subcontract expenses. The sum of all such bid-item direct costs gives the estimated total direct cost of the entire project ($398,975). To this are added the job overhead, small tools, tax, markup, and the premium charge for the performance and payment surety bonds, giving the total price of $566,516. In Figure 3.8, the 15 percent markup includes an allowance for office overhead. Dividing the total project bid by the total direct project cost gives a factor of 1.4199. By multiplying the total direct cost of each bid item by this factor, the total amount of that bid item is obtained. Dividing the total bid cost of each bid item by its quantity gives the bid unit price. Bid unit prices customarily are rounded off to even figures.

The unit prices just computed have been obtained on the basis that each bid item includes its own direct cost plus its pro rata share of the project overhead, small tools, tax, markup, and bond. If these unit prices are now entered without change onto the bid form, this is called a *balanced bid*. For several reasons, a contractor may raise the prices on certain

Recap Sheet
Bid Date: April 25, 20XX

Job: Highway Bridge
Estimator: GAS

Item No.	Bid Item	Unit	Estimated Quantity	Labor Cost	Equipment Cost	Material Cost	Subcontract Cost	Direct Cost	Bid Total	Unit
1	Excavation, unclassified	cy	1,667	$3,569	$929	$0	$0	$4,497	$6,386	$3.83
2	Excavation, structural	cy	120	$3,731	$555	$0	$0	$4,286	$6,086	$50.72
3	Backfill, compacted	cy	340	$3,591	$663	$0	$0	$4,254	$6,041	$17.77
4	Piling, steel	lf	2,240	$21,390	$11,176	$72,279	$0	$104,845	$148,872	$66.46
5	Concrete, footings	cy	120	$3,661	$685	$12,186	$0	$16,532	$23,475	$195.62
6	Concrete, abutments	cy	280	$37,459	$7,104	$30,542	$0	$75,105	$106,644	$380.87
7	Concrete, deck slab, 10 in.	sy	200	$11,745	$1,790	$8,020	$0	$21,555	$30,606	$153.03
8	Steel, reinforcing	lb	90,000	$0	$0	$0	$66,240	$66,240	$94,056	$1.045
9	Steel, structural	lb	65,500	$4,220	$1,680	$73,688	$0	$79,588	$113,009	$1.725
10	Bearing plates	lb	3,200	$1,943	$420	$3,250	$0	$5,613	$7,970	$2.49
11	Guardrail	lf	120	$1,664	$420	$5,826	$0	$7,910	$11,231	$93.59
12	Paint	ls	job	$0	$0	$0	$8,550	$8,550	$12,140	$12,140.39
	Totals			$92,974	$25,422	$205,790	$74,790	$398,975	$566,516	

Factor = $566,516 / $398,975 = 1.4199

Job overhead	$69,868
	$468,844
Small tools (5% of labor)	$4,649
	$473,492
Tax 3%	$14,205
	$487,697
Markup 15%	$73,155
	$560,852
Bonds	$5,665
Total Project Bid	$566,516

Figure 3.8
Highway bridge, recap sheet

bid items and decrease the prices on others proportionately so that the bid amount for the total job remains unaffected. This is called an *unbalanced bid*. It is assumed here that the contractor will submit a balanced bid. Likewise, the unit prices determined in Figure 3.8 are now entered in column 5 of Figure 3.9, the unit-price schedule of the bid form. The total estimated amounts in column 6 of Figure 3.9 are obtained by multiplying the unit prices just entered (column 5) by the estimated quantities (column 4). Because of the rounding off of unit prices, the project bid total is slightly different on the unit-price schedule of the bid form ($566,483.90; Figure 3.9) from what it is on the recap sheet ($566,516; Figure 3.8).

When unit-price proposals are submitted to the owner, the low bidder is determined on the basis of the total estimated amount. Consequently, for contract award purposes, the amount of $566,483.90 in Figure 3.9 is treated just as a lump-sum bid. In cases of error in multiplication or addition by the contractor in obtaining the estimated amounts in column 6 of Figure 3.9, it is usual for the unit prices to control and the corrected total sum to govern.

3.24 Cost Models

The project estimate is the most detailed and complete cost model of the project; however many other cost models are used by the project manager to track and manage costs throughout the project. They include:

- ❑ Project budget used as a basis for the cost-control system.
- ❑ Project cost breakdown or schedule of values defining how the client will be billed throughout the project (Chapter 12, "Project Financial Management").

		Unit Price Schedule			
Item No. [1]	Description [2]	Unit [3]	Estimated Quantity [4]	Unit Price [5]	Estimated Amount [6]
1	Excavation, unclassified	cy	1,667	$3.83	$6,384.61
2	Excavation, structural	cy	120	$50.72	$6,086.40
3	Backfill, compacted	cy	340	$17.77	$6,041.80
4	Piling, steel	lf	2,240	$66.46	$148,870.40
5	Concrete, footings	cy	120	$195.62	$23,474.40
6	Concrete, abutments	cy	280	$380.87	$106,643.60
7	Concrete, deck slab, 10 in.	sy	200	$153.03	$30,606.00
8	Steel, reinforcing	lb	90,000	$1.045	$94,050.00
9	Steel, structural	lb	65,500	$1.725	$112,987.50
10	Bearing plates	lb	3,200	$2.49	$7,968.00
11	Guardrail	lf	120	$93.59	$11,230.80
12	Paint	ls	job	$12,140.39	$12,140.39
				Total Estimated Amount =	$566,483.90

Figure 3.9
Highway bridge, completed bid form

- Progress estimates used as a basis for monthly billings.
- Change order estimates evaluating the cost of additional (or eliminated) work.
- Project cash flow used to track the flow of money into and out of the project (Chapter 12, "Project Financial Management").
- Project completed estimate, a key document for the postproject review that defines how much the project actually cost and how the money was spent.

These cost models will be elaborated upon later in this chapter or later in the book.

3.25 Project Budget

When the contractor is recognized as the successful bidder, the estimate must be restructured into a more suitable format for subsequent cost control of the actual construction work. This involves the preparation of the control budget or the project budget, which is the detailed schedule of expenses that the project manager will use for cost-control purposes during the construction phase. Converting the project estimate developed as a basis to bid or negotiate the job to the project budget used to actually do the work is an excellent way for the project manager to become very familiar with the project estimate and to take ownership for the project estimate. During this process, the project manager may work with the estimator to resolve any discrepancies between figures in the project estimate and the project manager's determination of what is required by the construction documents.

Figure 3.10a is the control budget for the highway bridge project. The work quantities and prices contained in this figure have been extracted from the bid-item summary sheets previously discussed and presented. In Chapter 11, "Project Cost System," where cost accounting will be addressed, the actual construction expenses of the highway bridge will be compared with the programmed costs contained in the project budget in Figure 3.10a. As is discussed in Chapter 11, "Project Cost System," unit prices are especially useful for making quick and meaningful comparisons of actual and budgeted expense for both labor and equipment. Unit costs of materials are not especially significant except for estimating purposes and are not shown in Figure 3.10a.

Figure 3.10b shows the control budget for the commercial building project. This is a summary estimate of costs without the normal breakdown into labor, material, and equipment, modified from the original estimate to reflect anticipated field expenditures. Even so, it is much more extensive than the bridge estimate, since even a simple building involves a far greater variety of work items than a simple bridge. Because of the

Project Budget

Job: Highway Bridge
Estimator: GAS

Cost Code	Work Type	Quantity	Unit	Direct Labor Cost	Labor Unit Cost	Equipment Cost	Equipment Unit Cost	Material Cost
	General Requirements							
01500.00	Move in	ls	1	$8,112	$8,112	$3,204	$3,204	$0
01700.00	Clean up	ls	1	$4,248	$4,248	$1,240	$1,240	$0
			Subtotals	$12,360		$4,444		$0
	Sitework							
32220.10	Excavation, unclassied	1,667	cy	$2,643	$1.59	$929	$0.56	$0
32222.10	Excavation, structural	120	cy	$2,985	$24.88	$555	$4.63	$0
32226.10	Backfill, compacted	340	cy	$2,873	$8.45	$663	$1.95	$0
31350.00	Piledriving rig, mobilization & demobilization	job	ls	$6,528	$6,528.00	$5,448	$5,448.00	$300
31361.10	Piling, steel, driving	2,240	lf	$8,224	$3.67	$5,728	$2.56	$71,979
			Subtotals	$23,253		$13,323		$72,279
	Concrete							
03150.10	Footing forms, fabricate	360	sf	$936	$2.60	$0	$0.00	$392
03150.20	Abutment forms, prefabricate	1,810	sf	$3,548	$1.96	$0	$0.00	$2,153
03157.10	Footing forms, place	720	sf	$360	$0.50	$0	$0.00	$202
03159.10	Footing forms, strip	720	sf	$158	$0.22	$0	$0.00	$0
03157.20	Abutment forms, place	3,620	sf	$7,312	$2.02	$2,520	$0.70	$1,014
03159.20	Abutment forms, strip	3,620	sf	$3,511	$0.97	$840	$0.23	$0
03157.30	Deck forms, place	1,800	sf	$3,190	$1.77	$420	$0.23	$1,890
03159.30	Deck forms, strip	1,800	sf	$1,460	$0.81	$210	$0.12	$0
03251.30	Concrete, deck, saw joints	60	lf	$360	$6.00	$110	$1.83	$0
03311.10	Concrete, footings, place	120	cy	$1,200	$10.00	$685	$5.71	$11,592
03311.20	Concrete, abutments, place	280	cy	$9,088	$32.46	$3,744	$13.37	$27,048
03345.30	Concrete, deck, place & screed	200	sy	$1,200	$6.00	$980	$4.90	$5,410
03346.20	Concrete, deck, finish	1,800	sf	$2,250	$1.25	$70	$0.04	$630
03370.20	Concrete, abutments, rub	1,960	sf	$3,762	$1.92	$0	$0.00	$235
03370.20	Concrete, abutments, curing	3,820	sf	$244	$0.06	$0	$0.00	$92
03370.30	Concrete, deck, curing	1,800	sf	$108	$0.06	$0	$0.00	$90
			Subtotals	$38,689		$9,579		$50,747
	Metals							
05120.00	Steel, structural, place	65,500	lb	$2,911	$0.04	$1,680	$0.03	$73,688
05520.00	Guardrail	120	lf	$1,216	$10.13	$420	$3.50	$5,826
05812.00	Bearing plates	3,200	lb	$1,340	$0.42	$420	$0.13	$3,250
			Subtotals	$5,467		$2,520		$82,764

Figure 3.10a
Highway bridge, project budget

Project Budget		
Work Type	**Estimated Cost**	**Subtotals**
Utilities & Sitework		$292,357
Mobilize	$23,780	
Utilities	$61,000	
Site Electrical	$30,000	
Sitework	$48,500	
Paving	$93,221	
Landscape	$25,681	
Irrigation	$10,175	
Structure		$265,585
Drilled Piers	$14,000	
Grade Beams and Slabs	$82,747	
Structural Steel	$139,500	
Miscellaneous Steel	$29,338	
Envelope		$336,789
Roofing	$83,797	
Dampproofing	$6,000	
Insulation	$7,000	
Masonry	$60,580	
Aluminum Frames	$19,463	
Storefront & Glazing	$104,149	
Plaster	$55,800	
Finishes		$338,892
Framing	$60,000	
Gypsum Drywall	$120,000	
Millwork	$28,797	
Wood Doors	$10,331	
Ceilings	$15,000	
Ceramic Tile	$23,500	
Flooring	$36,900	
Paint	$36,000	
Signage	$2,651	
Toilet Partitions	$2,487	
Toilet Accessories	$3,226	
Systems		$437,971
Plumbing	$88,685	
HVAC	$99,986	
Electrical	$187,500	
Fire Protection	$42,300	
Fire Alarm	$19,500	
Closeout		$20,000
	$20,000	
Project Total		**$1,691,594**

Figure 3.10b
Commercial building project budget

complexity of the example building project, detailed cost analysis will not be provided.

3.26 Progress Estimate

The project manager may or may not be involved in developing the project estimate. However, the project manager will assume responsibility for the estimate once the project has been handed off from the estimator. As previously indicated, once the project has been awarded to the company, the project estimate will be restructured to form the basis of the cost-control system for the project. Then, as the project progresses, typically on a monthly basis, a new estimate will be made of work that has been completed during the month. This will be used as a basis for billing, which has a critical influence on the cash flow of the project, as well as the company cash flow. This progress estimate is the responsibility of the project manager. Accurately determining the cost of work completed is critical. If the amount is underestimated, the contractor will not recover costs actually incurred during the month and the project cash flow will be negative, requiring the company to finance part of the work. If the amount is overestimated, it is likely that the monthly pay request will be rejected and returned for resubmittal of a correct amount, or simply turned down for the month. As a result, again, the company ends up financing even more of the work. This will be further elaborated upon in Chapter 12, "Project Financial Management."

3.27 Estimating for Changes

A fundamental characteristic of construction projects is that they are constantly changing. There are many causes for changes, including differing site conditions, changes in owner requirements, or recognizing ways to improve the design. The source of changes can be the site itself, the owner, the designer, or even the contractor. Unless the change is the result of contractor error, the cost of changes will normally be borne by the project owner. Unless the contractor is at fault, any change should result in fair compensation to the contractor. This requires a detailed and defensible estimate by the contractor of the actual cost of the change, including its impacts. It also requires an estimate of any additional time required by the change. Some contracts on projects that are highly sensitive to completion and/or milestone dates are written such that there will be no additional time provided for changes. In that situation, the cost of the change should reflect the cost to expedite work so that any delay caused by the change can be offset by such tactics as working overtime or adding workers in order to maintain the original dates.

Change order estimating is a highly specialized skill required of any project manager. The project manager must be able to anticipate any exceptional costs, such as restocking small quantities of material, special material handling, or disturbances due to reassignment of work in the field, all superimposed on the standard unit costs for the work in question. Detailed information is required from the field supervisor about actual working conditions and how the change will disturb ongoing operations.

The change order estimate must be prepared on a timely basis so as not to disturb either ongoing operations or project cash flow. Change work cannot be executed until it has been signed off, which often requires an estimate of the cost of the change. Therefore, an estimate is often required prior to starting the change work, which generally involves an ongoing operation. The change work will incur an added and typically immediate cost to the project that can only be recuperated after a formal modification to the contract has been negotiated and signed off. Estimates for change work are often subject to a high level of scrutiny by the owner and designer and so must be accurate and defensible. Though a contested change order may well be paid in full eventually, delays in collecting the payment will have a negative impact on the project cash flow.

Much work in the heavy civil area is based on unit costs, which makes the estimation of the cost of changes much easier. If the volume of the changed work can be accurately estimated, the adjustment for more or less work is easily accounted for, since the price is equal to quantity times unit cost. Where this becomes a problem in unit cost work is that the deduct unit cost typically is less than the add unit cost, so anticipating changes in unit cost work requires two sets of costs: one for additional units and the other for a reduction in units.

Key Points and Questions

Key Points
- The project manager works with many cost models throughout the project.
- Preliminary estimates provide a quick approximation of the project cost.
- The project manager may not be involved in developing the project estimate but becomes responsible for it when the project is handed off.
- Progress and change estimates are developed throughout the project by the project manager.

Review Questions and Problems

1. What are the primary ways in which project managers use estimates throughout the project?
2. Which type of preliminary estimate would be the best in each of the following situations, and why did you choose it?

a. Decide whether it is feasible to proceed with constructing a new small office building.

b. Establish the budget for a bond issue to build three new elementary schools.

c. Evaluate which of three alternate designs best fits the budget for a new clinic.

3. Why is an accurate quantity survey essential to the project estimate?
4. What is the best source for labor and equipment unit prices that will be used in developing an estimate?
5. If, halfway through the year, you determine that the projected volume of work for the year will be significantly below the targeted annual volume, what adjustment should be made to the percentage markup for company overhead added on to project estimates?
6. What impact will an inaccurate progress estimate have on cash flow?
7. What are some of the reasons change order work costs more than work in the original scope?
8. What is the intended use for each of the following cost models?

a. Project budget

b. Project cost breakdown or schedule of values

c. Progress estimates

d. Change order estimates

e. Project cash flow

f. Final estimate

4 Project Planning

4.1 Introduction

Project planning is central to project management. Planning takes place at all stages. When owners want a new facility, they put together a plan to acquire what they need. The plan is typically very simple in concept (though it may be quite complex in execution), consisting of a few steps:

- ❑ Identify the need as clearly as possible.
- ❑ Determine a budget and completion date.
- ❑ Bring together a team that can design and build the facility.
- ❑ Monitor the process throughout the project's schedule.

Successive participants in the process—designers, contractors, specialty contractors, material suppliers—all plan, often with much greater detail but with limited scope, in order to execute their part of the project.

This chapter begins a series of three chapters primarily focused on time control, but related to all the elements of project management. Planning, the subject of this chapter, is essential to any task, whether it be management oriented or focused on execution in the field. The product of the plan is often a schedule (Chapter 5, "Project Scheduling Concepts"), which demonstrates its utility through the numerous ways in which it is applied (Chapter 6, "Production Planning"). As readers move through the book, they will see that the planning process, resulting in the project schedule, is what ties all of the elements of project management together.

Learning objectives for this chapter include:
- Understand the critical importance of planning.
- Be able to develop a workable plan for a construction project.
- Learn the steps in developing a precedence network.
- Be able to determine the level of detail required for the particular plan being developed.

4.2 CPM Procedure

Construction time control is a difficult, time-consuming, and arduous management function. Project managers work within an extremely complex and shifting time frame, and they need a management tool that will enable them to manipulate large numbers of job activities and complicated sequential relationships in a simple and understandable fashion. The Critical Path Method (CPM) is just such an expedient and constitutes the basis for the ensuing treatment of project time control. This method applies equally well to all construction work, large and small, intricate and straightforward.

The management techniques of CPM are based on a graphical project model called a network. This network presents in diagrammatic form those job activities that must be carried out and their mutual time dependencies. A diagram of this type is a simple and effective medium for communicating complex job interdependencies. It serves as a basis for the calculation of work schedules and provides a mechanism for controlling project time as the work progresses.

CPM is a three-phase procedure consisting of planning, scheduling, and time monitoring or controlling. Planning construction operations involves the determination of what must be done, how it is to be performed, and the sequential order in which it will be carried out. Scheduling determines calendar dates for the start and completion of project components. Time monitoring is the process of comparing actual job progress with the programmed schedule.

4.3 Planning Phase

Planning is the process of devising of workable scheme of operations that, when put into action, will accomplish an established objective. The most time-consuming and difficult aspect of the job management system—planning—is also the most important. It requires an intimate knowledge of construction methods combined with the ability to visualize discrete work elements and to establish their mutual interdependencies. If planning were to be the only job analysis made, the time would be well spent. CPM planning involves a depth and thoroughness of study that gives the construction team an invaluable understanding and appreciation of job requirements.

Construction planning, as well as scheduling, must be done by people who are experienced in and thoroughly familiar with the type of fieldwork involved. Significant learning takes place during the planning phase of a project. Therefore, the people doing the planning are in the best position to manage the work. The project network and the management data obtained from it will be realistic and useful only if the job plan is produced and updated by those who understand the job to be done, the ways in which it can be accomplished, and the job-site conditions. Those executing the work in the field are most likely to be committed to ensuring that the work is done according to the plan if they participate in the planning process.

To construct the job network, information must be sought from many sources. Guidance from key personnel involved with the project, such as estimators, the project manager, the site superintendent, and the field engineer, can be obtained from a planning meeting or perhaps a series of meetings. The network serves as a medium whereby the job plan can be reviewed, criticized, modified, and improved. As problems arise, consultations with individuals can clear up specific questions. The important point here is the need for full group participation in the development of the network, and collective views must be solicited.

Participation by key subcontractors and suppliers is also vital to the development of a workable plan. Normally, the prime contractor sets the general timing reference for the overall project. Individual subcontractors then review the portions of the plan relevant to their work and help develop additional details pertaining to their operations. An important side effect is that this procedure brings subcontractors and the prime contractor together to discuss the project. Problems are detected early, and steps toward their solutions are started well in advance.

It must be recognized that the project plan represents the best thinking available at the time it is conceived and implemented. However, no such scheme is ever perfect, and the need for change is inevitable as the work progresses. Insight and greater job knowledge are acquired as the project evolves. This increased cognizance necessarily results in corrections, refinements, and improvements to the operational plan. The project program must be viewed as a dynamic device that is continuously modified to reflect the progressively more precise thinking of the field management team.

Construction planning may be said to consist of five steps:

1. A determination of the general approach to the project.
2. Breakdown of the project into job steps or "activities" that must be performed.
3. Ascertainment of the sequential relationships among these activities.

4. Graphic presentation of this planning information in the form of a network.
5. Endorsement by the project team.

Two different planning methodologies are presented in this chapter: beginning-to-end planning and top-down planning. Beginning-to-end planning breaks the job into steps or activities, starting with mobilization of the project, and proceeds step-by-step through the project to completion. This method presumes some level of detail from the beginning or starts with limited detail and adds detail as planning proceeds.

Top-down planning, sometimes referred to as work breakdown structure, starts with the overall project, breaking it into its major pieces, then breaking the major pieces into their component pieces. This process continues until the pieces are of sufficient detail to satisfy the complexity of the project. Both methods arrive at the same result: job activities that can be used to form a graphical logic diagram.

4.4 Job Activities

The segments into which a project is subdivided for planning purposes are called *activities*. An activity is a single continuous work step that has a recognizable beginning and end and requires time for its accomplishment. The extent to which a project is subdivided depends on a number of practical considerations, but these eight are suggested as guidelines for use when activities are being identified:

1. *By area of responsibility*, where work items done by the general contractor and each of its subcontractors are separated.
2. *By category of work* as distinguished by craft or crew requirements.
3. *By category of work* as distinguished by equipment requirements.
4. *By category of work* as distinguished by materials such as concrete, timber, or steel.
5. *By distinct structural elements* such as footings, walls, beams, columns, or slabs.
6. *By location on the project* when different times or different crews will be involved.
7. *With regard to the owner's breakdown* of the work for bidding or payment purposes.
8. *With regard to the contractor's breakdown* for estimating and cost accounting purposes.

The activities used may represent relatively large segments of a project or may be limited to small steps. For example, a reinforced concrete wall may be only a single activity, or it may be broken down into erect outside forms, tie reinforcing steel, erect inside forms and bulkheads, pour concrete, strip forms, and cure. Trial and error together with experience are the best guides regarding the level of detail needed. What is suitable for one project may not be appropriate for another. Too little detail will limit planning and control effectiveness. Too much will inundate the project manager with voluminous data that tend to obscure the significant factors and needlessly increase the cost of the management system. As a rule of thumb, the project manager should schedule to the same level of detail as he plans to exercise management control.

Not only is network detail a function of the individual project; it is also highly variable depending on who will be using the information. For example, the project general contractor might show the installation of a containment vessel on a nuclear power plant as a single activity. However, the subcontractor responsible for moving and installing this enormously heavy vessel may require a complete planning network to accomplish the task. Planning detail also varies with the level of project management involved. This matter is further discussed in Sections 4.16 and 4.17.

4.5 Job Logic

Job logic refers to the determined order in which the activities are to be accomplished in the field. The start of some activities obviously depends on the completion of others. A concrete wall cannot be poured until the forms are up and the reinforcing steel has been tied. Yet many activities are independent of one another and can proceed concurrently. Much job logic follows from well-established work sequences that are usual and standard in the trade. Nevertheless, for a project of any consequence, there is always more than one general approach, and no unique order of procedure exists. It is the planner's responsibility to winnow the workable choices and select the most suitable alternatives. At times, doing this may require extensive studies, including the preparation of a separate network for each different approach to the work. Herein lies one of the major strengths of the CPM planning method: It is a versatile and powerful planning tool that is of great value in the time evaluation of alternative construction procedures.

Showing the job activities and their order of sequence (logic) in pictorial form produces the project network, which is a graphical display of the proposed job plan. In general, job logic is not developed to any extent ahead of the network. Rather, job logic evolves in a natural fashion as the job is discussed and the network diagram progresses.

4.6 Restraints

To be realistic, a job plan must reflect the practical restraints or limitations of one sort or another that apply to most job activities. Such restraints stem from a number of practical considerations. The restrictions of job logic itself might be called physical restraints. Placing forms and reinforcing steel might be thought of as restraints to pouring concrete. Such normal restrictions arise from the necessary order in which construction operations are physically accomplished and are simply a part of job logic previously discussed. However, there are other kinds of restraints that need to be recognized.

Reinforcing steel, for example, cannot be tied and placed until the steel is available on the site. Steel availability, however, depends on the approval of shop drawings, steel fabrication, and its delivery to the job site. Consequently, the start of an activity that involves the placing of reinforcing steel is restrained by the necessary preliminary actions of shop drawing approval, steel fabrication, and delivery. Practical limitations of this sort on the start of some job activities are called *resource restraints,* or more particularly in this case, *material restraints.* Another common example of a job restraint is an equipment restraint, where a given job activity cannot start until a certain piece of construction equipment becomes available. Other instances of job restraints are availability of special labor skills or crews, delivery of owner-provided materials, receipt of completed project design documents, and turnover of owner-provided sites or facilities. At times there may be safety restraints, especially on the sequencing of structural operations on multistory buildings. The recognition and consideration of job restraints is an important part of job planning. Failure to consider such restraints can be disastrous to an otherwise adequate job plan.

Some restraints are shown as time-consuming activities. For example, the preparation of shop drawings and material fabrication and delivery are material restraints that require time to accomplish and are depicted as activities on project networks. Restraints are also shown in the form of dependencies between activities. If the same crane is required by two activities, an equipment restraint is imposed by having the start of one activity depend on the finish of the other. These types of restraints can be handled in two different ways. Since historically these networks were prepared and all calculations made by hand, such restraints were entered into the network as logical relationships. This method of "hard-wiring" restraints makes the process of updating and changing the schedule more arduous because the reordering of resource priorities on the project must always be accompanied by the rewiring of all such restraint relationships. In recent years, sophisticated computer scheduling programs have enabled the scheduler to present these restraints as resource utilization requirements with priority heuristics to determine which activity gets the use of limited resources first in the event of competing needs. This method of handling restraints is significantly more powerful and saves schedule update time on projects where changes occur frequently.

4.7 Beginning-to-End Planning

For many practitioners, projects are visualized in the order in which they will be built. Therefore, in planning a project, it is natural to start with mobilization and proceed step-by-step through the project to final inspection. This method of planning from beginning to end typically requires a predetermination of who will be using the network and the level of detail they will require.

In the case of the highway bridge, a project team would probably begin with the field operations, starting with "Move in," followed by "Excavation of abutment #1," then "Pile driving abutment #1." Soon it would become obvious that a series of procurement activities would be required to provide the materials necessary for construction. "Order & deliver piles," "Prepare & approve shop drawings for footing rebar," and other procurement activities would be added.

For the commercial building, field operations would begin with "Mobilization," that is, moving onto the job site to begin setting up temporary services and facilities. This would be followed by "Site work" to prepare the site for constructing the facility, followed by "Structural excavation for the foundations" and "Forming and reinforcing for the foundations." At this point, procurement becomes a restraint, requiring completing the fabrication of reinforcing steel for the foundations and delivery of such to the job site before reinforcing steel can be placed.

Although intuitive, beginning-to-end planning has some drawbacks. Some projects are so complex that it is difficult to visualize the best way to sequence the construction operations. Team members may not agree on the level of detail required. There is also a tendency to use greater detail in the earlier stages of the job and to begin to generalize in the later stages. In other instances, the planning team may leave out entire groups of vital operations in their forecast and hence disastrously underestimate the time required for completion of the project. For these reasons and others, top-down planning has significant advantages and is preferred by the industry.

4.8 Top-Down Planning and the Work Breakdown Structure

Top-down planning starts with an examination of the project from an overall perspective. In the case of the heavy civil example project, it would be viewed first as a dam project. The dam project is made up of a number of major segments: the earth dam, highway bridge, and pipeline relocation. The earth dam is made up of a number of segments including borrow pit development, haul road construction, river diversion, and others.

For the commercial building, the overall project might be considered in terms of site preparation, building construction, and tenant improvements. Depending on where the site is, site preparation might include infrastructure development, demolition of existing structures, and/or preparation of the building pad. For the example commercial building project, we will consider that the building site has already been developed with the infrastructure (roads and utilities) to support commercial development. The project will begin with mobilization and site work to prepare the site for the specific building designed for this site. Tenant improvements, including such items as installation of cubical partitions and office furniture, will begin after our construction project has been completed.

Top-down planning is often accomplished using a project outline called a work breakdown structure (WBS), which can be made manually or with the use of a computer outliner in the scheduling software, a spreadsheet, or a word processing program. In a collaborative environment, the team will often use a whiteboard to accumulate the consensus decisions or a sticky note process that allows each participant to place their activities on the board in relation to other activities. Regardless of the size or complexity of a project, it generally can be broken down into its major components. These principal job elements form the first level of the WBS.

In the case of the highway bridge, the six major components are:

1. Procurement
2. Field mobilization and site work
3. Pile foundations
4. Concrete abutments and wing walls
5. Deck
6. Finishing operations

These are the six components shown in the highway bridge general time schedule in Figure 3.5a. In turn, each of these major project segments can be broken down into its subcomponents to form the second level of the WBS. For example:

4. Concrete abutments and wing walls
    4.1 Abutment #1
    4.2 Abutment #2

Similarly, each of these can be further subdivided.

4.1 Abutment #1
    4.1.1 Forms and rebar abutment #1
    4.1.2 Pour abutment #1
    4.1.3 Strip and cure abutment #1
    4.1.4 Backfill abutment #1
    4.1.5 Rub concrete abutment #1

A similar analysis can be made for the commercial building project. Figure 3.5b gives the general time schedule for the commercial building project and has the following components:

1. Utilities and Site Work
2. Structure
3. Envelope
4. Finishes
5. Systems
6. Closeout

In turn, each of these major project segments can be broken down into its subcomponents to form the second level of the WBS. For example, for the structure:

2. Structure
    2.1 Drilled Piers
    2.2 Grade Beams and Slabs
    2.3 Structural Steel
    2.4 Miscellaneous Steel

Each of these can be further broken down, as, for example, the grade beams and slabs:

2.2 Grade Beams and Slabs
    2.2.1 Final grade for slabs
    2.2.2 Excavate grade beams
    2.2.3 Place reinforcing steel
    2.2.4 Pour concrete

It should be noted that the soil conditions are such that the grade beams can be formed directly in the earth and wood or metal forms are not required. Earth forming the grade beams allows them to be cast monolithically with the slabs, as indicated on the drawings.

A complete, high-level WBS for each of the two example projects is provided on the companion website. Chart 4.1a represents the example bridge project outline, and Chart 4.1b represents the example commercial building project outline.

The WBS has a number of other applications in addition to facilitating the planning process. In Chapter 7, "Managing Time," the WBS is used as the basis of a hierarchy of reports and bar charts, and in Chapter 11, "Project Cost System," it is used for cost reporting.

It should be apparent from the discussion that, as planning moves forward, different project teams would break down the work to reveal further detail. For example, in the heavy civil project, the earth dam construction team would continue to subdivide the WBS on the dam while the highway bridge team worked on the bridge. For the building project, the general contractor would provide a high-level WBS for the entire project, while each of the contractors responsible for specialty areas, such as the concrete contractor or the electrical contractor, would develop a detailed breakdown of their area. This process of developing more detail in the WBS will continue to include the construction crafts that will plan their weekly work schedules based on their continued breakdown of the work.

The project team that will oversee the work for a given area should develop the WBS structure for that area. Collaboration works very well in this process, allowing the preparation of the WBS to move along efficiently with the contribution of specialists from each of the various specialty areas. As can be seen from the examples just cited, the WBS develops quite naturally from basic project components through a number of levels to activity-sized segments of the project.

When the WBS is fairly complete, the arrangement of the activities and their logical dependencies constitute the next step. Here again, it is the responsibility of the project management team to determine the necessary order of activity accomplishment. It is during this type of planning meeting that the project management team begins to visualize the project as a whole and the manner in which the individual pieces fit together. For example, with the bridge project, one team member may suggest that one of the abutments needs to be backfilled before the deck girders are delivered so that the crane can unload the girders from the truck and put them directly into place. Another member may propose that both abutments be backfilled before the deck concreting operations start so that the concrete trucks can reach both ends of the bridge deck. This would enable much of the concrete to be placed directly from the trucks. The person drawing the network, probably using a computer and a projector, will record each of the suggestions made. In such a process, the planning team can review the information continually and make informed decisions regarding the project construction plan. In addition to generating the planning information, the project management group begins to think and act as a team.

4.9 Precedence Notation

Two symbolic conventions have traditionally been used to draw construction networks. The convention first used depicted activities as connected arrows separated by circles or nodes. This method, known as arrow notation or activity on arrow (AOA) notation, has largely been phased out with the growth in popularity of the more flexible precedence notation or activity on node (AON), where activities are depicted by boxes and linked together with arrows. Precedence diagrams have several important advantages over arrow diagrams, simplicity being the primary one.

Other advantages of precedence diagrams are discussed in Section 4.12. Because of these advantages, precedence notation will be emphasized in this text. However, since older schedules are still encountered, examples of AOA networks will be found on the companion website to enable the interested reader to be able read an AOA schedule when it is encountered.

In drawing a precedence network, each time-consuming activity is portrayed by a rectangular box. The dependencies between activities are indicated by dependency or sequence lines connecting one activity to another. The identity of the activity and a considerable amount of other information pertaining to it are entered into its rectangular box. This matter will be further developed as the discussion progresses.

4.10 Precedence Diagram

The preparation of a realistic precedence diagram requires time, effort, and experience with the type of construction involved. The management data extracted from the network will be no better than the diagram itself. The diagram is the key to the entire time control process. When the network is first being developed, the planner must concentrate on job logic. The only consideration at this stage is to establish a complete and accurate picture of activity dependencies and interrelationships. Restraints that can be recognized at this point should be included. The time durations of the individual activities are not of concern during the planning stage. Additionally, it is assumed that the labor and equipment availability is sufficient to cover demands. Where advanced recognition of restraints and resource conflicts are apparent, the situation should be noted for further consideration later. Matters of activity times and resource conflicts will be considered in detail later during project scheduling.

To illustrate the mechanics of network diagramming, consider a simple project, such as the construction of a pile-supported concrete footing. The sequence of operations will be excavation, building of footing forms, and the procurement of piles and reinforcing steel. These opening activities can begin immediately and can proceed independently of one another. After the excavation has been completed and the piles are delivered, pile driving can start. Fine grading will follow the driving of piles, but the forms cannot be set until the piles are driven and the forms have been built. The reinforcing steel cannot be placed until fine grading, form setting, and rebar procurement have all been carried to completion. Concrete will be poured after the rebar has been placed, and stripping of the forms will terminate this sequence. In elementary form, this is the kind of information generated as the planning group analyzes a project. The precedence diagram describing the prescribed sequence of activities is shown in Figure 4.1.

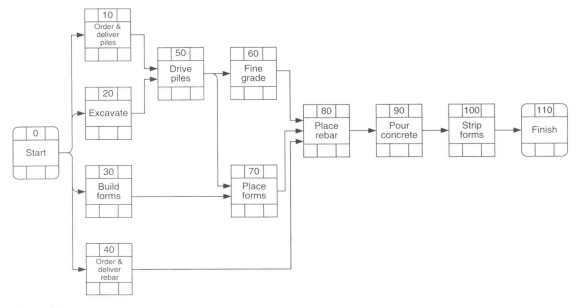

Figure 4.1
Concrete footing, precedence diagram

Each activity in the network must be preceded either by the start of the project or by the completion of a previous activity. Each path through the network must be continuous with no gaps, discontinuities, or dangling (detached or dead-end) activities. Consequently, all activities must have at least one activity that follows, except the activity that terminates the project. It is standard practice, as well as being a requirement of some computer programs, that precedence diagrams start with a single opening activity and conclude with a single closing activity. In this regard, "Start" and "Finish" activities appear in precedence diagrams throughout this text. These two zero time figures, referred to as milestones, are shown herein with rounded corners to distinguish them from the usual time-consuming activities.

The numbering of network activities, as shown in Figure 4.1, is not standard practice for hand-drawn networks but is generally required for those developed using computers. Activity numbers are used in this text for the purpose of easy and convenient activity identification. When network activities are numbered, each activity should have a unique numerical designation, with the numbering proceeding generally from project start to project finish. Leaving gaps in the activity numbers is desirable so that numbers are available for subsequent refinements and revisions. In this text, the network activities are given identifying names and are numbered by multiples of 5 or 10.

4.11 Network Format

A horizontal diagram format has become standard in the construction industry. The general synthesis of a network is from start to finish, from project beginning on the left to project completion on the right. The sequential relationship of one activity to another is designated by the dependency lines between them. In the usual precedence diagram, the length of the lines between activities has no significance because they indicate only the dependency of one activity on another. Arrowheads are not always shown on the dependency lines because of the obvious left-to-right flow of time. However, arrowheads are shown herein on precedence diagrams for additional clarity.

During initial diagram development, the network is sketched emphasizing activity relationships rather than the appearance or style of the diagram. Corrections and revisions are plentiful. The rough-and-ready appearance of the first version of the diagram is of no concern; its completeness and accuracy are. The finished diagram can always be put into a more tidy form at a later date.

To identify activities, we recommend writing out the description of each activity rather than using symbols. For instance, "PR" might be used for "Place rebar" or "PC" for "Pour concrete." Even mnemonic codes such as this make a diagram difficult to read, causing the user to spend a considerable amount of time consulting the symbol listing to identify activities. Networks are much more intelligible and useful when each activity is clearly identified on the diagram. In cases where activity descriptions become inconveniently long, some of the vital information, such as work location, type of operation, or responsibility, can be coded into the activity number rather than stated explicitly in the description.

When a working precedence diagram is being prepared, the scale and spacing of the activities deserve attention. If the scale is too big and activities are widely spaced, the resulting network is likely to become so large that it is unmanageable. Yet a small scale and overly compact makeup renders the diagram difficult to read and inhibits corrections and modifications. With experience and observation of the work of others, practitioners will soon learn to adjust the scale and structure of their diagrams to the scope and complexity of the project involved. It must not be forgotten that the network is intended to be an everyday tool, used and consulted by a variety of people. Emphasis here is not on drafting elegance but on contriving the most realistic, intelligible, and flexible form of network possible.

Dependency lines that go backward from one activity to another should not be used and are not possible with most computer programs. "Backward," in this instance, means going from right to left on the diagram, against the established direction of time flow. Backward sequence lines are confusing and increase the chance of unintentional logical loops being included in the network. A logical loop involves the impossible requirement that activity A be followed by activity B, with activity B needing to be accomplished before activity A. Logical loops may be inadvertently included in large and complex networks if backwardly directed sequence lines are permitted. Crossovers occur when one dependency line must cross over another to satisfy job logic. Careful layout will minimize the number of crossovers, but usually some cannot be avoided.

Color coding the diagram can be very useful at times. Various colors can be used to indicate different trades, work classifications, major job segments, and work that is subcontracted. Prefixes and suffixes can also be used as part of the activity numbering system to designate physical location, trade, or responsibility.

4.12 Lag Relationships

Figure 4.1 has been drawn on the customary basis that a given activity cannot start until *all* of those activities immediately preceding it have been completed. Also inherent in the notation used in the figure is that an activity can start once all of its immediately preceding activities have been finished. In the figure, activity 60, "Fine grade," cannot start until activity 50, "Drive piles," has been finished, and the start of activity 90, "Pour concrete," must await completion of activity 80, "Place rebar." In addition, by way of example, activity 80, "Place rebar," can start immediately once activity 60, "Fine grade," activity 70, "Place forms," and activity 40, "Order and deliver rebar," have all been finished. Under these conditions, it is not possible to have the finish of one activity overlap the start of a following activity. Where such a condition potentially exists, the activity must be further subdivided.

There are cases, however, where there may be a delay between the completion of one activity and the start of a following activity, or there is a need to show that one activity will overlap another in some fashion. Precedence diagrams can be made to show a variety of such conditions using lag relationships. This is, in fact, one of the primary advantages of precedence over arrow diagrams. The concept of lags is developed more completely in Section 5.21.

4.13 Precedence Diagram for the Example Projects

Getting started on a large project can be overwhelming, and a general job plan of limited size can be useful as a means of getting started. Once a general job plan has been put together, it is an excellent framework on which to amplify the network to the level of detail desired. Some practitioners favor starting job planning by compiling a relatively short list of major project operations arranged in chronological order. For the highway bridge, described in Section 3.5, this initial list could be:

A. Procurement
B. Field mobilization and site work

C. Pile foundations

D. Concrete abutments and wing walls

E. Deck

F. Finishing operations

For the commercial building described in Section 3.5, the preliminary list is easily developed from the first-level plan created in Section 4.8. The activity Mobilize has been added as a convenient starting point for the project.

A. Mobilize

B. Utilities and Site Work

C. Structure

D. Envelope

E. Finishes

F. Systems

G. Closeout

These lists can be used to prepare a preliminary job plan, such as the one shown in Figure 4.2a for the bridge or 4.2b for the building. The major operations in the preceding lists are much the same as, and could be identical to, those previously identified by the bar chart in Figures 3.5a and 3.5b, when a preliminary time analysis of the project was made during the estimating period. A general job plan like those shown in Figures 4.2a and 4.2b is useful in the sense that it places the entire project in perspective. It is profitable for the planner to establish a general frame of reference before beginning to struggle with the intricacies of detailed job planning.

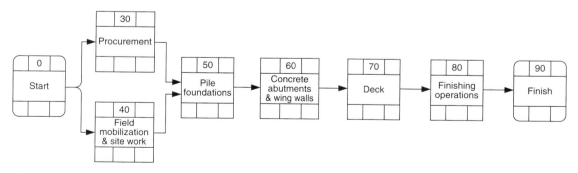

Figure 4.2a
Highway bridge, general job plan

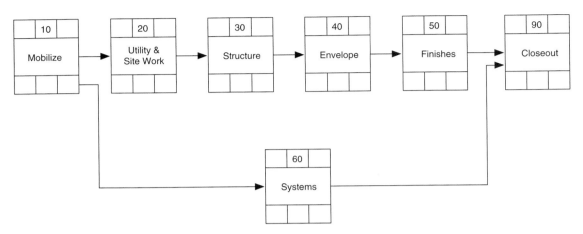

Figure 4.2b
Commercial building, general job plan

All of the steps in job planning have now been reviewed. Providing that the simple rules of precedence diagramming are followed, it should be possible to develop a working diagram for the example projects. A company prebid conference has established the general ground rules of procedure for the projects.

For the bridge, one abutment at a time will be constructed for reasons of equipment limitations and the economy of form reuse. Abutment #1 will be constructed first, with its footing being poured initially, followed by the breast and wing walls. As excavation, pile driving, forming, and pouring are completed on abutment #1, these operations move over to abutment #2. Only one set of footing forms and abutment wall forms will be made, these being used first on abutment #1 and then moved over to abutment #2.

As soon as the steel girders have been delivered, the concrete abutments stripped, and abutment #1 backfilled, the steel girders will be placed. Forms and reinforcing steel for the deck slab can follow. Abutment #2 must be backfilled before the deck can be poured. Concrete patching and rubbing can be started as the abutments are stripped and must be finished before painting can start. Job cleanup and final inspection complete the project.

The full precedence diagram that results from the highway bridge logic just established is found on the companion website as Chart 4.2a. Most of the dependencies on the diagram are normal dependencies in that they are the natural result of the physical nature of the activities themselves. However, the diagram also includes some resource restraints. When such restraints have been firmly established, it is advisable to include them in the first draft of the network because they can be significant planning

4.13 Precedence Diagram for the Example Projects

factors that have a major effect on the project plan and schedule. By way of explanation, refer to activities 10, 20, and 50 in the diagram, using activity 20 as a typical example.

Activity 20, "Prepare & approve S/D footing rebar," in Chart 4.2a, is a consequence of a construction contract requirement that the prime contractor and its subcontractors submit shop drawings (S/D) and other descriptive information concerning project materials and machinery to the owner or project design professional for their approval. This must be done before these materials can be fabricated and provided to the job site. Shop drawing approval and material fabrication and delivery are shown separately on the precedence diagram. This is because the delivery of reinforcing steel and other materials to the project typically is quoted by the vendor as requiring a stipulated period of time after the general contractor returns approved shop drawings to the vendor.

With regard to material restraints such as activity 30 in Chart 4.2a, an additional matter may have to be considered. For example, if the piles are sold to the contractor "FOB trucks, job site"—and this is typical—the contractor is responsible for unloading the vendor's trucks when they arrive on the job site. (FOB is an abbreviation for "freight on board" and refers to the delivery point covered by the vendor's quote.) In such an instance, the planner may wish to emphasize this fact by including another activity, "Unload piles," at the end of activity 30. This simply serves as a reminder to all concerned that the contractor must make advance arrangements to have suitable unloading equipment and workmen available when delivery is made. Unloading activities have not been included in the precedence diagram, however, so that the discussion can concentrate on basics.

An equipment restraint has also been included in the precedence diagram. The driving of steel pilings for the two abutments requires an assembly of equipment units, including a large crane, which must be mobilized before pile driving can start. This is illustrated by activity 100, "Mobilize pile-driving rig," which precedes activity 110, "Drive piles, abutment #1." After the piles are driven for both abutments, the pile-driving rig is demobilized. As shown, this disassembly process must occur before activity 180, "Forms & rebar, abutment #1," can start because the crane used for pile driving is required to handle the forms and rebar for this abutment. Abutment #2 follows at a later date. Another resource constraint in the precedent diagram arises because the same concrete forms are going to be used for both abutments. As a result of this decision, activity 240, "Forms & rebar, abutment #2," cannot start until activity 220, "Strip & cure, abutment #1," has been completed. In a similar manner, the same forms will be used for both footings.

Similar logic has been used to develop the full precedence diagram for the commercial building example, which is also found on the companion website as Chart 4.2b.

4.14 Value of Precedence Network

Chart 4.2 is not only a lucid job model; it is also an effective tool useful for day-to-day direction and control of the work. Chart 4.2a is the job plan for the highway bridge project, and Chart 4.2b is the job plan for the commercial building project. Preparing the network has forced the job planners to think the job through completely from start to finish. Decisions have been made about equipment, construction methods, and sequence of operations. The field supervisory team now possesses a depth of knowledge about the project that can be obtained only through such a disciplined process of detailed job analysis.

The job plan, in the form of a precedence diagram, is comprehensive, detailed, and in a form that is easy to communicate to others. The network diagram is an expedient medium for communication between field and office forces. If for some reason the project manager or field superintendent must be changed during construction, the diagram can assist appreciably in effecting a smooth transition. The diagram makes job coordination with material dealers, subcontractors, owners, and architect-engineers a much easier matter. The orderly approach and analytical thinking that have gone into the network diagram have produced a job plan far superior to any form of bar chart or narrative analysis. Invariably, synthesis of the network results in improvements to original ideas and a sharpening of the entire approach to the project.

4.15 Repetitive Operations

Some kinds of construction projects involve a long series of repetitive operations. Transmission lines, highways, pipelines, multistory buildings, and tract housing are familiar examples of construction jobs that entail several parallel strings of continuing operations. One segment of the heavy civil example project described in Section 2.18 is the relocation of five miles of natural gas pipeline. This is a good example of a construction operation that will involve repetitive operations that proceed simultaneously. The repetitive operation analysis would be appropriate for a high-rise building, but since our building example is a single story, discussion of repetitive operations will be left to the pipeline segment of the heavy civil project.

For purposes of discussing the planning of such a project, the major repetitive segments of the pipeline relocation have been identified as:

Locate & clear	Lay pipe
Excavate	Test
String pipe	Backfill

4.15 Repetitive Operations

The basic plan for this job is shown in Figure 4.3, with all the operations being sequential except for excavate and string pipe, which are done concurrently. However, no pipeline contractor would proceed in such a single-file manner unless the pipeline was very short in length or some special circumstance applied. Rather, location and clearing work would get well under way. Excavation, together with pipe stringing, would then start, and pipe laying would proceed fairly closely behind. Pressure testing of the pipe and backfilling would complete the sequence. After the project got "strung out" along the right-of-way, all these operations would move ahead sequentially, one stage following the next.

All this makes it clear that more detail than is included in Figure 4.3 will be necessary if the job plan is to be useful. The way to accomplish this is to divide the pipeline into arbitrary but typical repeating sections. The length of the repeating section chosen can be quite variable, depending on the length of the pipeline, terrain, contract provisions, and other job conditions. For discussion purposes, suppose the contractor decides that a mile-long section of pipeline represents a fairly typical unit of work. This would be exclusive of river, railroad, and highway crossings, which are frequently done in advance of the main pipeline and which may require their own planning and scheduling studies as separate operations. Figure 4.4 indicates how the basic plan can be broken down into mile-long units.

The notation used in Figure 4.4 is one where the individual activity box indicates only the section of the pipeline involved. The work categories listed at the left margin of the figure apply to the horizontal string of activities at each successive level. The logic of the figure shows that after a mile of the pipeline right-of-way has been located and cleared, both excavation and pipe stringing start. These latter two activities proceed simultaneously, since one does not depend on the progress of the other. After excavation and pipe stringing have proceeded one mile, pipe-laying starts. Testing and backfill begin in the order shown. Once a work phase is started, it will proceed more or less continuously until its completion. This type of linear

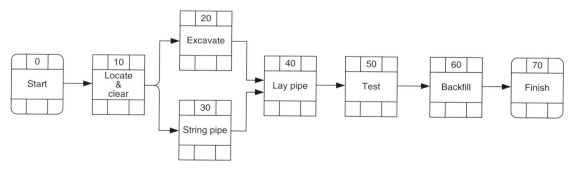

Figure 4.3
Pipeline relocation, basic plan

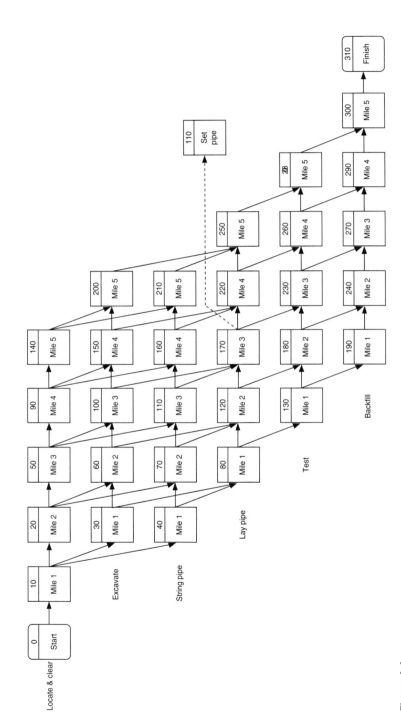

Figure 4.4
Pipeline relocation, precedence diagram planning

4.16 Network Interfaces

Often, different portions of the same project are planned separately from one another. However, as frequently happens, the individual construction plans are not truly independent of one another, and the two networks are actually related in some way. Thus, the two networks must interface with each other so that mutual dependencies are transmitted properly from one plan to the other. The term *interface* refers to the dependency between activities of two different networks. This dependency can be indicated on the networks by dashed sequence lines between the affected activities.

The pipeline relocation will be used to illustrate the workings of an interface. Assume that this job includes a stream crossing at the end of the third mile of pipeline. The construction of the crossing structure will be separate from that of the pipeline, but the two must be correlated so the crossing structure is ready for pipe by the time the pipeline construction has progressed to the crossing location. In this instance, planning of the pipeline itself and of the crossing structure proceed independently of one another. Figure 4.4 is the planning network for the pipeline work. The crossing is to be effected by a suspension-type structure, and Figure 4.5 is its construction plan.

The required job logic of having the crossing structure ready to accept pipe by the time the pipeline has reached the crossing site can be imposed by having activity 110 of Figure 4.5 immediately follow activity 170 of Figure 4.4. This is shown in both figures by the dashed dependency line. Ensuring that the crossing structure is started soon enough for the desired meshing of the two networks is now a matter of scheduling and will be treated in Chapter 5, "Project Scheduling Concepts."

4.17 Master Network

As illustrated by Figure 2.1a, the total heavy civil example project consists of several major subprojects, each being relatively self-contained and independent of the others. The planning networks of two of these segments of the heavy civil example project, the highway bridge and the pipeline relocation, have been developed. It is now easy to visualize that similar planning diagrams must be developed for the river diversion, haul roads, borrow development, earth dam, and other major project elements. These separate networks would be used for the detailed time scheduling and daily field management of the several components of the heavy civil example project. In a general sense, each of the major project segments is constructed and managed separately.

The planning for each subproject is done in considerable detail. This fact is obvious from the nature of the networks developed for the highway bridge

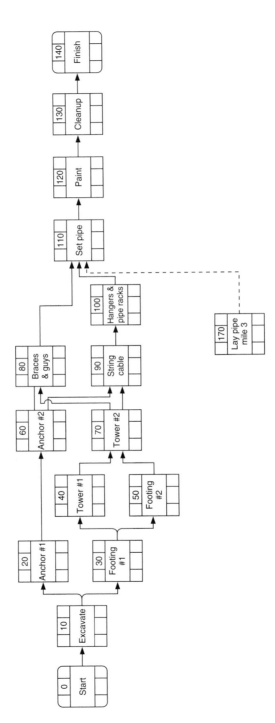

Figure 4.5
Pipeline crossing structure, precedence diagram planning

and the pipeline relocation. Although this level of detail is necessary for the day-to-day field control of construction operations, it is overwhelming for an owner or project manager who wants to keep abreast of overall site operations in a more general way. The amount of detail required by managers decreases with their increased span of authority. A field manager requires information and data that are tailored to the level of his responsibilities.

On the heavy civil example project, a master planning network would be prepared that encompassed and included every subproject. The level of detail of this diagram would be relatively gross, with each activity representing substantial segments of the fieldwork. The highway bridge and the pipeline relocation would each appear as a small cluster of associated activities. It is likely that the highway bridge would appear in the master diagram as the activities shown in Figure 4.2a, and the pipeline relocation would be as presented in Figure 4.3. The master job plan concerns itself essentially with the big picture, the broad aspects of the major job segments and how they relate to each other. Keeping the master network free of excessive detail is necessary to the production of an overall plan that can be comprehended and implemented by those who must apply it.

4.18 Subnetworks

The detailed planning diagrams of the highway bridge (Chart 4.2a on the companion website) and the pipeline relocation (Figure 4.4) are spoken of as subnetworks of the master network. As has been discussed, lower-level field management uses them for everyday direction and control of the work. It would be possible, of course, to draw an overall project network that would combine the detailed planning networks of all the subprojects into one enormous diagram. However, such an all-inclusive network would be so large and cumbersome as to be virtually useless to anyone concerned with the heavy civil example project. The detailed planning of an extensive construction contract such as the heavy civil example project is accomplished primarily through the medium of many subnetworks, with their interdependencies indicated by appropriate interfaces. In this way, management personnel can concentrate on their own localized operations. Each manager can monitor the work according to his own plan without the distraction of having to wade through masses of information irrelevant to him and his responsibilities. It is also entirely possible that certain activities of the highway bridge network (Chart 4.2a on the companion website) or of the pipeline relocation network (Figure 4.4) might require further expansion into more detailed subnetworks. To ensure their timely accomplishment, it is sometimes desirable to subject certain critical activities to further detailed planning study. To illustrate, activity 100, "Mobilize pile-driving rig," of Chart 4.2a might be expanded into its own planning subnetwork, as shown by Figure 4.6.

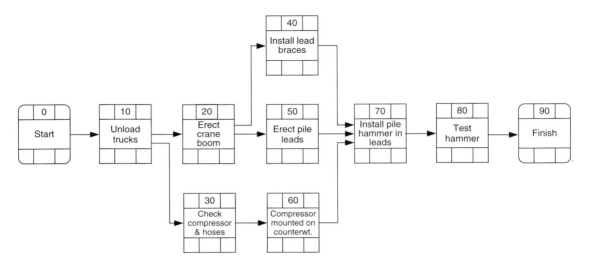

Figure 4.6
Mobilize pile-driving rig, precedence diagram planning

4.19 Computer Applications for Planning

Computers have become an invaluable tool that is indispensible as construction professionals plan, schedule, and control projects. In the succeeding chapters, computer applications for each phase of construction management will be discussed. Because of rapidly changing computer technology and software sophistication, only the fundamentals of computer applications are appropriate for discussion.

Earlier in this chapter, it was stated that planning is the most important phase of project management, and it is also the most difficult and time consuming. This is because the knowledge, experience, and insight of the project team must be brought together to identify a plan that is both complex and uncertain. For years, the process was one of holding a planning meeting, drawing the network on paper, holding another planning meeting, and revising the network drawing again and again until the team was satisfied or ran out of time.

Computers have had a major impact on the planning process, enabling networks to be developed on the computer screen and then revised in real time as the planning progresses. Though the initial team planning is still often done on a whiteboard, the outcome of this planning is quickly moved into a digital environment. One person, using the computer, can record ideas as the plan is being developed. Each addition to the plan can be seen, understood, and if necessary criticized by the other team members. In this way, the skills of the team are combined and a dynamic plan is developed in much less time than was previously required in a manual environment. With networked computers, it is now possible to hold the planning meeting in a virtual environment in which all participants can be in their own office on their own computer. This facilitates planning for large and complex projects, with expertise being drawn from anywhere in the world. Computer-based network development offers significant added flexibility over manual drafting. Individual activities or whole groups of activities can be moved around on the screen. Individual parts of the network can be created independently and then combined with the main network. Multiple approaches to a particular planning problem can be created, evaluated, and compared, with the best solution used in the final plan.

Perhaps the best part of this type of planning is that it involves each member of the project team in a highly collaborative planning process. The resulting plan is perceived to be the team's plan rather than one provided by an individual or the planning department. Everyone on the team now has a stake in making this plan work. This perception alone causes each phase of the project management system described in the succeeding chapters to contribute to a successful project.

Key Points and Questions

Key Points

- Planning is of critical importance and underlies all aspects of the project.
- Planning is best performed as a team activity involving key stakeholders, rather than by a single individual who then imposes the plan on the other project participants.
- Precedence-based critical path scheduling is a powerful and versatile tool for project managers and supervisors, enabling them to better understand complex projects and providing more opportunity to control time, cost, and resources.

Review Questions and Problems

1. Explain the three phases of the Critical Path Method.
2. What are the five steps in developing the plan?
3. Why is inclusion of subcontractors and other stakeholders a key part of the planning process?
4. Discuss the drawbacks to beginning-to-end planning?
5. Explain why various stakeholders are interested in differing levels of detail on their particular plan.
6. What are the advantages of top-down planning over beginning-to-end planning?
7. Select a specialty discipline of interest to you for the building example. Develop a general precedence-based plan for that discipline.

5 Project Scheduling Concepts

5.1 Introduction

Chapter 4, "Project Planning," dealt with the procedures followed in developing a project plan and describing that plan as a Critical Path Method (CPM) network diagram. Once this network diagram has been developed, the time management system enters the next phase: that of work scheduling. Thus far, all planning effort has been directed at defining the work to be accomplished and the sequence of activities, that is, the order in which that work must be carried out. Time relating to overall project construction duration or the time required to complete individual activities has not been factored into the plan. This chapter is concerned with the time scheduling of construction projects. Once calculations have established the time-based characteristics of the project and each activity, subsequent chapters will apply this knowledge to manage and control the project.

This chapter will begin with determination of durations for individual activities. The bridge project is used as the basis for the discussion of the calculation process. However, the same calculation process is used for the commercial building project, with the resulting network found on the companion website. The chapter will then describe the calculation process for project times. New terminology for scheduling is introduced, including early and late start and finish, float, critical activities, and lag time. The schedule based on project days is then converted to calendar dates.

Learning objectives for this chapter include:

❏ Be able to determine activity durations.
❏ Understand how network times are calculated.
❏ Become familiar with scheduling terminology.
❏ Review ways in which scheduling information can be displayed.

5.2 Scheduling Procedure

As has been stated previously, the CPM was developed especially for the planning and scheduling of construction operations and is the procedure used throughout this text. However, it is of interest to note that a somewhat different procedure, Program Evaluation and Review Technique (PERT), has been devised for application to the scheduling of research and development projects. Because research is generally highly exploratory in nature, historical experience and background are rarely good measures for future time estimates, and make reasonably accurate time estimates difficult to establish. As a result, researchers have developed PERT as a method of statistically evaluating project duration over a time-sensitive domain. Although basically similar, CPM and PERT differ in several important respects concerning the estimation of activity time durations. Though PERT can provide some understanding regarding probability and project duration even on construction projects, it will not be addressed in this text, as its primary value is in estimation of durations for projects with high levels of uncertainty in terms of activity durations. In this chapter and throughout the book, only the CPM scheduling method is developed and applied.

A project schedule is a projected timetable of construction operations that will serve as the principal guideline for project execution. Several steps are involved in devising an efficient and workable job schedule. The following eight steps are offered as a procedural guide:

1. *Estimate the time* required to carry out each network activity.
2. *Compute the time period required for the overall project completion* using these time estimates.
3. *Establish time intervals* within which each activity must start and finish to satisfy the completion date requirement.
4. *Identify those activities* whose expedient execution is crucial to timely project completion.
5. *Shorten the project duration at the least possible cost* if the project completion date will not meet the contract or other requirements.
6. *Adjust the start and finish times of selected activities to minimize resource conflicts and smooth out demands for manpower and equipment* using surplus or float times that most activities possess.

7. *Make a working project schedule* that shows anticipated calendar dates for the start and finish of each network activity.
8. *Record the assumptions* made and the plan's vital boundary conditions. These will become an integral aspect of the completed baseline project schedule.

This chapter discusses the first four steps just described. The remainder are presented in subsequent chapters. The highway bridge and pipeline relocation, together with the commercial building project, will be used to present the several facets of construction scheduling.

5.3 Activity Times

CPM customarily uses a single time estimate for each network activity. In construction, each activity is "deterministic" in the sense that similar or identical work has been performed many times before. Such prior experience enables the contractor to estimate with reasonable accuracy the time required to carry out each job operation. Single time estimates for network activities will be used exclusively in this book.

In the construction industry, activity durations are customarily expressed in terms of working days, although other time units, such as hours, shifts, or calendar weeks, are sometimes used. The main criterion is that the unit chosen be harmonious with the project and the management procedures being used. Once a time unit is selected, it must be used consistently throughout the network. More will be said about this later in the chapter. Working days will be used herein as the standard time unit for activity duration estimates and scheduling computations.

5.4 Rules for Estimating Activity Durations

The true worth of a project work schedule and the confidence with which it can be applied depend almost entirely on the job logic as contained in the planning network and the accuracy with which the individual activity times can be estimated. As a general rule, time estimates can be made more reliably for activities of limited scope than for those of larger extent. In itself, this is not necessarily an argument for the use of great detail in job planning. Nevertheless, when considerable uncertainty surrounds the duration of a given activity, sometimes it can be helpful to subdivide that activity into smaller elements.

Six important rules apply to the estimation of activity durations:

1. *Evaluate activities one at a time, independently of all others.* For a given activity, assume that materials, labor, equipment, and other needs will be available when required. If there is a fundamental reason to believe

that this will not be true, then the use of a preceding restraint may be in order.

2. *For each activity, assume a normal level of manpower and/or equipment.* Exactly what "normal" is in this context is difficult to define. Most activities require only a single crew of workers or a standard spread of equipment. Based on experience, conventional crew sizes and equipment spreads have emerged as being efficient and economical. In short, a normal level is about optimum insofar as expedient completion and minimum costs are concerned. A normal level may be dictated by the availability of labor and equipment. If shortages are anticipated, this factor must be taken into account. However, conflicting demands among concurrent activities for workers or equipment will be temporarily ignored. At this stage, such conflicts are only matters of conjecture, and they will be investigated in detail during a later stage of scheduling.

3. *If time units of working days are being used, assume a normal workday.* Do not consider overtime or multiple shifts unless this is typical or a part of the standard workday. Around-the-clock operations are normal in most tunnel work, for example, and overtime is extensively used on highway jobs during the summer months to beat the approaching cold weather. Some labor contracts guarantee overtime work as a part of the usual workday or workweek. In these cases, the extra hours are normal and should be considered.

4. *Concentrate on estimating the duration of the individual activity and ignore all other time considerations.* In particular, the completion date of the project must be put entirely out of mind. Otherwise, there is apt to be an effort made, consciously or unconsciously, to fit the activities within the total time available. This is one of the serious drawbacks of the bar chart as a planning and scheduling device. Most contractors will admit that the average bar chart is made up primarily by adjusting the individual work items to fit within an overall time requirement. The only consideration pertinent to estimating an activity duration is how much time is required to accomplish that activity, and that activity alone.

5. *Use consistent time units throughout.* When using the working day as a time unit, it must be remembered that weekends and holidays are not included. Certain job activities, such as concrete curing or systems testing, continue during nonworking days. To some extent, allowances can be made for this. For example, curing periods of seven days will involve only five working days. In most cases, however, such corrections and conversions cannot be exact and must be based on the scheduler's best judgment. When using a computer program to facilitate schedule calculations, check to see whether the program supports multiple

calendars and use the method suggested by the software developer. Associated with the use of consistent time units is the matter of conversion of time periods from one base to another. For example, vendors invariably express delivery times in terms of calendar days. If the delivery of a pump is given as 30 days by the vendor, this will translate into approximately 21 working days. Again, most computer programs handle these problems with multiple calendars. Manual calculations require the scheduler to make these adjustments to the durations by hand.

6. *Assume normal weather conditions in estimating the duration needed to accomplish each activity.* Some operations are sensitive to the effects of weather and may not be performed at all or will take longer to complete if necessary climatic conditions are absent. In general, such activities should be estimated assuming the existence of conducive weather. Using historical weather data for the site location, operation-specific calendars can be developed to account for the seasonal variations that weather will have on these activities. This process is discussed further in Section 5.20.

5.5 Estimating Activity Durations

It is important that someone experienced in and familiar with the type of work involved be responsible when the activity durations are being estimated. With respect to work done by subcontractors, it is good practice to solicit input from them concerning the times required for those activities for which they are responsible. Subcontractors usually are in the best position to render judgments concerning the times required for the accomplishment their work.

One effective way of estimating an activity duration is to compute it by applying a crew or equipment production rate to the total number of units of work to be done. For illustrative purposes, the determination of time estimates for two of the activities on the highway bridge will be discussed. First, consider activity 110, "Drive piles, abutment #1," as it appears in Chart 4.2a. The summary sheet for Bid Item No. 4 shows that a pile-driving production rate of 70 linear feet per hour was used when the cost of the highway bridge was estimated. Each abutment involves the driving of twenty-eight 40-foot-long piles. Dividing the total lineal footage of 1,120 feet by 70 gives 16 hours, or two working days.

$$28 \text{ piles} \times 40 \text{ ft/pile} = 1.140 \text{ ft}$$

$$1{,}140 \text{ ft} \div 70 \text{ ft/hr} = 16.3 \text{ hrs}$$

In addition to the actual pile driving, most of one day will be required to prepare the templates, cut off pile heads, and move the equipment. Thus, the time estimate for this particular activity will be three days. In a similar manner for activity 200, "Pour abutment #1," Figure 3.6 shows that the concrete for this abutment will be poured at the rate of 8.75 cubic yards per hour. This production rate can be used to compute the time required to pour the abutment. Each abutment contains 140 cubic yards of concrete, and dividing this by 10 cubic yards per hour gives 14 hours, or about two working days.

$$140 \text{ cy} \div 8.75 \text{ cy/hr} = 16 \text{ hrs}$$

Another approach in determining activity times is to assume a crew size and use the estimated labor unit cost rather than a production rate. To illustrate this procedure, Bid Item No. 3 shows the unit labor cost for compacted backfill to be $8.45 per cubic yard. Activities 280 and 310 each include 170 cubic yards of compacted backfill. Thus, each of these activities has a direct labor cost of:

$$170 \text{ cy} \times \$8.45/\text{cy} = \$1,436.50$$

Assume a crew of three laborers with a daily labor cost of:

$$3 \text{ laborers} \times 8 \text{ hrs/day} \times \$22.00/\text{hr} = \$528/\text{day}$$

Dividing $1,436.50 by $528/day gives 2.7 days, or approximately three days. Hence, the estimated duration of activities 280 and 310 will each be three working days.

When estimating activity times, one special circumstance must be kept in mind: the case where the same work item is repeated several times during the construction period. For example, successive job activities may involve repetitions of essentially identical concrete forming. The time performance on such work will improve considerably during the first few cycles. This learning-curve effect causes the later items to be accomplished in less time than the first ones. The basic proposition of the learning-curve phenomenon is that skill and productivity in performing the same work improve with experience and practice and therefore should be reflected in the network durations.

Time estimates of surprising accuracy often can be made informally. Experienced construction supervisors have an almost uncanny ability to give off-the-cuff time estimates that usually prove to be reasonably close. This may seem to be an almost casual approach to such an important matter, but experience shows that it has its place, especially when checking against time values obtained by presumably more exact means. Input from field superintendents is valuable and desirable, but it would be a mistake to allow them to make all the duration estimates in such an informal fashion. If for no other reason, they are human and their time estimates are apt to be generous, so their chances of staying on schedule later on are commensurately improved.

Activity durations customarily are expressed in terms of full working days because, in most cases, to do otherwise is to assume a fictitious degree of accuracy. If an activity time is less than one working day, the activity concerned may be too small for practical job scheduling and control. Refer to Charts 5.1a and b on the companion website for the logic that was developed in Chapter 4, "Project Planning," for both the bridge and the commercial building. To each logic diagram has been added the estimated durations for each activity. Each activity duration, in terms of working days, is shown in the lower, central part of the activity box. Each activity is also given an identifying number, located in the upper, central part of the activity box. The other numerical values shown with the activities and the contingency activity at the right end of the figures will be discussed later.

5.6 Time Contingency

When applying the CPM procedure to a construction project, it is assumed that individual activity durations are deterministic in the sense that they can be estimated relatively accurately and that their actual durations will have only relatively minor variances from the estimated values. An estimated activity duration is the time required for its usual accomplishment and does not include any allowance for random or unusual happenings. An estimated or likely completion time for the entire project can be computed by using such estimates of individual activity times. The actual project completion time probably will vary from this estimated value for a number of reasons that cannot be entirely predicted or quantified by the planning team. When the time estimate is made for an activity, it is based on the assumption that "normal" conditions will prevail during its accomplishment. Although normal conditions are difficult to define, the concept is accurate enough to recognize that many possibilities of "abnormal" occurrences can substantially increase the actual construction time. Concessions for abnormal or random delays are accounted for in a number of ways, the simplest being a contingency allowance. Where these occurrences can be recognized as activity specific, the contingency may be added in the form of an uncertainty variance to the activity duration. This concept is fundamental to the PERT method previously mentioned.

In most cases, though, a contingency allowance cannot be applied to individual activities to account for general project delays, such as those caused by fires, accidents, equipment breakdowns, labor problems, late material deliveries, damage to material shipments, unanticipated site difficulties, and the like. Often, it is impossible to predict which activities may be affected and by how much. As a result, a general allowance for such time contingencies normally is added to the overall project duration or at the end of specific construction sequences.

5.7 Project Weather Delays

Probably the most common example of project delay is that caused by inclement weather. It is important that the probable effect of adverse weather be reflected in the final project time schedule. The usual basis for making time estimates is on the assumption that construction operations will proceed on every working day. However, it is obvious that there can be a profound difference in the time required for excavation, depending on whether the work is to be accomplished during dry or wet weather. Similarly, snow, extreme temperatures, and high winds can substantially affect the times required to do certain types of construction work.

How allowances for time lost because of inclement weather are handled depends to a great extent on the type of work involved. When most highway, heavy, and utility work are affected by the weather, normally the entire project is shut down. On buildings and other work that can be protected from the weather, allowances for time lost are commonly applied only to those groups of activities susceptible to weather delay. Seldom is a job of this type completely shut down by bad weather after it is enclosed. Some parts of the job may be at a standstill while others can proceed. Although one or more weather contingencies can be used in a network, a preferred solution is to account for weather using activity calendars, a subject that is covered in Section 5.19.

5.8 Network Time Computations

The scheduling of detailed activities in a network generally is done by computer, with times expressed in terms of calendar dates or expired working days. However, the optimum use of this information for project time control purposes requires the user to have a thorough understanding of the computations and the true significance of the data generated. Time values generated by a "black box" with no insight into the process cannot be used in optimal fashion to achieve time management purposes. For this reason, the manual computation of activity times is discussed in detail in the sections that follow. When these calculations are made by hand, they normally are performed directly on the network itself. When making the initial study with manual computations, activity times usually are expressed in terms of expired working days. Likewise, the start of the work being planned customarily is taken to be at time zero.

After a time duration has been estimated for each activity, some simple step-by-step computations are performed. The purpose of these calculations is to determine (1) the overall project completion time and (2) the time brackets within which each activity must be accomplished to meet the completion date. The network calculations involve only additions and subtractions and can be made in different ways, although the data produced are comparable in all cases. The usual procedure is to calculate what are referred to as activity times. Activity times play a fundamental role in project scheduling, and their determination is treated in this chapter.

The calculation of activity times involves the determination of four limiting times for each network activity. The "early start" (ES) or "earliest start" of an activity is the earliest time that the activity can possibly start, allowing for the times required to complete the preceding activities. The "early finish" (EF) or "earliest finish" of an activity is the earliest possible time that it can be completed, and is determined by adding that activity's duration to its early start time. The "late finish" (LF) or "latest finish" of an activity is the very latest that it can be finished and allow the entire project to be completed by a designated time or date. The "late start" (LS) or "latest start" of an activity is the latest possible time that it can be started if the project target completion date is to be met and is obtained by subtracting the activity's duration from its latest finish time.

The computation of activity times can be performed manually or by computer. When calculations are made by hand, they normally are performed directly on the network itself. When the computer is used, activity times are shown on the computer-generated network or in tabular form. When making manual computations, it is normal to use project days, working in terms of expired working days. Commensurately, the start of the project is customarily taken to be time zero (the end of project day zero). When computations are done by computer, calendar dates often are used so that early and late starts are in the morning of the calendar day and early and late finishes are in the afternoon of the calendar day.

5.9 Early Activity Times

The highway bridge network, Chart 5.1a on the companion website, is used here to discuss the manual calculation of activity times directly on a precedence diagram. The complete network precedence diagram for the commercial building is provided as Chart 5.1b. Precedence diagrams are exceptionally convenient for the manual calculation of activity times and afford an excellent basis for describing how such calculations are done. This section discusses the computation of the early-start (ES) and early-finish (EF) times. The determination of late activity times is described subsequently. The early time computations proceed from project start to project finish and from left to right in illustrated networks; this process is termed the *forward pass*. The basic assumption for the computation of early activity times is that every activity will start as early as possible. That is to say, each activity will start just as soon as the last of its predecessors is finished.

The ES value of each activity is determined first, with the EF time then being obtained by adding the activity duration to the ES time. To assist the reader in understanding how the calculations proceed, small sections of the bridge are used in Figure 5.1a to illustrate the numerical procedures. Reference to the network shows that activity 0 is the initial activity. Its earliest possible start is, therefore, zero elapsed time. As explained by the sample activity shown, the ES of each activity is entered in the upper left of its activity box. The value of zero is entered at the upper left of activity 0 in Figure 5.1a. The EF of an activity is obtained by adding the activity duration to its ES value. Activity 0 has a duration of zero. Hence, the EF of activity 0 is its ES of zero added to its duration of zero, or a value of zero. EF values are entered into the upper right of activity boxes, and Figure 5.1a shows the EF value of activity 0 to be zero. Activity 0 calculations are trivial, but the use of a single opening activity is customary with precedence diagrams.

The network shows that activities 10, 20, 30, 40, and 50 can all start after activity 0 has been completed. In going forward through the network, the earliest these five activities can start is obviously controlled by the EF of the preceding activity. Since activity 0 has an EF equal to zero, then each of the five activities that follow can start as early as time zero. Figure 5.1b shows that activities 30 and 40 have ES values of zero. Zeros have been entered, therefore, in the upper left of activities 30 and 40. This is typical for activities 10, 20, and 50 as well. The EF of activity 30 is its ES of zero plus its duration of 15, or a value of 15. Likewise, the EF value of activity 40 is zero plus 3, or a value of 3.

Continuing into the network, Figure 5.1c shows that activities 80 and 90 and 100 cannot start until activity 40 has been completed. Activity 40 is referred to as a burst activity, which is an activity that is followed by two or more activities. The earliest that activity 40 can finish is at the end of the third day. Using activities 80 and 90 as examples, it is seen that the earliest these two activities can be started is day 3. The EF of activity 80 will be its ES value of 3 plus its duration of 3, or a value of 6. In like fashion, activity 90 will also have an EF value of 6. These values have been entered into the activity boxes in Figure 5.1c.

The bridge network and 5.1d show that activity 130 cannot start until both activities 70 and 110 are finished. The EF values of activities 70 and 110 are 12 and 18, respectively. Because the ES of activity 130 depends on the completion of both activities, it follows that the finish of activity 110, not that of activity 70, actually controls the start of activity 130 and that activity 130 will have an early start of 18. Activity 130 is an example of a merge activity—an activity whose start depends on the completion of two or more

5.9 Early Activity Times

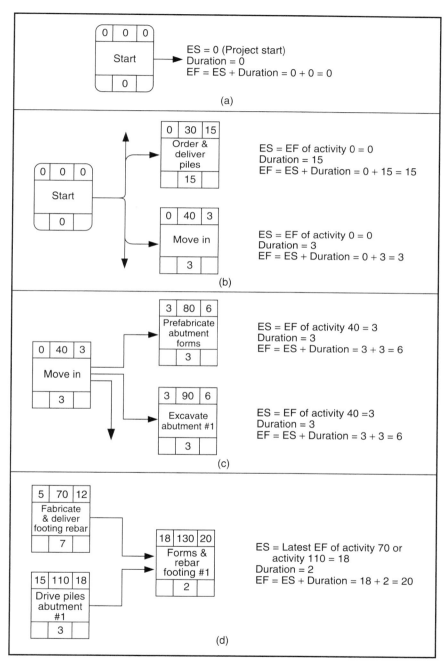

Figure 5.1
Highway bridge, forward-pass calculations

preceding activities. The rule for this and other merge activities is that the earliest possible start time of such an activity is equal to the latest (or largest) of the EF values of the immediately preceding activities.

The forward-pass calculations consist only of repeated applications of the few simple rules just discussed. Working methodically in step-by-step fashion, the computations in the bridge network proceed from activity to activity until the end of the network is reached. In similar fashion, the forward pass can be executed through the commercial building network as well. The reader is reminded that the figures and calculations shown in Figure 5.1 were for explanatory purposes only. The actual forward-pass calculations would have commenced with activity 0 in the network, with the ES and EF values being entered onto the activity boxes as the calculations progressed. As is now obvious, the calculations are elementary and, with practice, can be performed rapidly. Even so, when several hundred activities are involved, the manual development of activity times can be tedious, time consuming, and subject to error. Moreover, any time the schedule is updated or there is a change to either activity logic or durations, the entire set of calculations must be repeated. It soon becomes apparent that the calculation of CPM schedules is ideal for a computer, but extremely tedious for a person.

5.10 Project Duration

Reference to the bridge network (Chart 5.1a on the companion website) will disclose that the early-finish time for the last work activity (400) is 64 elapsed working days. For the job logic established and the activity durations estimated, it will require 64 working days to reach the end of the project, provided that each activity is started as soon as possible (or at its ES time). Thus, if a competent job of planning has been done, if activity durations have been accurately estimated, and if everything goes well in the field, project completion can be anticipated in 64 working days or about $7/5 \times 64 \approx 90$ calendar days. In this regard, any labor holidays that occur on regular workdays must be added to the 90 calendar days just obtained. As will be seen in Section 5.17, the construction period for the highway bridge will be during the months of June into September. During this time, the holidays of Independence Day and Labor Day will occur. Therefore, the construction period will require approximately 92 calendar days.

The matter of contingency must again be considered at this point. Some provision must be made for general project delays caused by a variety of troubles, oversights, difficulties, and job casualties. At this stage of project scheduling, many contractors will plan on a time overrun of 5 to 10 percent and add this to the overall projected time requirement for the entire work. The percentage actually added must be based on a contractor's judgment and experience. In the bridge network (Chart 5.1a), a contingency of six working days has been added to the diagram in the form of the final contingency activity 410. Adding an overall contingency of six working days gives a probable job duration of 70 working days. To the contractor's way of thinking, 70 working days represents a more realistic estimate of actual project duration than does the value of 64. If everything goes as planned, the job probably will be finished in about 64 working days. However, if the usual difficulties arise, the contractor has allowed for a 70-working-day construction period.

Whether the contractor chooses to add in a contingency allowance or not, now is the time to compare the computed project duration with any established project time requirement. Again, the highway bridge will be used as an example. Assuming a contingency of six working days, the estimated project duration is:

$$7/5 \times (64+6) + 2 \text{ labor holidays} \approx 100 \text{ calendar days}$$

This project duration is compared with the completion date established by the example project master schedule or by the time provision in the construction contract. If the highway bridge must be completed in 90 calendar days, then the contractor will have to consider ways in which to shorten its time duration. If a construction period of 100 calendar days is permissible, the contractor will feel reasonably confident that it will be able to meet this requirement and no action to shorten the work is required. How to go about decreasing a project's duration is the subject of Chapter 7, "Managing Time."

The probable project duration of 100 calendar days, or approximately 14 weeks, is a valuable piece of information. For the first time the contractor has an estimate of overall project duration that it can rely on with considerable trust. The difference in confidence level between the value of the 14 weeks that has just been obtained and the 15 weeks derived earlier from the bar chart in Figure 3.5a should be apparent.

5.11 Late Activity Times

For purposes of discussion in the remainder of this chapter, it is assumed that a project duration of 100 calendar days, or 70 working days, for the highway bridge is acceptable. Unless there is some mitigating circumstance, there is no point in the contractor's attempting to rush the job, and certainly there is nothing to be gained by deliberately allowing the work to drag along. The normal activity durations used as a scheduling basis represent efficient and economical operation. Shorter activity times usually will require expensive expediting actions. Longer activity times suggest a relaxed attitude and increased costs of production. Certainly, job overhead expense increases with the duration of the construction period.

Having established that 70 working days is satisfactory, job calculations now turn around on this value, and a second series of calculations is performed to find the late-start (LS) and the late-finish (LF) times for each activity. These calculations, called the backward pass, start at the project end and proceed backward through the network, going from right to left in the bridge network on the companion website. The late activity times to be computed are the latest times at which the several activities on the highway bridge can be started and finished with project completion still achievable in 70 working days. The supposition during the backward pass is that each activity finishes as late as possible without delaying project completion. The LF value of each activity is obtained first and is entered into the lower right portion of the activity box. The LS, in each case, is obtained by subtracting the activity duration from the LF value. The late-start time is then shown at the lower left. Small sections of the bridge network on the companion website are used in Figure 5.2 to help illustrate the numerical procedures.

The backward pass through the bridge network is begun by giving activity 420 an LF time of 70. Figure 5.2a shows the value of 70 entered in the lower right of activity 420. The LS of an activity is obtained by subtracting the activity duration from its LF value. Activity 420 has a duration of zero. Hence, the LS of activity 420 is its LF of 70 minus its duration of zero, or a value of 70. This value of 70 has been entered at the lower left of activity 420 in Figure 5.2a. Again, this is a trivial calculation, but it is customary to end precedence diagrams with a single closing activity. Continuing with Figure 5.2a, activity 410 immediately precedes activity 420. In working backward through the network as shown in Figure 5.2a, the latest that activity 410 can finish obviously is controlled by the LS of its succeeding activity, 420. If activity 420 must start no later than day 70, then activity 410 must finish no later than that same day. Consequently, activity 410 has an LF time equal to the LS of the activity following (420), or a value of 70. With a duration of 6, it has an LS value of 70 – 6, or 64. These values are shown on activity 410 in Figure 5.2a.

The bridge network (Chart 5.1b) and Figure 5.2b show that activity 400 immediately precedes activity 410. The latest that activity 400 can finish is controlled by the LS of its succeeding activity, 410. If activity 410 must start no later than day 64, then activity 400 must finish no later than that same day. Consequently, activity 400 has an LF time equal to the LS of the activity following (410), or a value of 64. With a duration of 1, it has an LS value of 64 – 1, or 63. These values are shown on activity 400 in Figure 5.2b.

The bridge network (Chart 5.1a) and Figure 5.2c disclose that activity 390 is preceded by three activities: 360, 370, and 380. The LF of each of these activities is set equal to the LS of activity 390, or day 60. Subtracting the activity durations from their LF values yields the LS times. The LS times for activities 360, 370, and 380 are, correspondingly, equal to 57, 55, and 59. These values have been entered in Figure 5.2c.

Some explanation is needed when the backward pass reaches a burst activity (one that has more than one activity immediately following it). In

5.11 Late Activity Times

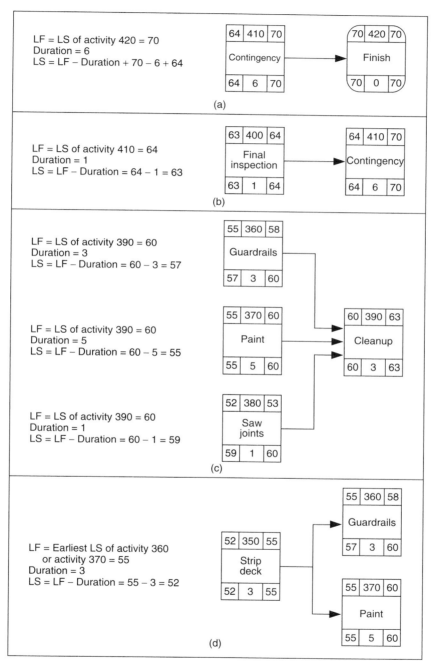

Figure 5.2
Highway bridge, backward-pass calculations

the bridge network on the companion website, activity 350 would be the first such activity reached during the backward pass, and this activity would be followed immediately by activities 360 and 370. To obtain the late finish of activity 350, the late starts of the immediately succeeding activities are noted. These are obtained from the bridge network as 57 for activity 360 and 55 for activity 370 and are entered into Figure 5.2d. Keeping in mind that activity 350 must be finished before either activity 360 or 370 can begin, it is logical that activity 350 must be finished no later than day 55. If it is finished any later than this, the entire project will be delayed by the same amount. The rule for this and other burst activities is that the LF value for such an activity is equal to the earliest (or smallest) of the LS times of the activities that follow.

The backward-pass computations proceed from activity to activity until the start of the project is reached. All that is involved is repetition of the rules just discussed. In making actual network calculations, of course, the LF and LS values are calculated directly on the bridge network, with the times being entered into the activity boxes as they are obtained. The reader is encouraged to verify all of the early and late activity times shown in the networks for the bridge and the building on the companion website as a test of how well the rules of computational procedure have been mastered.

5.12 Total Float

Examination of the activity times appearing in the bridge network (Chart 5.1a) discloses that the early- and late-start times (also early- and late-finish times) are the same for certain activities and not for others. The significance of this fact is that there is leeway in the scheduling of some activities and none at all in the scheduling of others. This leeway is a measure of the time available for a given activity above and beyond its estimated duration. This extra time is called *float,* two classifications of which are in general usage: total float and free float.

The total float of an activity is obtained by subtracting its ES time from its LS time. Subtracting the EF from the LF gives the same result. Once the activity times have been computed on a precedence diagram, values of total float are easily computed and may be noted on the network if desired. This has not been done on the bridge network on the companion website, however, in an attempt to keep the figure as simple as possible. Referring to the bridge network, the total float for a given activity is found as the difference between the two times at the left of the activity box or between the two at the right. The same value is obtained in either case. An activity with zero total float has no spare time and is, therefore, one of the operations that controls project completion time. For this reason, activities with zero total float are called critical activities. The second of the common float types, free float, is discussed in Section 5.14.

5.13 Critical Path

In a precedence diagram, a critical activity is quickly identified as one whose two start times (ES and LS) at the left of the activity box are equal. Also equal are the two finish times (EF and LF) at the right of the activity box. Inspection of the activities in the bridge network on the companion website (Chart 5.1a) discloses that 18 activities have total float values of zero. Plotting these on the figure discloses that these 18 activities form a continuous path from project beginning to project end; this chain of critical activities is called the critical path. The critical path usually is indicated on the diagram in some distinctive way, such as with red or bold lines, which are used in the bridge network on the companion website.

Inspection of the network diagram for the bridge network shows that numerous paths exist between the start and end of the diagram. These paths do not represent alternate choices through the network. Rather, each of these paths must be traversed during the actual construction process. If the time durations of the activities forming a continuous path were to be added for each of the many possible routes through the network, a number of different totals would be obtained. The largest of these totals is the critical or minimum time for overall project completion. Each path must be traveled, so the longest of these paths determines the length of time necessary to complete *all* of the activities in accordance with the established project logic.

If the total times for all of the network paths on the bridge network were to be obtained, it would be found that the longest path is the critical path already identified using zero total floats and that its total time duration is 70 days. Consequently, it is possible to locate the critical path of any network by merely determining the longest path. However, usually this procedure is not practical. The critical path normally is found by means of zero total float values. It needs to be pointed out that the scheduler can be badly fooled by attempting to prejudge which activities will be critical or to locate the critical path by inspection. Critical activities are not necessarily the most difficult or those that seem to be the most important job elements. Although there is only one critical path in the bridge network on the companion website, more than one such path is always a possibility in network diagrams. One path can branch out into a number of paths, or several paths can combine into one. In any event, the critical path or paths must consist of an unbroken chain of activities from start to finish of the diagram. There must be at least one such critical path, and it cannot be intermittent. A break in the path indicates an error in the computations. On the highway bridge, 15 of the 40 activities (exclusive of start, contingency, and finish), or about 38 percent, are critical. This is considerably higher than is the case for most construction networks because of the small size of the highway bridge. In larger diagrams, critical activities generally constitute 20 percent or less of the total.

Any delay in a critical activity automatically lengthens the critical path. Because the length of the critical path determines project duration, any delay in the finish date of a critical activity, for whatever reason, automatically prolongs project completion by the same amount. Thus, identification of the critical activities is an important aspect of job scheduling because it pinpoints those job areas that must be closely monitored at all times if the project is to be kept on schedule.

5.14 Free Float

Free float is another category of spare time. The free float of an activity is found by subtracting its EF time from the earliest of the ES times of the activities directly following. To illustrate how free floats are computed, consider activity 260 in the bridge network, which shows the EF time of activity 260 to be 35. Activity 320 is the only following activity. In this simple case, the free float of activity 260 is the difference between the ES of activity 320 and the EF of activity 260, 43 − 35 = 8 days. When an activity has more than one following activity, the following activity with the earliest ES time controls. An example is activity 90 with an EF time of 6. It is followed by activities 110 and 120 with ES times of 15 and 6, respectively. The free float of activity 90 is therefore 6 − 6 = 0 days. Another example could be activity 270, which has an EF time of 37. The earliest ES date of the following activities, 360 and 370, is 55. Thus, activity 270 has a free float value of 55 − 37 = 18 days. As an alternate statement of procedure, the free float of a given activity can be obtained by subtracting its EF (upper right) from the smallest of the ES values (upper left) of those activities immediately following.

The free float of an activity is the amount by which the completion of that activity can be deferred without delaying the early start of the following activities or affecting any other activity in the network. To illustrate, activity 270, which has a free float of 18 days, can have its completion delayed by up to 18 days because of late start, extended duration, or any combination thereof, without affecting any other network activity.

5.15 Activity Time Information

Information concerning activity times and float values can be presented in three ways: on the network diagram itself, in tabular format, and in the form of bar charts. Using the precedence diagram of the highway bridge as a basis, the preceding sections presented the manual computation of activity times and floats. When such computations are made, they are almost always performed directly on the network. This procedure is faster, more convenient, and conducive to greater accuracy.

To serve a variety of purposes, activity times and floats can be presented in table form, as has been done in Figures 5.3a for the highway bridge and

5.15 Activity Time Information

Activity (Bold type denotes critical activities) (1)	Activity Number (2)	Activity Arrow (3)	Duration (Working Days) (4)	Earliest Start (ES) (5)	Earliest Finish (EF) (6)	Latest Start (LS) (7)	Latest Finish (LF) (8)	Float Total (TF) (9)	Float Free (FF) (10)
Start	0		0	0	0	0	0	0	0
Prepare & approve S/D abutment & deck rebar	**10**	**10–50**	**10**	**0**	**10**	**0**	**10**	**0**	**0**
Prepare & approve S/D footing rebar	20	10–60	5	0	5	9	14	9	0
Order & deliver piles	30	10–70	15	0	15	3	18	3	0
Move in	40	10–20	3	0	3	12	15	12	0
Prepare & approve S/D girders	50	10–40	10	0	10	8	18	8	0
Fabricate & deliver abutment & deck rebar	**60**	**50–160**	**15**	**10**	**25**	**10**	**25**	**0**	**0**
Fabricate & deliver footing rebar	70	60–100	7	5	12	14	21	9	6
Prefabricate abutment forms	80	20–160	3	3	6	22	25	19	19
Excavate abutment #1	90	20–30	3	3	6	15	18	12	0
Mobilize pile-driving rig	100	20–70	2	3	5	16	18	13	10
Drive piles-abutment #1	110	70–80	3	15	18	18	21	3	0
Excavate abutment #2	120	30–90	2	6	8	19	21	13	10
Forms & rebar footing #1	130	100–110	2	18	20	21	23	3	0
Drive piles abutment #2	140	90–140	3	18	21	21	24	3	0
Pour footing #1	150	110–120	1	20	21	23	24	3	0
Demobilize pile-driving rig	160	140–160	1	21	22	24	25	3	3
Strip footing #1	170	120–130	1	21	22	24	25	3	0
Forms & rebar abutment #1	**180**	**160–180**	**4**	**25**	**29**	**25**	**29**	**0**	**0**
Forms & rebar footing #2	190	150–170	2	22	24	30	32	8	0
Pour abutment #1	**200**	**180–200**	**2**	**29**	**31**	**29**	**31**	**0**	**0**
Pour footing #2	210	170–190	1	24	25	32	33	8	0
Strip & cure abutment #1	**220**	**200–210**	**3**	**31**	**34**	**31**	**34**	**0**	**0**
Strip footing #2	230	190–220	1	25	26	33	34	8	8
Forms & rebar abutment #2	**240**	**220–230**	**4**	**34**	**38**	**34**	**38**	**0**	**0**
Pour abutment #2	**250**	**230–240**	**2**	**38**	**40**	**38**	**40**	**0**	**0**
Fabricate & deliver girders	260	40–260	25	10	35	18	43	8	8
Rub concrete abutment #1	270	210–300	3	34	37	52	55	18	18
Backfill abutment #1	280	210–260	3	34	37	40	43	6	6
Strip & cure abutment #2	**290**	**240–250**	**3**	**40**	**43**	**40**	**43**	**0**	**0**
Rub concrete abutment #2	300	250–300	3	43	46	52	55	9	9
Backfill abutment #2	310	250–280	3	43	46	46	49	3	3
Set girders	**320**	**260–270**	**2**	**43**	**45**	**43**	**45**	**0**	**0**
Deck forms & rebar	**330**	**270–280**	**4**	**45**	**49**	**45**	**49**	**0**	**0**
Pour & cure deck	**340**	**280–290**	**3**	**49**	**52**	**49**	**52**	**0**	**0**
Strip deck	**350**	**290–300**	**3**	**52**	**55**	**52**	**55**	**0**	**0**
Guardrails	360	300–310	3	55	58	57	60	2	2
Paint	**370**	**300–320**	**5**	**55**	**60**	**55**	**60**	**0**	**0**
Saw joints	380	290–320	1	52	53	59	60	7	7
Cleanup	**390**	**320–330**	**3**	**60**	**63**	**60**	**63**	**0**	**0**
Final inspection	**400**	**330–340**	**1**	**63**	**64**	**63**	**64**	**0**	**0**
Contingency	**410**	**340–350**	**6**	**64**	**70**	**64**	**70**	**0**	**0**
Finish	**420**		**0**	**70**	**70**	**70**	**70**	**0**	**0**

Figure 5.3a
Highway bridge, activity times

5.3b for the commercial building. Such a table of values serves to collect pertinent time information of the project into a form that is useful and convenient. In the figures, column 2 lists the activity numbers as they appear in the precedence diagram of the bridge network or the commercial building network, with the critical activities appearing in boldface type. Column 3 in the highway bridge table provides the activity arrow designations for reference to the AOA diagram provided on the companion website.

1	2	3	4	5	6	7	8	9
Activity (BF type denotes critical activities)	Act #	Dur	ES	EF	LS	LF	TF	FF
Start	1	0	0	0	0	0	0	0
Mobilize	2	5	0	5	25	30	25	25
Utilities	3	45	0	45	80	125	20	80
Site Work	4	65	0	65	0	65	0	0
Site Electrical	5	15	65	80	205	220	140	140
Site Paving, Start	6	45	65	110	125	170	60	50
Site Paving, Finish	7	20	160	180	170	190	10	0
Irrigation	8	15	180	195	190	205	0	10
Landscape	9	15	195	200	205	220	10	10
Drilled Piers	10	10	20	30	20	30	0	0
Grade Beams & Slabs	11	30	30	60	30	60	0	0
Structural Erection	12	15	60	75	60	75	0	0
Miscellaneous Steel	13	20	75	95	125	145	50	10
Roofing	14	20	75	95	85	105	10	0
Framing	15	30	75	105	75	105	0	0
Dampproof	16	25	105	130	115	140	10	0
Masonry	17	30	130	160	140	170	10	0
Insulation	18	30	105	135	105	135	0	0
Gypsum Drywall	19	30	135	165	135	165	0	0
Aluminum Frames	20	30	105	135	120	150	15	0
Wood Doors	21	15	135	150	150	165	15	15
Storefront & Glazing	22	15	135	150	170	185	35	15
Plaster	23	20	105	120	145	165	40	40
Ceilings	24	20	165	185	185	205	20	0
Ceramic Tile	25	20	165	185	185	205	20	0
Flooring	26	15	185	200	205	220	20	20
Paint	27	35	165	200	165	200	0	0
Millwork	28	20	200	220	200	220	0	0
Signage	29	15	200	215	205	220	10	0
Toilet Partitions	30	15	185	200	205	220	20	20
Toilet Accessories	31	15	185	200	205	220	20	20
HVAC Rough-in	32	30	95	125	175	205	80	75
HVAC Finish	33	15	200	215	205	220	5	5
Plumbing under Slab	34	35	35	70	75	110	40	35
Plumbing in Partitions	35	25	105	130	110	135	5	5
Plumbing Fixtures	36	15	200	215	205	220	5	5
Fire Protection Rough-in	37	15	105	120	190	205	85	80
Fire Protection Finish	38	15	200	215	205	220	5	5
Electrical under Slab	39	30	35	65	75	105	40	40
Electrical in Partitions	40	30	105	135	105	135	0	0
Electrical Finish	41	50	150	200	170	220	20	20
Fire Alarm Rough-in	42	20	95	115	190	210	100	90
Fire Alarm Finish	43	10	205	215	210	220	5	5
Closeout	44	5	220	225	220	225	0	0

Figure 5.3b
Commercial building, activity times

5.16 Float Paths

Activity time data presented in bar chart form are widely used for project time control purposes during the construction process. Project time information in the form of bar charts is discussed in Section 5.29.

All paths through the diagram in the bridge network, except the critical path, have summations of activity times less than 70 working days and are called float paths. The longest float path, or the one with the least float, can be determined by referring to the total float values of Figure 5.3a and noting that one activity (360) has a total float value of only two days. This indicates that the next longest path through the network is 70 − 2 = 68 working days. Reference to the bridge network will show that this path is the same as the critical path except that activity 360 is substituted for activity 370. This path is said to have a float value of 2. The total float values of 3 in Figure 5.3a indicate that the next longest path through the network totals 67 days. As a matter of fact, there are two different paths through the network, each with a cumulative time total of 67 days or with float times of 3. Float paths can be located by linking activities with the same total floats together with critical activities as needed to form a continuous chain through the entire diagram.

When beginning a backward pass through a construction network, an established target duration for the project may be used for the "turn around" rather than the EF of the terminal activity. In such a case, it is possible for all paths through the network to be float paths. To illustrate, suppose that the backward pass through the highway bridge network on the companion website had started with the LF of terminal activity 420 equal to 74 rather than 70. The effect of this on the values shown in Figure 5.3a would be to increase all of the late activity times (columns 6 and 7) and all of the total float values by the constant amount of 4. In this case, there would be no activities with zero total floats. Nevertheless, there is still a critical path: the float path with a minimum and constant total float value of 4. This path would be the same as the original critical path. The only difference is that each critical activity would now have a total float of 4 rather than zero. In a similar manner, if 67 had been used as the LF of terminal activity 420, all late activity times and total floats would be reduced by 3. In this case, the critical path would have a constant total float of −3 and would be the same critical path as that originally found.

The concept of shared float is of importance. The notion of float on construction projects is often misunderstood and, if not monitored and controlled, can become a source of conflict. Although the delay of an activity from its early-start position by a duration equal to less than its total float will not impact the overall project duration, it consumes float from each of the paths of which it is a member. Therefore, the use of float by one activity reduces the amount of float available to other activities. Activity total float is shared by every activity on its corresponding paths. Consequently, no single activity has unilateral ownership of available float. Rather, it must be shared, managed, and distributed as a fundamental project resource.

5.17 Early-Start Schedule

After the network calculations have been completed, the resulting activity times are used to prepare various forms of calendar date schedules that are used for project time management. One of these is a schedule of activities based on their early-start and finish times. This time schedule is called an early-start schedule or a normal schedule. The subject of preparing field operation schedules is discussed in Chapter 7, "Managing Time." The treatment of early-start schedules in this chapter is limited to introducing the general concept of project schedules and explaining how expired working days are converted to calendar dates.

Activity times obtained by manual calculations are expressed in terms of expired working days. For purposes of project monitoring and control, it is necessary to convert these times to calendar dates on which each activity is expected to start and finish. This is done easily with the aid of a calendar on which the working days are numbered consecutively, starting with number one on the anticipated start date and skipping weekends and holidays (July 5 and September 6 in our case). Figure 5.4a is the conversion calendar for the highway bridge, assuming that the starting date is to be Monday, June 14. Calendar dates appear in the upper left-hand corner of each box, and working days are circled.

The reader is reminded at this point that each major portion of the example projects is described and scheduled using its own unique planning network. When making manual time computations, each of the networks begins at zero expired working days. It is at this stage of conversion of expired working days to calendar dates that the time relationships among the various networks are established. There is a different calendar similar to Figure 5.4a or Figure 5.4b for each network; from this calendar, its unique calendar date time schedule is obtained.

When making up a job calendar, the true meaning of elapsed working days must be kept in mind. To illustrate, the early start of activity 180 in Figure 5.3a is 25. This means this activity can start after the expiration of 25 working days, so the starting date of activity 180 will be the morning of calendar date numbered 26. From Figure 5.4a, working day 26 equates to the calendar date of July 20. There is no such adjustment for early-finish dates. In the case of activity 180, its early-finish time is 29, which indicates that it is finished by the end of the twenty-ninth working day. Hence, from Figure 5.4a, the early-finish date of that activity will be the afternoon of July 23.

5.18 Tabular Time Schedules   **119**

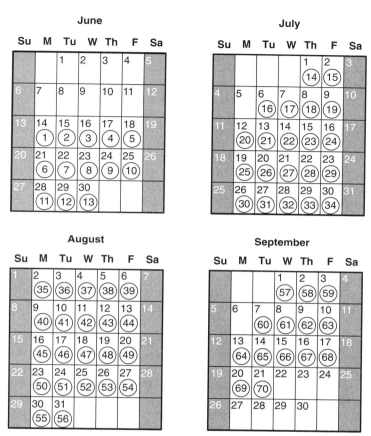

Figure 5.4a
Highway bridge, conversion calendar

It has been mentioned previously that calendar date information concerning network activities can be presented in different ways, depending on how the data will be used. Chart 5.2 on the companion website shows one way in which computers can enter the early-start and early-finish dates of the individual activities on the precedence diagram for the highway bridge and the building.

5.18 Tabular Time Schedules

When the computed activity times are converted to calendar dates, this schedule information often is presented in a tabular format. The table in Figure 5.5a is the early-start schedule for the highway bridge, with the activities listed in the order of their starting dates. Figure 5.5b provides

5 Project Scheduling Concepts

December
Su	M	Tu	W	Th	F	Sa
1	2	3	4	5	6	7
8	9	10	11	12	13 0	14
15	16 1	17 2	18 3	19 4	20 5	21
22	23 6	24 7	25	26 8	27 9	28
29	30 10	31 11				

January
Su	M	Tu	W	Th	F	Sa
			1	2 12	3 13	4
5	6 14	7 15	8 16	9 17	10 18	11
12	13 19	14 20	15 21	16 22	17 23	18
19	20 24	21 25	22 26	23 27	24 28	25
26	27 29	28 30	29 31	30 32	31 33	

February
Su	M	Tu	W	Th	F	Sa
						1
2	3 34	4 35	5 36	6 37	7 38	8
9	10 39	11 40	12 41	13 42	14 43	15
16	17 44	18 45	19 46	20 47	21 48	22
23	24 49	25 50	26 51	27 52	28 53	

March
Su	M	Tu	W	Th	F	Sa
						1
2	3 54	4 55	5 56	6 57	7 58	8
9	10 59	11 60	12 61	13 62	14 63	15
16	17 64	18 65	19 66	20 67	21 68	22
23	24 69	25 70	26 71	27 72	28 73	29
30	31 74					

April
Su	M	Tu	W	Th	F	Sa
		1 75	2 76	3 77	4 78	5
6	7 79	8 80	9 81	10 82	11 83	12
13	14 84	15 85	16 86	17 87	18 88	19
20	21 89	22 90	23 91	24 92	25 93	26
27	28 94	29 95	30 96			

May
Su	M	Tu	W	Th	F	Sa
				1 97	2 98	3
4	5 99	6 100	7 101	8 102	9 103	10
11	12 104	13 105	14 106	15 107	16 108	17
18	19 109	20 110	21 111	22 112	23 113	24
25	26	27 114	28 115	29 116	30 117	31

June
Su	M	Tu	W	Th	F	Sa
1	2 118	3 119	4 120	5 121	6 122	7
8	9 123	10 124	11 125	12 126	13 127	14
15	16 128	17 129	18 130	19 131	20 132	21
22	23 133	24 134	25 135	26 136	27 137	28
29	30 138					

July
Su	M	Tu	W	Th	F	Sa
		1 139	2 140	3 141	4	5
6	7 142	8 143	9 144	10 145	11 146	12
13	14 147	15 148	16 149	17 150	18 151	19
20	21 152	22 153	23 154	24 155	25 156	26
27	28 157	29 158	30 159	31 160		

August
Su	M	Tu	W	Th	F	Sa
					1 161	2
3	4 162	5 163	6 164	7 165	8 166	9
10	11 167	12 168	13 169	14 170	15 171	16
17	18 172	19 173	20 174	21 175	22 176	23
24	25 177	26 178	27 179	28 180	29 181	30
31						

September
Su	M	Tu	W	Th	F	Sa
	1 182	2 183	3 184	4 185	5	6
7	8 186	9 187	10 188	11 189	12 190	13
14	15 191	16 192	17 193	18 194	19 195	20
21	22 196	23 197	24 198	25 199	26 200	27
28	29 201	30 202				

October
Su	M	Tu	W	Th	F	Sa
			1 203	2 204	3 205	4
5	6 206	7 207	8 208	9 209	10 210	11
12	13 211	14 212	15 213	16 214	17 215	18
19	20 216	21 217	22 218	23 219	24 220	25
26	27 221	28 222	29 223	30 224	31 225	

November
Su	M	Tu	W	Th	F	Sa
						1
2	3	4	5	6	7	8
9	10	11	12	13	14	15
16	17	18	19	20	21	22
23	24	25	26	27	28	29
30						

Figure 5.4b
Commercial building, conversion calendar

5.18 Tabular Time Schedules

Project Calendar				
Activity (Bold type denotes critical activities)	Activity Number	Duration (Working Days)	Scheduled Starting Date A.M.	Scheduled Completion Date P.M.
Prepare & approve S/D, abutment & deck rebar	**10**	**10**	**June-14**	**June-25**
Prepare & approve S/D, footing rebar	20	5	June-14	June-18
Order & deliver piles	30	15	June-14	July-2
Move in	40	3	June-14	June-16
Prepare & approve S/D, girders	50	10	June-14	June-25
Prefabricate abutment forms	80	3	June-17	June-21
Excavate abutment #1	90	3	June-17	June-21
Mobilize pile-driving rig	100	2	June-17	June-18
Fabricate & deliver footing rebar	70	7	June-21	June-29
Excavate abutment #2	120	2	June-22	June-23
Fabricate & deliver abutment & deck rebar	**60**	**15**	**June-28**	**July-19**
Fabricate & deliver girders	260	25	June-28	August-2
Drive piles, abutment #1	110	3	July-6	July-8
Forms & rebar, footing #1	130	2	July-9	July-12
Drive piles, abutment #2	140	3	July-9	July-13
Pour footing #1	150	1	July-13	July-13
Demobilize pile-driving rig	160	1	July-14	July-14
Strip footing #1	170	1	July-14	July-14
Forms & rebar, footing #2	190	2	July-15	July-16
Pour footing #2	210	1	July-19	July-19
Forms & rebar, abutment #1	**180**	**4**	**July-20**	**July-23**
Strip footing #2	230	1	December-21	July-20
Pour abutment #1	**200**	**2**	**July-26**	**July-27**
Strip & cure, abutment #1	**220**	**3**	**July-28**	**July-30**
Forms & rebar, abutment #2	**240**	**4**	**August-2**	**August-5**
Rub concrete, abutment #1	270	3	August-2	August-4
Backfill abutment #1	280	3	August-2	August-14
Pour abutment #2	**250**	**2**	**August-6**	**August-9**
Strip & cure, abutment #2	**290**	**3**	**August-10**	**August-12**
Rub concrete, abutment #2	300	3	August-13	August-17
Backfill abutment #2	310	3	August-13	August-17
Set girders	**320**	**2**	**August-13**	**August-16**
Deck forms & rebar	**330**	**4**	**August-17**	**August-20**
Pour & cure deck	**340**	**3**	**August-23**	**August-25**
Strip deck	**350**	**3**	**August-26**	**August-30**
Saw joints	380	1	August-26	August-26
Guardrails	360	3	August-31	September-2
Paint	**370**	**5**	**August-31**	**September-7**
Cleanup	**390**	**3**	**September-8**	**September-10**
Final inspection	**400**	**1**	**September-13**	**September-13**
Contingency	**410**	**6**	**September-14**	**September-21**

Figure 5.5a
Highway bridge, early-start schedule

ID	Task Name	Start	Finish	Late Start	Late Finish	Free Slack	Total Slack
1	Start	Mon 12/16/13	Mon 12/16/13	Mon 12/16/13	Mon 12/16/13	0 days	0 days
2	Mobilize	Mon 12/16/13	Fri 12/20/13	Mon 12/8/14	Fri 12/12/14	60 days	255 days
3	Utilities	Mon 12/16/13	Tue 2/18/14	Wed 4/9/14	Wed 6/11/14	20 days	80 days
4	Site Work	Mon 12/16/13	Tue 3/18/14	Mon 12/16/13	Tue 3/18/14	0 days	0 days
5	Site Electrical	Wed 3/19/14	Tue 4/8/14	Mon 10/6/14	Fri 10/24/14	140 days	140 days
6	Site Paving, Start	Wed 3/19/14	Tue 5/20/14	Thu 6/12/14	Thu 8/14/14	50 days	60 days
7	Site Paving, Finish	Fri 8/1/14	Thu 8/28/14	Fri 8/15/14	Fri 9/12/14	0 days	10 days
8	Irrigation	Fri 8/29/14	Fri 9/19/14	Mon 9/15/14	Fri 10/3/14	0 days	10 days
9	Landscape	Mon 9/22/14	Fri 10/10/14	Mon 10/6/14	Fri 10/24/14	10 days	10 days
10	Drilled Piers	Wed 1/15/14	Tue 1/28/14	Wed 1/15/14	Tue 1/28/14	0 days	0 days
11	Grade Beams & Slabs	Wed 1/29/14	Tue 3/11/14	Wed 1/29/14	Tue 3/11/14	0 days	0 days
12	Structural Erection	Wed 3/12/14	Tue 4/1/14	Wed 3/12/14	Tue 4/1/14	0 days	0 days
13	Miscellaneous Steel	Wed 4/2/14	Tue 4/29/14	Thu 6/12/14	Thu 7/10/14	10 days	50 days
14	Roofing	Wed 4/2/14	Tue 4/29/14	Wed 4/16/14	Tue 5/13/14	0 days	10 days
15	Framing	Wed 4/2/14	Tue 5/13/14	Wed 4/2/14	Tue 5/13/14	0 days	0 days
16	Dampproof	Wed 5/14/14	Wed 6/18/14	Thu 5/29/14	Wed 7/2/14	0 days	10 days
17	Masonry	Thu 6/19/14	Thu 7/31/14	Thu 7/3/14	Thu 8/14/14	0 days	10 days
18	Insulation	Wed 5/14/14	Wed 6/25/14	Wed 5/14/14	Wed 6/25/14	0 days	0 days
19	Gypsum Drywall	Thu 6/26/14	Thu 8/7/14	Thu 6/26/14	Thu 8/7/14	0 days	0 days
20	Aluminum Frames	Wed 5/14/14	Wed 6/25/14	Thu 6/5/14	Thu 7/17/14	0 days	15 days
21	Wood Doors	Thu 6/26/14	Thu 7/17/14	Fri 7/18/14	Thu 8/7/14	15 days	15 days
22	Storefront, Glazing	Thu 6/26/14	Thu 7/17/14	Fri 8/15/14	Fri 9/5/14	15 days	35 days
23	Plaster	Wed 5/14/14	Wed 6/11/14	Fri 7/11/14	Thu 8/7/14	40 days	40 days
24	Ceilings	Fri 8/8/14	Fri 9/5/14	Mon 9/8/14	Fri 10/3/14	0 days	20 days
25	Ceramic Tile	Fri 8/8/14	Fri 9/5/14	Mon 9/8/14	Fri 10/3/14	0 days	20 days
26	Flooring	Mon 9/8/14	Fri 9/26/14	Mon 10/6/14	Fri 10/24/14	20 days	20 days
27	Paint	Fri 8/8/14	Fri 9/26/14	Fri 8/8/14	Fri 9/26/14	0 days	0 days
28	Millwork	Mon 9/29/14	Fri 10/24/14	Mon 9/29/14	Fri 10/24/14	0 days	0 days
29	Signage	Mon 9/29/14	Fri 10/17/14	Mon 10/13/14	Fri 10/31/14	0 days	10 days
30	Toilet Partitions	Mon 9/8/14	Fri 9/26/14	Mon 10/6/14	Fri 10/24/14	20 days	20 days
31	Toilet Accesseries	Mon 9/8/14	Fri 9/26/14	Mon 10/6/14	Fri 10/24/14	20 days	20 days
32	HVAV Rough in	Wed 4/30/14	Wed 6/11/14	Fri 8/22/14	Fri 10/3/14	75 days	80 days
33	HVAC Finish	Mon 9/29/14	Fri 10/17/14	Mon 10/6/14	Fri 10/24/14	5 days	5 days
34	Plumbing, under Slab	Wed 2/5/14	Tue 3/25/14	Wed 4/2/14	Tue 5/20/14	35 days	40 days
35	Plumbing in Partitions	Wed 5/14/14	Wed 6/18/14	Wed 5/21/14	Wed 6/25/14	5 days	5 days
36	Plumbing Fixtures	Mon 9/29/14	Fri 10/17/14	Mon 10/6/14	Fri 10/24/14	5 days	5 days
37	Fire Protection Rough in	Wed 5/14/14	Wed 6/4/14	Mon 9/15/14	Fri 10/3/14	80 days	85 days
38	Fire Protection Finish	Mon 9/29/14	Fri 10/17/14	Mon 10/6/14	Fri 10/24/14	5 days	5 days
39	Electrical, under Slab	Wed 2/5/14	Tue 3/18/14	Wed 4/2/14	Tue 5/13/14	40 days	40 days
40	Electrical, RI Partitions	Wed 5/14/14	Wed 6/25/14	Wed 5/14/14	Wed 6/25/14	0 days	0 days
41	Electrical Finish	Fri 7/18/14	Fri 9/26/14	Fri 8/15/14	Fri 10/24/14	20 days	20 days
42	Fire Alarm Rough in	Wed 4/30/14	Wed 5/28/14	Mon 9/22/14	Fri 10/17/14	90 days	100 days
43	Fire Alarm Finish	Mon 10/6/14	Fri 10/17/14	Mon 10/13/14	Fri 10/24/14	5 days	5 days
44	Closeout	Mon 10/27/14	Fri 10/31/14	Mon 10/27/14	Fri 10/31/14	0 days	0 days

Figure 5.5b
Commercial building, early-start schedule

the comparable information for the commercial building. These calendar dates often are referred to as scheduled or expected dates. Not only is this sort of operational schedule useful to the contractor; it also can be used to satisfy the usual contract requirement of providing the owner and architect-engineer with a projected timetable of construction operations.

Whether the contractor prepares activity schedule data in the form of network diagram information or activity timetables depends on the use for which the information is intended. It is important to note that although tabular reports provide activity numbers, descriptions, schedule dates, and float information, they do not reflect project logic. Tabular reports, such as bar charts, communicate only basic information concerning individual activities. They do not communicate the sequence of activities and, therefore, are not diagnostic tools. Only network diagrams have this capability. This fact means that the form of time control information provided to a member of the project management team must be selected to meet the demands and responsibilities of that position.

5.19 Activities and Calendar Dates

Until elapsed working days have been converted to calendar dates, there is no accurate way to associate activities with calendar times. In the case of short-duration work, such as the highway bridge, this has not been a problem because it was recognized from the beginning that the work would be done during the summer months. However, on projects requiring many months or years, it is important to associate general classes of work with the seasons of the year during which the work will be performed. The general time schedule developed during project cost estimating has provided guidance in this regard. Nevertheless, this preliminary construction schedule is at best approximate and may be altered during project planning. Consequently, the first version of the working job calendar must be examined with the objective of comparing activities with the weather expected during their accomplishment. This perusal might well disclose some activities that should be expedited or delayed to avoid cold or wet weather, spring runoff, or other seasonal hazards. It may reveal a need for cold-weather operations hitherto unanticipated. It can be a good guide for the final inclusion of weather contingency allowances.

5.20 Calendars for Weather

Estimating and presenting the effect of weather on construction processes is a difficult task and requires both knowledge of the geographic location in which the work is to be performed and specific experience with the kind of construction operations considered. Several methods are currently in use for estimating these effects and adding them to the project plan. As discussed, a contingency may be added to the schedule to account for probable weather delays. This contingency can be either added to specific activities or combined generally and placed at the closure of a string of operations.

A preferred method for presenting weather effects is through the use of weather calendars. By determining the expected number of days lost to weather per calendar month and then removing those days randomly from the work calendar, the scheduler may indirectly extend the duration of the affected activities by the number of days removed from the calendar. Using this same technique, the scheduler may prepare various levels of weather-affected calendars and assign the individual activities to them based on the operation's sensitivity to the environment. As an example, consider two activities on the highway bridge project that have differing sensitivities to weather: activity 90, "Excavate abutment #1," and activity 370, "Paint." While abutment excavation may continue until two inches of rain has fallen within a 24-hour period, the more sensitive painting operation must be suspended if more than one inch of rain falls. Historical weather records for this project location indicate that five days in January receive more than one inch of rainfall and three days receive more than two inches. In response, the scheduler can prepare two separate weather calendars: one for highly sensitive operations such as painting and another for less sensitive operations such as abutment excavation. Activity 90 then would be assigned to the first calendar, while activity 370 would be assigned to the second. Consequently, each of these operations would be halted during the days removed from their respective weather calendars. This would cause the activity to take a commensurately longer period to complete.

This method of accounting for weather effects is particularly powerful, as it automatically adjusts for large changes in the schedule. If an activity is originally scheduled for performance during the winter and later is rescheduled for summer, the seasonal effect of weather is accounted for and commensurately adjusted without the scheduler's intervention. Additionally, this method withstands the close scrutiny of owners and public authorities that frequently are concerned with the amount of contingency added to construction schedules and the computational methods supporting them. A more detailed handling of weather delays is included in Section 9.10.

5.21 Lags between Activities

All previous discussions of project planning and scheduling have been on the basis of two important assumptions. One is that an activity cannot start until *all* the *immediately* preceding activities have been completed. The other is that once all preceding activities have been finished, the following activity can immediately start. Although these assumptions are more or less true for most network relationships, there are instances where they are

5.21 Lags between Activities

not. Consequently, a more flexible notation convention has been developed that can be used to show these more complex activity relationships. Such a system involves the use of lags between activities.

Figure 5.6 presents several examples of the use of lags for the highway bridge. Unfortunately, nomenclature in this area is not standardized, and "lead times" often are used to describe exactly the same precedence relationships. This is merely a matter of which activity is taken as the reference. A successor "lags" a predecessor, but a predecessor "leads" a successor. In this text, only lag times are used. Lag time can be designated on a dependency line with a positive, negative, or zero value. If no time is designated, it is assumed to be zero. In effect, a negative lag time is a lead time.

It is not intended that the activities and activity times shown in Figure 5.6 relate or pertain to the highway bridge. These are just general examples designed to illustrate each particular time relationship. The activity times shown in the figure, diagrams 1 through 8, illustrate how the forward-pass and backward-pass computations are performed between the two affected activity boxes.

Diagram 1. This figure shows that the succeeding activity 190 can start no earlier than the completion of activity 180. Here, no value of lag time is indicated, which means a value of zero, or that there is no lag between the finish of activity 180 and the beginning of activity 190. This figure indicates that the wall concrete can be poured immediately after the wall forms have been completed with no delay in between. This is the usual form of activity dependency and is the only one that has been used in the network diagrams presented thus far in this text.

Diagram 2. This figure indicates that activity 200 cannot start until three days after activity 190 has been completed. This condition is reflected in the time calculations by making the early start of activity 200 equal to the early finish of activity 190, plus the time delay of three days, or a value of 46.

Diagram 3. The dependency arrow in this figure shows that activity 120 can start no earlier than the start of activity 110. Since there is no lag value shown (hence a value of zero), activity 120 can start immediately after the start of activity 110. In this case, the ES of activity 120 is equal to the ES of activity 110.

Diagram 4. This figure shows that after one day of stripping wall forms, the cement masons can start patching and rubbing the wall surfaces. The ES of activity 210 is made equal to the ES value of activity 200, plus one day of lag, for a value of 47.

Diagram 5. Here the finish of activity 270 occurs immediately after, but only after, the completion of activity 260. In this case, the EF of activity 270 is made the same as the EF of activity 260.

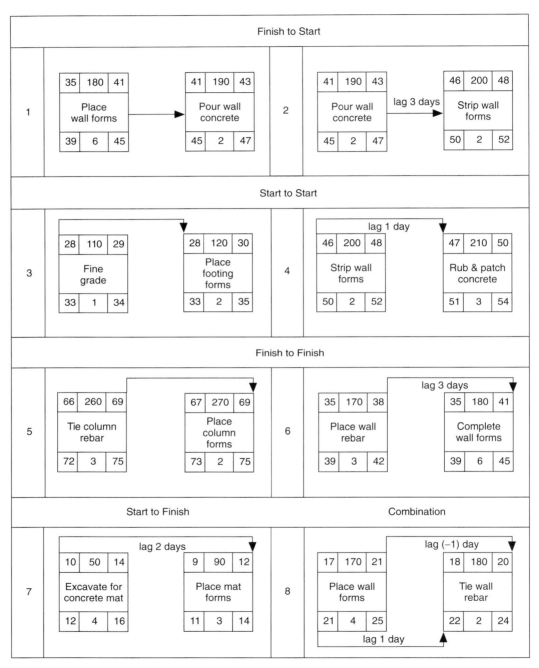

Figure 5.6
Lag relationships, precedence notation

Diagram 6. This diagram shows that the finish of activity 180 follows the completion of activity 170 by three days. In the time calculations, activity 180 will have an EF equal to the EF value of activity 170, plus the three days lag, for a value of 41.

Diagram 7. The start-to-finish dependency shown here indicates that the finish of activity 90 is achieved two days after the start of activity 50. Two days following the start of activity 50 will be its ES (10 expired working days) plus the delay of two days, or a value of 12 for the EF of activity 90.

Diagram 8. The combination of lags shown indicates these conditions: The start-to-start lag of one day indicates that tying reinforcing steel can start one day after the wall forming has begun. The finish-to-finish dependency of −1 (note that this lag is minus and could be described as a lead time of +1) indicates that the placing of wall forms cannot finish until one day after completion of the reinforcing steel placement.

Any construction network involves many simplifications and approximations of reality. Certainly, among these is the usual assumption that an instantaneous transition occurs between a completed activity and those that immediately follow it. It is undoubtedly true that most sequential transitions between successive activities on actual construction jobs involve either some overlapping or some delay between the finish of one and the start of the next. Absolute precision in making up project networks would necessitate the widespread usage of lag relationships, which would complicate the planning and scheduling process substantially with little gain in management efficacy. For this reason, lag times are not used extensively except where the time effects are substantial or for special construction types. The use of lags is especially convenient when working with long strings of simultaneous and repetitive operations, which is demonstrated with a pipeline relocation example.

5.22 Pipeline Scheduling Computations

Network computations for projects that involve repetitive operations proceed in precisely the same manner as for any other project. Figure 5.7, which depicts the same job logic as Figure 4.4, illustrates the determination of activity times for the pipeline relocation discussed in Section 4.15. In actual practice, the contractor undoubtedly would have the crews and equipment in better balance than indicated by the activity times used in Figure 5.7. In other words, the times to accomplish a mile of location and clearing, a mile of excavation, and a mile of the other operations would be about the same. This fact obviously would help to prevent the undesirable situation of having one operation unduly limit another with attendant

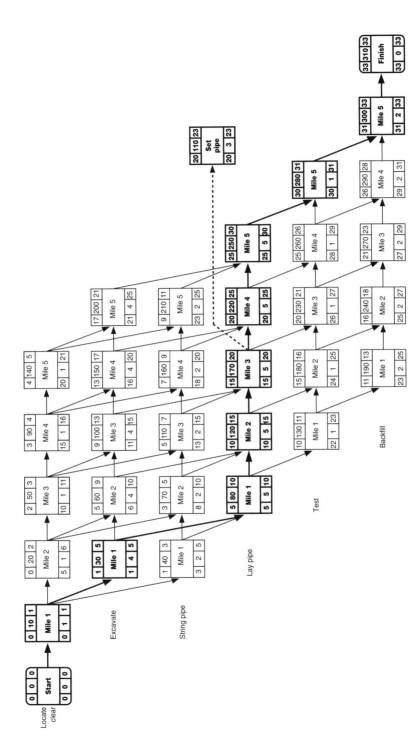

Figure 5.7
Pipeline relocation, precedence diagram time computations

128

wasted time and loss of operational efficiency. The activity times used in the figure were selected for purposes of illustrating the generality of the procedure. The bold lines in the figure constitute the critical path, located by those activities with zero total float.

5.23 Pipeline Summary Diagram

Summary diagrams for repetitive operation projects are of particular importance. For example, suppose the pipeline in Figure 5.7 were 20 miles in length rather than 5. A detailed figure like Figure 5.7 for the entire project length would be impossibly large. Figure 5.8 is a condensation of Figure 5.7 obtained by using lag notation. Figure 5.8 shows all 5 miles of each individual operation as a single activity box and is of a type that lends itself well to long strings of repeated operations. For example, activity 10 in this figure represents the locating and clearing of the entire 5 miles of right-of-way and has a duration of 5 days. In a similar manner, activity 20 represents all the excavation and has a total time duration of 20 days. The other four job operations are shown in a similar manner. As explained previously, excavation and string pipe can proceed simultaneously with one another following location and clearing.

To discuss the activity time computations with regard to Figure 5.8, reference is made to activity 40, which may be considered typical.

1. *Computation of ES.* The ES of activity 40 is computed twice. One possible value is the ES of activity 20 added to the delay of 4, giving a value of 5. The other is the ES of activity 30 added to the delay of 2, or a value of 3. Since activity 40 cannot start until four days after activity 20 has started and two days after activity 30 has started, it is most severely restrained by activity 20 and, therefore, has an ES of 5.
2. *Computation of EF.* The EF of activity 40 is computed three times. First, the addition of the EF of activity 20 to the delay of 5 gives a value of 26. Second, adding the delay of 5 to the EF of activity 30 yields 16. Third, adding the ES of activity 40 to its duration of 25 gives 30. The largest value (30) is the EF value.
3. *Computation of LF.* The LF of activity 40 has only one possible value. This is computed by subtracting the lag of one day from the LF of activity 50, giving a value of 30.
4. *Computation of LS.* The LS of activity 40 has two possible values. First, the duration of activity 40 is 25. Subtracting this from its EF value of 30 yields 5. Second, the LS of activity 50 is 22. Subtracting the five-day lag from this gives 17. The LS of activity 40 is the smaller of 5 and 17, or a value of 5.

The use of such a summary diagram can sacrifice much of the internal logic. Reference is made to the location of the critical path in Figure 5.8,

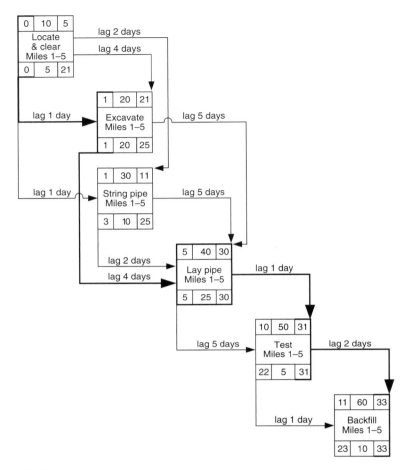

Figure 5.8
Pipeline relocation, summary precedence diagram

which does not reveal the same level of detail as does Figure 5.7. To illustrate this point, it should be noted in Figure 5.8 that summary activities 10 and 20 have equal values of ES and LS. However, there are different values of EF and late LF. These values indicate that part, but not all, of the "Locate & clear" sequence (activity 10) and of the "Excavate" sequence (activity 20) are critical. In themselves, these summary activities do not indicate how much of each sequence is critical. The bold line in Figure 5.8 merely shows that parts of these two work sequences are critical. Further study will disclose that only the first day of activity 10 and the first four days of activity 20 are critical. The computed early and late times for activity 40 indicate that the entire "Lay pipe" sequence is critical, which is indicated by a bold line around the entire activity box.

5.24 Interface Computations

In Section 4.16, a pipeline crossing structure network that interfaces with the pipeline relocation network was discussed. The network for the pipeline crossing structure, which is located at the end of mile 3 of the pipeline relocation right-of-way, is shown in Figure 5.9. The interfacing activities between these two networks are activity 110 (Figure 5.9) and activity 170 (Figure 5.7). It has been decided that the crossing structure must be ready to accept pipe by the time the pipe-laying operation of the main pipeline reaches it. The interrelationship between the two interfacing activities is established in this manner. Reference to Figure 5.7 shows that activity 170 has an EF = LF = 20, which means that the third mile of the pipeline will be laid and in place by elapsed working day 20. This simply says that the crossing structure must be able to accept pipe (activity 110 in Figure 5.9) by no later than 20 elapsed working days. Thus, the LS for activity 110 in Figure 5.9 is set equal to 20. When this is done, the time schedules for both the pipeline and crossing structure are meshed together and are on the same basis.

The LS value of 20 has been entered for activity 110 in Figure 5.9 and the forward pass completed from that point. A backward pass through the crossing network is now performed, showing that the LS value of the first activity (10) is 3. This value signifies that the construction of the crossing structure must be started no later than the beginning of the fourth day (time = 3) if the crossing is to be ready for pipe when it is needed. This follows from the fact that the same time scale is being used for both the pipeline relocation and the crossing structure; that is, zero time is the same for both. Making a forward pass up to activity 110, using an ES of activity 10 in Figure 5.9 equal to zero, yields the ES and EF times shown.

The total float values for each activity have also been determined and are shown in Figure 5.9. As can be seen, there is a continuous path through the network from activity 10 to activity 110, with a constant total float of 3. This float path is the critical path for that project. The significance of the float value of 3 is that, if the crossing structure is started at the same time as the main pipeline (time = 0), the longest path through the network has 3 days of spare time associated with it. If everything goes reasonably smoothly, the crossing structure should be ready for pipe in 17 working days, or 3 days ahead of the arrival of the pipeline itself. If troubles develop on the crossing structure, a general time contingency of 3 days has been provided.

If the EF of activity 170 in Figure 5.7 had a value of 16 rather than 20, all the LS and LF values in Figure 5.9 would be four days less than those shown. The LS of activity 10 would have been 21, indicating that the crossing structure would have to be started one day before the main pipeline if the crossing is to be ready on time.

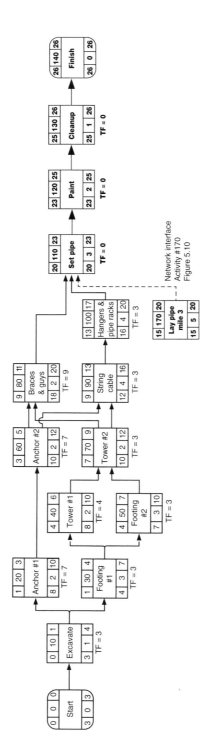

Figure 5.9
Pipeline crossing structure, precedence diagram time computations

5.25 Hammock Activity

A useful concept with regard to drawing project diagrams is the hammock activity, one that extends from one activity to another but that has no estimated time duration of its own. It is an activity in the usual sense because it is time consuming and requires resources, but its duration is controlled not by its own nature but by the two activities between which it spans. Its ES and LS times are determined by the predecessor activity, and its EF and LF times are dictated by the successor activity. Common examples of hammock activities are dewatering and haul road maintenance, construction operations whose time spans are not self-imposed but are dictated by other job factors.

As a specific illustration, suppose construction of the highway bridge were to involve diversion of the stream during the construction and backfilling of the abutments. Reference to Chart 5.1a the highway bridge precedence diagram time computations on the companion website shows that the stream must be diverted prior to the start of activity 90, "Excavate abutment #1," and can be removed only after the completion of activity 310, "Backfill abutment #2." Figure 5.10 illustrates how hammock activity 35, "Maintain stream diversion," would span the time interval between activity 25, "Divert stream," and activity 275, "Remove stream diversion." Activity 35 has no intrinsic time requirement of its own; it is determined entirely by the finish of activity 25 and the start of activity 275. According to the current plan and schedule, the duration of activity 35 can vary anywhere from 34 to 43 working days (values obtained from the start and finish times of this activity in Figure 5.10).

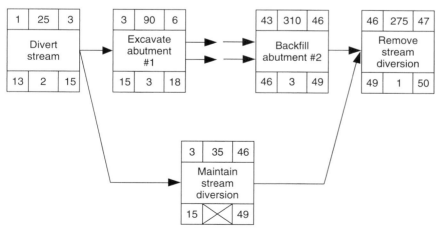

Figure 5.10
Highway bridge, hammock activity

Job overhead is a good example of a hammock activity. The cost of job overhead is related directly to the duration of the project. In Chapter 11, "Project Cost System," project costs are assigned to network activities and the notion of a hammock activity to account for job overhead costs becomes important.

5.26 Milestones

Milestones are points in time that have been identified as being important intermediate reference points during the accomplishment of the work. Milestone events can include dates imposed by the owner for finishing certain tasks as well as target dates set by the contractor for completing certain segments of the work. A milestone is usually the scheduled date for either the start or the completion of some difficult or important aspect of the project. On large projects, contractors frequently establish a series of milestones extending throughout the project and use these as reference points for project monitoring.

Events, or points in time, do not appear as such on the customary precedence diagram. Nevertheless, milestones can be shown on a project network, should project management desire this. The usual convention is to show an event as a zero time duration box at the appropriate location within the diagram. If some distinctive geometric figure is preferred, circles, ovals, triangles, or other shapes can be used. Any information pertaining to a milestone and considered to be useful may be entered. To illustrate how a milestone can be indicated on a precedence network, suppose that a target date for the highway bridge job is when the abutments have been completed and backfilled so that water flow down the ravine being spanned by the bridge is no longer a potential problem. Reference to the highway bridge precedence diagram time computations on the companion website shows that this point in time is reached with the completion of activity 310. A symbolic convention commonly used for depicting milestones on precedence diagrams is shown in Charts 5.1a and 5.2 on the companion website as the start and finish activities.

5.27 Time-Scaled Networks

The original project network is arranged to show its activities in the general order of their accomplishment but is not plotted to a time scale. A time-scaled network has some distinct advantages over the regular diagram for certain applications because it provides a graphical portrayal of the time relationships among activities as well as their sequential order. Such a plot enables one to determine immediately which activities are scheduled to be in process at any point in time and to detect quickly where problem areas exist.

When drawing a time-scaled diagram, two time scales can be used: working days and calendar dates. One scale is, of course, immediately convertible to the other. Chart 5.3a on the companion website is the early-start schedule of the highway bridge plotted to a working-day scale with the calendar months also indicated. Chart 5.3b on the companion website is the early-start schedule of the building plotted to a working-day scale with the calendar months also indicated. These are the same network diagrams as Charts 5.1a and 5.1b. In each case, the two portray the same logic and convey the same scheduling information.

Chart 5.3(a or b) is obtained by plotting ES and EF values for each activity. Anchored at its left end by its ES time, each activity is visualized as being stretched to the right so that its horizontal length is equal to its estimated time duration. Each activity is shown as a one-dimensional line rather than as a two-dimensional box. Vertical solid lines indicate sequential dependence of one activity on another. When an activity has an early-finish time that precedes the earliest start of activities following, the time interval between the two is, by definition, the free float of the activity. Free floats are shown as horizontal dashed lines in Chart 5.3(a and b), and a time-scaled plot of this type automatically yields to scale the free float of each activity. When an activity has no free float, no dashed extension to the right of that activity appears. The horizontal dashed lines also represent total float for groups or strings of activities, a topic that is discussed in Section 5.28.

The circles shown in Chart 5.3a are entered to clarify where one activity ends and another begins as well as to indicate sequential dependencies. The numerical values in the various circles are the ES and EF times in terms of expired working days for the respective activities.

Time-scaled diagrams are very convenient devices for checking daily project needs of labor and equipment and for the advanced detection of conflicting demands among activities for the same resource. Chapter 8, "Resource Management," discusses this subject in detail. A network plotted to a time scale is also useful for project financial management applications and for the monitoring of field progress. Summary diagrams for management, owners, and architect-engineers frequently are drawn to a time scale to enhance the diagram's ease of comprehension and ready application to job checking and evaluation. Milestones can be indicated on time-scaled networks using any desired distinctive symbol. For example, the triangle in Chart 5.3a at the finish of activity 310 indicates the previously discussed point in time when the abutments have been completed and backfilled.

5.28 Nature and Significance of Floats

A project network plotted to a time scale illustrates and clarifies the nature of free and total floats. In Chart 5.3(a and b), the critical path can be visualized as a rigid and unyielding frame extending through the diagram.

The horizontal dashed lines (float) can be regarded as elastic connections between activities that can be shortened or elongated. The vertical solid lines represent dependencies among activities. Under circumstances to be discussed in later chapters, activities can be rescheduled, provided the correct dependency relationships are maintained and there is sufficient float present to accommodate the change.

To illustrate the nature of free float, consider activity 310 in Chart 5.3a. This activity has a duration of three days and a free float of three days. Chart 5.3a clearly shows that a total of six working days are available to accomplish this activity. Within its two time boundaries (day 43 and day 49), this activity can be delayed in starting, have its duration increased, or a combination of the two, without disturbing any other activity. For all practical purposes, activity 310 can be treated as an abacus bead three units in length that can be moved back and forth on a wire six units long. Therefore, the free float of an activity is extra time associated with the activity that can be used or consumed without affecting the early-start time of any succeeding activity.

The nature of total float can also be understood from a study of Chart 5.3a. For example, activity 310 also has a total float of three days. If the completion of this activity were to be delayed by three days, it would become a critical activity and a new loop would materialize on the critical path. It follows from this demonstration that when free float is used for a given activity, total float is also used in the same amount.

In most cases, however, the nature of total float is considerably more involved than the example just given. The problem with total float is that strings or aggregations of activities usually share it. To maintain the necessary logic dependencies when using total float, one must often move whole groups of activities as a unit. The time-scaled plot is a convenient basis for doing this, but the process can become involved at times.

To illustrate a more complex case, reference is made to Chart 5.3a and activities 30, 110, 130, 150, and 170, each of which has a total float of three days. These five activities, as a group, have a combined flexibility of three days. If three days are lost in achieving completion of any of the activities of this string, the three days of flexibility are gone, the three days of total float for *all five* activities are consumed, and a new branch on the critical path is formed by activities 30-110-130-150-170. Chart 5.1a shows that activity 190 must follow activity 170. Consequently, when activity 170 moves three days to the right, Chart 5.3a reveals that activities 190, 210, and 230 do likewise. Thus, all floats of these three activities are reduced from 8 to 5, and the free float of activity 140 is increased by 3 to a value of 4.

This discussion shows that the total float of an activity is the length of time that the early finish of the activity can be delayed and not adversely affect project completion. If all of the total float of a given activity is consumed, accidentally or by design, the activity becomes critical and a new critical path, or branch thereon, is created in the network. In this new critical path, all the activities prior to the given activity must be completed by their EF times and all activities following will begin at their LS times.

5.29 Bar Charts

Bar or Gantt charts, briefly discussed in Section 3.10, present the project schedule plotted to a horizontal time scale. The bar chart has been a traditional management device for planning and scheduling construction projects. However, bar charts have serious and well-recognized shortcomings when used for the original development of project management information. For one thing, the interdependencies among activities are difficult to show and often are not reflected in the data generated. Additionally, the bar chart in itself does not provide a basis for ascertaining which activities are critical and which are floaters. Consequently, each activity receives the same consideration, with no indication of where management attention should be focused. The bar chart is not an adequate planning and scheduling tool because it does not portray a detailed, integrated, and complete plan of operations. Bar charts are completely ineffective for project shortening, resource management, and most of the other project management methods yet to be discussed.

However, the unsurpassed visual clarity of the bar chart makes it a very valuable medium for displaying job schedule information. It is immediately intelligible to people who have no knowledge of CPM or network diagrams. It affords an easy and convenient way to monitor job progress and record project advancement. For these reasons, bar charts continue to be widely used in the construction industry. The use of bar charts as project time management devices is discussed in Chapter 8, "Resource Management." While CPM networks are planning and diagnostic tools, bar charts are visual display devices.

Fortunately, it is possible to prepare bar charts on a more rational basis, avoiding their intrinsic weaknesses and incorporating the strengths and advantages of network analysis. This is possible simply by recognizing that a time-scaled diagram is an elaborate form of bar chart. Most computer programs will create a variety of bar charts based on the logic of the CPM network. Figure 5.11 is an early-start bar chart for the highway bridge presenting the programmed schedule for the entire project. For each activity, the shaded ovals extend from its ES to EF times (bolded ovals represent critical activities, and shaded ovals, noncritical activities). The white ovals that extend to the right of the noncritical activities represent the total float. Notice that no work is scheduled for weekends or holidays. Milestones can be indicated on bar charts and frequently play an important role in the monitoring of job progress. The milestone discussed in Section 5.26 associated with the completion of activity 310 is shown as a triangle in Figure 5.11.

Construction Progress Chart

Project _____ Highway Bridge _____ Job No. _____ 200008-05 _____

Activity	Activity number	June 14	21	28	July 5	12	19	26	August 2	9	16	23	30	September 6	13
Move in	40	▪▪▪													
Excavate abutment #1	90		▫▫▫▫▫	▫▫▫	▫▫▫										
Prefabricate abutment forms	80		▪▪▪	▫▫▫▫	▫▫▫▫	▫▫▫▫▫									
Excavate abutment #2	120		▫▫▫▫	▫▫▫▫	▫▫▫▫	▫▫									
Mobilize pile-driving rig	100		▪▪▪	▪▪▪	▫▫▫										
Drive piles abutment #1	110			▫▫▫▫▫	▪▪▪▪	▫▫									
Forms & rebar footing #1	130					▪▪▪▪									
Drive piles abutment #2	140					▪▪▪▪	▫								
Demobilize pile-driving rig	160					▫▫▫	▫								
Pour footing #1	150					▫▫▫	▫								
Strip footing #1	170						▪▪▪								
Forms & rebar footing #2	190						▫▫▫▫▫	▫▫▫							
Pour footing #2	210						▫▫▫▫	▫▫▫▫							
Forms & rebar abutment #1	**180**						▪▪▪▪	▪▪▪							
Strip footing #2	230						▫▫▫	▫▫▫▫▫							
Pour abutment #1	**200**							▪▪							
Strip & cure abutment #1	**220**								▪▪▪						
Backfill abutment #1	280								▫▫▫	▫▫▫▫					
Rub concrete abutment #1	270								▫▫▫	▫▫▫▫▫	▫▫▫▫▫				
Forms & rebar abutment #2	**240**								▪▪▪▪	▪▪▪▪					
Pour abutment #2	**250**									▪▪					
Strip & cure abutment #2	**290**									▪▪▪					
Rub concrete abutment #2	300										▫▫▫▫	▫▫▫▫▫	▫		
Backfill abutment #2	310										▫▫▫▫	▫▫▫▫▫	▫		
Abutments finished	315										△				
Set girders	**320**										▪▪				
Deck forms & rebar	**330**										▪▪▪▪	▪▪▪			
Pour & cure deck	**340**											▪▪▪	▪▪		
Strip deck	**350**												▪		
Saw joints	380												▫	▫▫▫▫	
Paint	**370**												▪▪	▪	
Guardrails	360												▫	▫	
Cleanup	**390**													▪▪▪	
Final inspection	**400**														▪

(Bold type denotes critical activities)

Figure 5.11
Highway bridge, bar chart schedule

Single activities may not always be the most desirable basis for the preparation of bar charts. Simpler diagrams with fewer bars and less detail may be more suitable for high-level management. In such cases, a bar chart can be prepared using larger segments of the project as a basis. In this regard, the concept of project outlines involving different degrees of work breakdown can be valuable. This matter was discussed in Section 4.8. Bar charts may or may not include job restraints, such as the time required for the preparation of shop drawings and for the fabrication and delivery of job materials. In general, such restraints are a function of job expediting (see Section 8.17), which is handled more or less separately from the field construction operations. Because of this fact, restraints that appear on the highway bridge network are not shown on the bar charts in Figure 5.11.

5.30 Computer Applications for Scheduling

This chapter has presented a comprehensive discussion of project scheduling, a process of establishing a calendar-date schedule for the field construction process. For reasons already explained, the scheduling procedure discussed herein has emphasized manual methods, with the objective of developing a thorough understanding of the procedures involved and the significance of the project time data generated.

Realistic job logic and accurate activity duration estimates should result from the project management team's knowledge, experience, intuition, and discerning judgment. It is important to note that the network logic and activity durations so developed provide a graphic and mathematical job model. This model is very powerful in that it allows a manager to look into the future and make project decisions based on information from the model. Like all models, the quality of the information obtained is directly proportional to the accuracy of the model itself. No amount of computing power can change this basic fact.

With regard to the mechanical process of project scheduling, however, computers enjoy a distinct advantage in their ability to make time computations accurately, with great rapidity, and to present this information in a variety of useful forms. As a result, computers are used universally for project scheduling purposes. Computers can provide the project team with an almost unlimited array of project data and graphic representations of network scheduling information. In addition to forward- and backward-pass calculations, the computer can convert expired working days into calendar dates. Different activities can be assigned to different calendars. For example, maintenance and support activities can be assigned to a weekend calendar while construction activities are assigned to a weekday schedule. Activity lags (Section 5.21) and hammock activities (Section 5.25) are easily incorporated into the project schedule, and a variety of activity sorts are available.

Calculations performed by computer can be displayed graphically in the form of project networks (or portions thereof) or in the form of tabular reports. These can be made to cover varying portions of the overall project and differing periods of time. Computers have the capacity to sort information in terms of specific activities, spans of time, physical location on the site, areas of responsibility, or other desired criteria. This capability eases the burden of getting the right information to the right person at the right time. Computer programs typically have the capability of reducing the network time schedule to bar chart form. These work well for reporting job progress to owners, design professionals, and others concerned with the construction schedule.

Computer programs used for project scheduling can differ substantially from one another in many important respects, and a wide variety of scheduling programs are available. For this reason, care must be exercised to select a program best suited to the specific management needs of a given project. Most of these programs will allow a scheduled time to be assigned for the completion of construction and will make the backward pass using this designated finish date rather than the value obtained during the forward pass. Scheduled dates also can be assigned to network milestones.

Chapter 6, "Production Planning," moves from viewing the project as a whole to planning the details that often are overlooked and can cause serious job delays.

Key Points and Questions

Key Points

- The true worth of a project schedule depends extensively on the job logic and the accuracy with which the individual activity times can be estimated.
- The optimum use of CPM information for project time control requires the user to have a thorough understanding of the computations and the true meaning of the data generated.
- The critical path determines the minimum time required to complete the project and also provides the information needed to expedite project completion.
- Float is a valuable resource that enables effective resource management.
- CPM-based computer programs enable extensive sorting of data that, in turn, facilitates more effective project management.

Review Questions and Problems

1. The CPM schedule provides a great deal of information that is useful to a construction project manager. What do you believe are the three most useful applications, and why do they top your list?

2. Since CPM calculations are best accomplished with the assistance of a computer, why is it important for project managers to know how CPM calculations are done?
3. Why are various types of CPM data useful? Give several examples.
4. Several examples of hammock activities have been given for the bridge project. Provide an example of a hammock activity (other than job overhead, which is equally applicable for both the bridge and the building) for the building project, together with the predecessor and successor activities.
5. Why is the bar chart not an adequate planning and scheduling tool?
6. Since the bar chart is ineffective as a planning and scheduling tool, why is it still widely used in the construction industry?
7. For the CPM Calculation Exercise found on the companion website, determine the ES, EF, LS, LF, FF, and TF for each activity. Identify the activities through which the critical path runs. How many project days will this project take?

6 Production Planning

6.1 Introduction

This chapter revisits the planning process. Chapter 4. "Project Planning" considered planning from a project point of view. Project planning of the example projects consisted of dividing the project into activities and establishing the logical relationships between them. This process, along with the project scheduling techniques of Chapter 5, "Project Scheduling Concepts," established what was going to be done on the project and when each activity was to be accomplished.

Production planning, however, is concerned with how these activities are going to be accomplished. If project planning is macroplanning, then production planning is microplanning. Production planning establishes the methods to be used, the assignment of personnel, the movement of material to the workface, and the process of assembling the pieces. Production planning begins well before the project is mobilized in the field and continues throughout the project until all field operations are closed out. The initial effort required to plan for production is equivalent to that required for project planning in Chapter 4, "Project Planning."

Before addressing the fundamentals of production planning, two additional topics of project planning must be discussed: the planning team and reengineering the project. This is a key point in the project where the project manager interfaces with the field supervisor. Many of the management responsibilities are handed off to the supervisor, who will take on the day-to-day responsibilities to execute the work in the field.

Learning objectives for this chapter include:
- ❏ Recognize planning as a key component of the construction process.
- ❏ Understand how lean production principles apply to construction.
- ❏ Review some key aspects of preconstruction planning.
- ❏ Become aware of how project management is changing with the advent of Building Information Modeling (BIM).

6.2 Planning Team

Once the contract is signed, there is very real pressure on the project team to get on with the work; success often is defined as making the dirt fly. Yet studies show conclusively that complete, thorough, and detailed planning both at the beginning of a project and carried on throughout the duration of the project is the key to successful completion. Not only does the project, as a whole, need to be thoroughly planned, but each activity must also be planned.

Often, the people assigned to a new project have never worked together. The planning process is the first opportunity to begin team building. Getting the team together in a planning session gives the members an opportunity to work together in an unstructured environment in which every idea is welcome and each individual's expertise can be called upon to identify future problems and solutions.

Many constructors at all levels find planning to be the hardest part of the construction process. Construction people are often more at ease when working with physical objects rather than with abstractions, and planning is an abstract process. There is a great deal of uncertainty at the beginning of a project. This causes frustration. There is the perception that starting work right away is the best way to come to grips with the uncertainties. In order to overcome these planning objections, several things can be done to make planning less abstract and more satisfying.

Computer technology can lend structure to the development of ideas, but computers must be regarded as no more than support tools to the very human process of planning. As discussed previously, the outliner in scheduling software or in word processing software can be used to break a project down into successive layers of finer detail. Using a computer, projector, and outlining software, the project team can quickly begin to contribute, improve, and agree upon how the project is to be organized. Once the outline is substantially complete, and using the same equipment, the team can start the process of moving the resulting breakdown to the network diagram and establishing the constructive logic between activities. With this process, each idea is placed on the screen, evaluated, and added to the project plan. Although not as physically permanent as concrete and steel, the computer images can be seen and provide a visual record of the planning process that will result in a physical schedule when completed.

When a computer is not used, there are many other ways to carry out successful planning meetings. Whiteboards and markers can be used, with different colored markers used for distinct operations or specialty areas. Sticky notes can also be used with participants making notes for their activities and placing them where they feel is the most appropriate location in the plan. As the planning evolves, these notes can be relocated in relation to other activities until the entire group feels that the required logic has been satisfied and that the most efficient sequencing for the project has been attained. Whatever method is selected for planning, it is essential that it be highly interactive and inclusive, involving each participant in the planning process.

These planning sessions need to have an accomplished facilitator who can move the process along and get all participants involved. After the meeting is completed, the facilitator can take a picture of the plan to make a permanent record, then develop a written or digital summary that can be distributed to all and can form the basis for the formal scheduling process that builds upon the plan.

There are a number of important elements in successful planning. First and foremost, the team must understand the project thoroughly. Everyone must be familiar with the type of work being planned and must fully understand the contract documents. All individuals must bring appropriate skills and experience to the team, and their abilities must complement each other. Sufficient time must be allotted for the planning process. Many successful project teams find it advisable to meet away from the office and construction site, thus avoiding interruptions. Sequestering the team for two or three days offsite allows the planning process to proceed uninterrupted through mealtimes and into the evening. If fieldwork must be started immediately, a temporary or mobilization team can be assigned to do the immediate tasks while the permanent team completes the planning function.

As the project plan develops, the team should bring others into the process. Vital suppliers and subcontractors should be included in meetings. When appropriate, the suppliers and subcontractors prepare specialty subschedules to cover their part of the work. These subschedules are presented to the project team and defended before they are added to the master plan. In this way, suppliers and subcontractors are made aware of the important part they play in the success of the project. By involving the broad cross-section of stakeholders in the planning process, commitment is gained across the construction team to the strategic plan as it evolves.

Before the plan is finalized, an analysis of possible risks should be made. This involves an evaluation of parts of the plan where things are most likely to go wrong. Questions such as "Where are we most vulnerable to weather damage?" or "Which material is most apt to have delivery or approval problems?" are used to identify problem areas. Once the areas of risk have been identified, plans need to be made to mitigate and manage these risks.

Having the project team work together to prepare a complete job plan strengthens buy-in and means that each team member will have developed a strong sense of dedication to the project's outcome. It also increases the likelihood that each participant understands what is expected of him in the execution of his component of the project. Success in the planning process strengthens the project team, aligns their objectives, and sets the project on course for a successful conclusion.

6.3 Reengineering the Project

There is often a better way to do things. Reengineering is the process of examining a project and its component parts in an effort to find improved or alternate ways to accomplish an operation. This reengineering process can be applied to any part of the project, from field operations, such as the way a piece of equipment is installed, to management procedures, such as the way change orders are handled. The key to successful reengineering is to identify alternative ways of doing things. A really great idea cannot be evaluated and implemented if it has never been suggested.

Using the highway bridge as an example, many of the major reengineering processes are beyond the responsibility of the contractor. For instance, a culvert might have been proposed in place of the bridge, with considerable cost savings. The location of the road may have been rerouted so that a bridge was unnecessary, but these decisions were made early in the project by the owner and/or design team and hence became the responsibility of the owner. In some project delivery systems, the contractor participates in early strategic planning of the project, during the design phase; however, in the more traditional design-bid-build approach, especially in the public sector, constructors are not brought into the project until after many critical decisions have been made.

In the heavy civil example project, the highway bridge contractor has the responsibility for choosing the means and methods to be used in the construction of the bridge. Improvements in these means and methods rest with the project team. In order to identify as many new ideas as possible, a brainstorming session is a very productive approach.

Brainstorming, or conceptual blockbusting technique, involves a free-format meeting of the project team where any idea is acceptable. This is a meeting where nothing is sacred and no one is right. Outrageous suggestions often trigger new and more practical ideas, and new ideas are what reengineering is all about. The leader of the meeting should encourage unconventional solutions, ask people to question the boundary conditions between operations and processes, and look at solutions used in other industries. Free, uninhibited thinking at this stage can provide new solutions to old problems and provide substantial savings in terms of time and money.

The brainstorming session might develop solutions that could save money and/or time or make the job safer, but would require some modification of the design and reallocation of the budget. For example, in building construction, there is a growing trend to prefabricate components either onsite or offsite for assembly in the field, rather than "stick building" everything at the workface. This will often provide a final product that is of higher quality, completed more quickly, at a substantial cost saving, with much safer execution. However, some details of the design often need to be modified to accommodate prefabrication. Budget reallocation may also be required, for example, when subassemblies are manufactured by suppliers resulting in the transfer of a great deal of the labor cost from the field contractor to the supplier. The project owner should be presented with such options, even if redesign is required. Cost savings can be shared among the various participants in the project, and some of the savings could be used to offset the additional cost of redesign or manufacture of subassemblies, where required.

6.4 Planning for Production

Production planning involves site layout, arrangement of utilities, and preparation of storage and prefabrication areas. It includes establishing traffic patterns for vehicles and material flow patterns from storage to prefabrication to final installation. This is a detailed study of exactly how the work is going to be accomplished. It is complicated, time consuming, and extremely important.

Traditional production planning has evolved in recent years to become a component of lean production with roots in the manufacturing industry. Lean production gained its name from its focus on squeezing inefficiencies out of the production process, thus making it lean. It developed from the Japanese automobile industry, specifically Toyota, but has evolved into broad application across the manufacturing industry and now is being adapted for application in the construction context.

Lean production addresses all aspects of the manufacturing process, including design, the supply chain, and the manufacturing component. Similarly, lean construction addresses all aspects of the construction process, including design, supply chain management, and field operations. It considers in detail many subcategories in these general areas, such as safety, quality, and efficiency of field operations. The brief treatment of lean construction in this text will focus on managing the construction process. For a much more comprehensive understanding of lean principles applied to construction, the reader is referred to the Lean Construction Institute at leanconstruction.org.

When focused on construction operations, the lean construction process begins with production planning prior to commencement of construction. Setup of the lean construction process is accomplished through a facilitated planning meeting as described in Section 6.2. Execution of the lean construction process continues throughout the project and provides a powerful means to better control production, safety, quality, cost, and schedule. Execution of the lean construction process throughout the field operations will be addressed in Chapter 10, "Project Coordination."

In the remainder of this chapter, production planning is broken down into its components and discussed.

6.5 Support Planning

One of the first things to consider is how equipment and materials will get to the site. For heavy civil projects, the route typically will involve public roads and bridges, both of which must be examined for alignment and capacity. Weight and dimension limits of roads and bridges are extremely important. Heavy or oversize loads may have to be routed far out of the way in order to get them to the site. Where bridges will not withstand the loads and alternate routes are unavailable, fords may have to be built around the bridge. Heavy loads may have to be divided at their source in order to get them to the site. There may be low utility lines and other height or width restrictions. Narrow roads or tight turns may cause problems. Identifying these restrictions for critical material deliveries in advance may save much-needed project time.

For building construction, similar problems must be resolved; however, since most building construction takes place in urban areas, additional questions need to be considered such as traffic and constrictions to urban thoroughfares. The planning should consider not only getting equipment and materials to the site but also offloading them from their conveyance in very tightly constricted areas. Supply chain management becomes a significant problem as construction projects increase in size and scope, in complexity, and in terms of schedules and budgets that continually become tighter.

A thorough investigation of onsite and offsite utilities is necessary. Such needs as the closest rail siding and dock facility may play an important part in getting materials to the site. The need for water, sewer, gas, high-voltage electricity, and three-phase electricity must be identified and sources found before starting work. Newer utility requirements may also include hazardous waste disposal facilities, groundwater testing facilities, silt retention basins, and wideband communication platforms for voice, video, and data transmissions.

A detailed drawing is required for efficient site layout. The number, size, and location of job-site buildings must be determined. It is important that the job-site office be located in such a way as to provide a secure entry to the site and have a good view of the work area. Other buildings include office space for inspectors and subcontractors, tool storage buildings, changing facilities for employees, and storage buildings for hazardous and flammable materials. A traffic pattern for vehicles needs to be established. Parking must be provided for employees, subcontractors, and visitors. Heavy trucks must have clear access to the storage and work areas.

Specific provisions must be planned for equipment maintenance and fueling, material laydown, material storage, and prefabrication. Fuel storage and waste storage also are important considerations. Figure 6.1a shows the resulting site plan for the highway bridge.

The site plan for the example building is shown in Figure 6.1b. Since the figure is highly compressed to be able to fit in this book, the reader is encouraged to find the site plan (Sheet A1.1) among the commercial building drawings provided on the companion website. It can be seen that this building is located on a relatively spacious site. This was done with the objective of providing adequate room to expand the building at a later date. The reader is encouraged to print this drawing from the companion

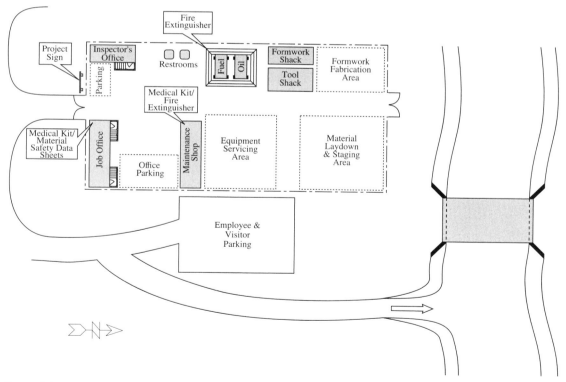

Figure 6.1a
Highway bridge, site layout

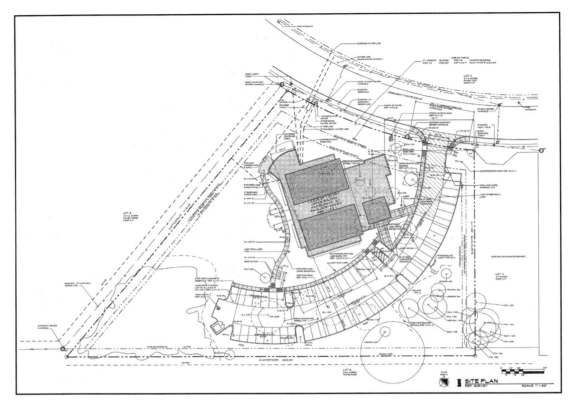

Figure 6.1b
Commercial building, site layout
The Arkitex Studio, Incorporated

website and sketch in the site plan for construction (see review question 4 at the end of the chapter). A fence should secure the boundaries and provide a primary and a secondary access and egress point. Space should be provided for offices for the general contractor and major specialty contractors (at a minimum, mechanical and electrical contractors). Show material storage (laydown) areas for construction materials for the general and specialty contractors.

6.6 Technical Problems

Many projects involve technical problems requiring significant advance planning. Problems on the highway bridge include dewatering of the foundation and placement of the crane for setting the deck girders. Most projects involve lifting of heavy materials, so crane capacities and coverage must be checked. Crane conflicts with power lines, structures, and other cranes often require preparation of detailed drawings. Getting materials and equipment into the structure often requires detailed "pathfinding" studies. Refineries, power plants, manufacturing facilities, and other projects may require complex computer studies in order to ensure that equipment and piping will fit in the location specified.

As previously indicated, construction of buildings in downtown areas requires detailed planning for unloading and storage of materials, traffic patterns, and accommodations for pedestrians. Deep excavations may require bracing, sheet piling, or other soil stabilization techniques. Complex projects on tight sites may require contractors to maintain separate laydown areas at a distance from the site, where materials can be received and prepared for delivery to the site on a just-in-time scheduling basis. In busy urban areas, providing materials to the site might require deliveries on off-peak hours or even at night.

Since just-in-time scheduling is a lean principle, it is becoming much more prevalent as the industry moves toward prefabrication of units to be assembled at the site rather than "stick-building" with basic materials. Extensive prefabrication is moving the industry toward much more commonality with manufacturing. Prefabrication demonstrates many benefits, including higher quality, lower cost, increased safety, and a much shorter schedule for site operations. Effective prefabrication and assembly is highly dependent on detailed and accurate planning.

Some projects have complex work sequences that require a detailed study to determine the exact sequence of steps to follow. Often, long-lead equipment deliveries require that certain parts of a project be left accessible until late in the project. More and more projects involve the use of, or disposal of, hazardous materials. Work with these materials requires notifying employees of the hazard and training workers in proper handling techniques and in the use of specific tools and safety equipment.

Early and complete identification of these technical challenges fosters proper planning. Although the problems remain technically difficult, proper planning gives project management time to consider alternative methods and bring in technical specialists when necessary.

6.7 Personnel Planning

The success of a project is largely dependent on the quality and morale of its craftspeople. Good planning at the beginning of the job helps ensure success in several ways. Craftspeople in proper number and with appropriate skill levels must be hired. It is important that consistent policies be established and applied to all staff. Job-site morale can be devastated when labor shortages require that new employees be given incentives not offered to earlier employees. Consistent wage rates and benefit packages are essential.

There is extensive human resource law to which all employers and employees are subject that covers such issues as equal opportunity, discrimination, and harassment. The project owner, construction manager, general contractor, and each specialty contractor should have their own company human resource policies, procedures, and programs. For a specific project, these should be consistent with one another and with federal, state, and local laws, resulting in a well-planned human resource program for that project. Supervisory people in all companies and at all levels need to be trained in the application and execution of the project human resource program.

Where there is a shortage of skilled people, training programs have to be implemented. Working with local technical-vocational schools or trade unions may help the situation. On-the-job training programs may require significant planning early in the project. When the shortage is severe, craftspeople may have to be brought in from other locations. In this case, transportation and housing may be required.

Projects in remote locations require camp facilities. Morale is greatly affected by the organization of the camp and the quality of the food. Poor camp facilities often result in low morale, poor production, and high personnel turnover. A highly qualified camp manager can be the most important person on the project management team.

6.8 Safety Planning

Construction work is inherently dangerous. Sharp objects, heavy loads, high places, and emphasis on production all provide a fertile environment for accidents. Good planning and the implementation of a comprehensive safety program are necessary for a safe project. Studies have shown that employee involvement in a safety program is essential. Establishment of a safety committee during the first week of the project is a good way to start a safety program.

The safety committee is mandated to implement and monitor the safety program. Some firms assign this duty to all the foremen on the project. They meet weekly, review the safety program, and conduct a safety inspection. In some construction companies, a different foreman is assigned as the safety inspector each week. This foreman inspects the project daily, correcting any safety problems observed and posting violations on the bulletin board. Typically, a foreman will strive harder for safety within his crew if another foreman is going to inspect his work area and report safety problems to his peers. Of course, a Monday morning toolbox safety program and incentives for perfect safety records are integral parts of safety planning.

Safety planning includes specific consideration regarding what to do in case of an accident. Staff with first aid certification should be identified. In case of a serious accident, a plan should be in place for making emergency calls, meeting the emergency medical people, and leading them to the victim. Emergency evacuation from high in a building or deep within a construction project requires forethought. Safety planning should consider fire and evacuation plans. Often, a fire is accompanied by a power failure, so lifts and electric cranes are not available. Similarly, if a project is subject to a flood, heavy rain, or a tsunami, plans should first consider evacuation of workers. Secondary plans should consider ways to minimize damage to the project. In all of these cases, waiting until the time of the disaster to assign people and decide what must be done will almost certainly reduce the success of the recovery effort.

Work in high places and activities involving heavy lifts need to be identified in advance and safety plans made. Plans for attaching barricades and lifelines and covering openings should be made in advance. Concrete forms need to be designed for safety, including safe work surfaces, railings, and ladders. These concrete-form safety considerations should apply to placing, moving, stripping, and cleaning operations as well as for placing concrete. Movement of large equipment or lifting heavy loads should be planned for after hours in order to minimize the number of people involved.

Planning should include a process for handling hazardous materials. Material Safety Data Sheets (MSDSs) should be provided with all material orders. These sheets are to be placed in a notebook and made available to employees. Workers are entitled to know when they are handling hazardous materials and to be aware of the dangers involved and the suggested handling procedures. Special attention should be given to work in enclosed spaces where chemicals may form toxic fumes. All chemical agents should be clearly marked and inspected regularly for condition.

Safety equipment needs to be stored in convenient locations where it is quickly accessible when operations call for its use. A regular inspection should be scheduled to make sure the safety equipment inventory is complete and in good condition. The availability of safety equipment, including personal protective equipment (PPE), needs to be advertised and its use made mandatory.

Because of the inherent dangers involved in construction, any comprehensive safety plan, together with the job human resource plan, should include methods of ensuring that employees are alert and substance free. This is best accomplished through well-planned substance policies and programs. Comprehensive substance abuse prevention programs should always be planned well in advance of construction and incorporated into company or project hiring and employment policy. This ensures that all employees joining the project are fully aware of the company's substance policies, testing methods, and actionable penalties. The policies should state the company's intention to operate a substance-free workplace, detail the testing methods to be applied, specify the kind of substances to be banned (such as alcohol), and explain any penalties and assistance programs. The substance-testing program should address compulsory testing prior to hiring, continuous random testing, and testing for cause. The employment application process should include mandatory substance testing. If need be, employees may be hired on a probationary contract pending the results of the test. This ensures that all employees join the project substance free. Random substance testing ensures that employees do not begin using banned substances during the project. Because some projects employ a large number of workers, random testing may only cycle through the job every six to eight months, reducing this type of testing as a deterrent. Therefore, a more active testing method may be needed to identify employee substance abuse at its onset. To meet this need, a testing policy should allow for immediate substance testing with just cause and compulsory testing following any accidents. Advanced planning and publication of the substance-testing policy will help deter potential users from applying for work and will maintain a substance-free and safe workplace.

6.9 Planning for Quality

In today's construction market, every project needs to have a formalized quality control and assurance system in order to be competitive. Contractors are expected to be able to produce high-quality products, and if they get a reputation for shoddy work, they will not be able to compete for top projects. The constant conflict between production and quality must be resolved for a successful project. Perhaps the most important part of a quality control program is the creation of a job-site culture where quality is asked for and expected. Craftspeople need to take pride in their work and expect their peers to do likewise. There are a number of ways to build this visibility and dedication to quality.

Incorporating lean construction principles is a good way to start. A lean project is built on a strong commitment to teamwork, where all stakeholders commit to work with one another to produce the highest-quality product for the project owner, while meeting the primary objectives of each individual participant. A lean project is based on honesty, integrity, accountability, and commitment, which are the basic principles for building a partnering relationship. Morale can be further enhanced by team building. An example of team building might be when the architect-engineer is called upon to design an inspiring project logo for use on hard hats and T-shirts. Another example might be when a safety award program is developed to encourage and recognize a high level of safety through out the project. Both of these help to build pride for all stakeholders in the project. Drawings and models of the project can be kept on display in the job office and shown to all new workers so they may better appreciate the construction process and the team effort going into the project. Some contractors plan a Saturday open house from time to time so that employees' families can visit the site and see the quality work being done.

It must be remembered that production control, including quality, safety, and efficiency, occurs at the workface. Craftspeople are in the strongest position to have a positive impact on these project characteristics. Management must appreciate that planning for construction quality results in turning the organizational chart upside down. In a lean production environment, each level of separation from the workface comes with a reduction in the manager's ability to directly affect quality, safety, and efficiency. This increases the manager's responsibility to support the workface activities at the highest level possible. Workers must be empowered through programs such as lean construction to search out better ways to get a quality job done safely and efficiently.

On the formal side, the project team needs to determine the required tests, standards, and metrics for the job. Procedures for identifying and dealing with substandard work must be established. Every job-site inspection, for whatever purpose, needs to have a quality and safety component included. Special attention is required to incorporate lean construction principles within subcontractor organizations. The subcontractor selection process needs to be based on prequalification for quality, safety, and team play alongside price considerations.

One last point: Quality does not apply only to the finished product. It applies to the process as well. Quality issues include such elements as the way projects are planned, identification of problems well in advance of construction, the rate at which change orders are handled, and the quality of employee training. Every phase of construction operation provides an opportunity for continuous improvement.

6.10 Material Ordering and Expediting

Major materials usually are ordered within days of signing a construction contract. The material delivery lead time determines the order in which purchase orders are prepared. Therefore, special attention must be given to determining which materials will require the longest lead time. Preparation of purchase orders must be coordinated with the project schedule such that timely delivery dates are specified on the order. Quantity and quality requirements also must be spelled out in the purchase order. The purchase order should state where and when the materials are to be delivered and notification requirements prior to delivery so that adequate preparation can be made for receiving, off-loading, inspecting, storing, or directly installing the material.

Simply placing a purchase order for materials provides little assurance that the materials will arrive on time. A systematic procurement process is the best insurance for on-time deliveries. In larger companies, this is generally the responsibility of a purchasing agent or a purchasing department. However, in many construction companies, procurement is the responsibility of the project manager. No matter who is responsible for procurement, all purchase orders need to include a delivery date based on the project schedule. Early deliveries can cause problems with storage and have an adverse effect on project cash flow. Late deliveries are worse.

Special attention is needed to ensure that submittals, such as shop drawings, are prepared, checked, submitted for approval, and returned to the vendor on a timely basis. Similarly, samples, tests, mill certificates, and other documents must be approved. The control system for these submittals must ensure prompt handling at every level. Once all approvals are in place, manufacturing can begin. Fabrication schedules must be such that finished materials are ready to be shipped, allowing sufficient time to get to the job site. Some materials and equipment require inspection during the fabrication process or testing prior to shipment. These items must be identified and a system must be in place to notify the proper inspecting or approving organizations.

While key materials and equipment are being fabricated, the expeditor should contact the supplier for assurance that shipping arrangements have been made. Often, space must be reserved in advance on common carriers. Bills of lading and insurance also have to be obtained in advance. Some contractors require all subcontractors to submit unpriced purchase orders to the expeditor. This allows the contractor to confirm that the material has been ordered and to assist with the shipping schedule when transportation disruptions occur. As little as possible is left to chance.

The submittal schedule and submittal log are two tools that support the procurement process. The submittal schedule should be based on the project schedule. It will define when each installed item must be available at the workface and from that can backtrack through the supply chain to determine critical dates for such tasks as submittals, approvals, fabrication, and shipping. The submittal log identifies each critical point in the procurement process, lists the associated dates, and then is used to track the process by entering dates that each critical task has been completed.

At the job site, the expeditor and project engineer meet regularly to determine the current status of materials and to identify problem areas. Any potential delay in manufacturing or shipping requires prompt action in order to avoid delaying the project. The longer the advance warning of a material delay, the more options there are to work around the problem.

6.11 Material Handling, Storage, and Protection

The arrival of material at the job site should never be a surprise. As stated in the previous section, the exact time of a delivery should be established in advance. Unexpected deliveries are a disruption to work at the site, and often the only person available at the time to accept the delivery is unqualified for the job. At the job site, a specific worker should be assigned to accept the delivery and to oversee unloading. This person needs to have a copy of the purchase order in hand when the material arrives. Unloading equipment should be available and a specific place assigned for storage. As the shipment is unloaded, it should be inspected for completeness and for damage. Any deviations from the purchase order need to be noted on the delivery ticket, and that information must be passed on to project management.

When warehousing is required, an inventory control system has to be established. Care must be taken that materials and equipment are issued to the specific installation for which they were ordered. Exceptions need to be approved and documented so that one construction operation does not use materials ordered and needed by another operation.

Whenever possible, the material or equipment should be moved directly from the carrier to the final installed location. If this is impossible or impractical, the material should be stored in a location that will minimize handling and handwork. It is estimated that handling, moving, and handwork account for as much as 30 percent of the cost of installation. With inadequate planning, material orders are too often incomplete or damaged and the problem is not noted until the day of installation. If proper arrangements are not made, waterproof materials may be stored in buildings while weather-sensitive materials wind up in the rain, or it is determined too late that a trench must go directly through the place where topsoil is stockpiled. Detailed planning regarding material handling and storage saves money and prevents unnecessary and frustrating delays.

At times, materials must be stored offsite. In this case, issues of title, insurance, and payment are involved. It must be determined when title changes from the supplier to the contractor and from the contractor to the owner. Whoever has title also has insurance responsibilities, and these must be coordinated.

It must also be determined when payment will be due, first from the contractor to the vendor, and then from the owner to the contractor. Sequencing billing and payments for materials is a key component of cash flow management, which will be addressed in Chapter 12, "Project Financial Management." Since materials can constitute half or more of the entire budget, especially in building construction projects, delays in payments can have a severe negative effect on cash flow because the contractor can end up financing materials for an extended period of time.

6.12 Equipment Planning

On some projects, especially in the heavy civil area, construction equipment constitutes a major portion of project cost. For that reason, planning the spread of construction equipment, the haul roads where the equipment will operate, and the maintenance facilities that will keep the equipment running are of utmost importance.

Selection of the most appropriate equipment often must be modified according to the equipment available. A cost analysis must be made of the impact of less appropriate equipment. Where additional equipment is required, a buy, lease, or rent analysis is necessary.

An efficient equipment maintenance plan must be put into place with regularly scheduled preventive maintenance. Layout for the fueling and maintenance yard must be coordinated with the other yard areas. This area must provide easy access for equipment and should be laid out with safety in mind. Planning must be done for efficient haul roads, matching grades to the capabilities of the equipment available, and a haul road maintenance plan must be devised.

6.13 The New Production Model

Although construction projects are typically one of a kind, similar construction operations are repeated on many projects. Most any building project and many heavy civil projects have concrete work. Most heavy civil projects have a significant portion of earthwork, as does any new building project. It is important to plan efficient field operations. Though each field operation can be studied individually as the project moves forward, many efficiencies with broader impact can be incorporated into the initial production planning process. For example, opportunities may be found for prefabrication, which can lead to efficiency through assembly-line techniques. On the highway bridge, it is determined that the abutment forms could be prefabricated and brought to the site ready for installation. In building construction where there is a great deal of repetition in projects such as office buildings, hotels, and multistory residential facilities, there are significant opportunities for prefabrication, resulting in field assembly of components rather than "stick building." In addition to saving time, prefabrication can often result in safer field operations with higher quality and lower costs.

Fundamental to planning efficient production methods is simplifying each step of the process. This starts with simplifying the drawings. Construction plans have traditionally been terse and diagrammatic. Dimensions are given once. Typical details are drawn with exceptions shown in a schedule or table. High rates of production require that the details be spelled out for each step of the work. Rather than have craftspeople search through the drawings for the location of inserts, weld plates, and dimensions, this information is typically shown on a shop drawing.

Significant changes are occurring to traditional construction drawings with the advent of Building Information Modeling (BIM). Instead of paper drawings, the design of most projects has moved to digital models based on advanced versions of computer-aided design (CAD) software. Currently, even though the design is in a digital format, the final product of the design is often still printed. But there is a growing trend toward digital-only documents. BIM has expanded the traditional two-dimensional CAD computer model to three dimensions, enabling the project to be rotated and moved so that it can be viewed from any point, internal or external. BIM then adds the fourth dimension of time so that sequencing of the construction operations can be studied and the schedule can be incorporated into the BIM model. Finally, BIM has the capability of adding what some refer to as the fifth dimension, the verbal definition of installed items. This includes specifications and other pertinent information, such as installation guidelines and production costs. BIM is taking on the role of becoming a repository for all information about the project from all stakeholders in a single large digital model.

Because of the integrated nature of the BIM model bringing together information from many sources into a single model, it is changing the very nature of the relationships among participants in a construction project. Traditionally, the architect-engineer provided the design for a project and retained the intellectual property rights to the design. The general contractor and various specialty contractors provided information on cost and schedule and protected proprietary information, especially related to costs. In a BIM environment, all of this information from the various participants in the project is being entered into the single BIM model, so information is shared. Even more significant, risk is shared. In this highly interrelated environment, a new contractual basis is required, leading to the emergence of a project delivery system that has become known as Integrated Project Delivery (IPD).

As BIM becomes more prevalent, it will change the way project managers and field supervisors do their jobs. Installation details that were accessed on shop drawings will be accessed on tablet computers and even smartphones. Planning and updating can be extended literally until the point of installation and adjusted based on real-time field conditions. This means the balance of responsibilities between the project manager and the field supervisor will evolve with more, higher-level decisions being made in real time in the field. Meanwhile, while project managers are delegating more decisions to the field, they are also working much more closely with the other stakeholders in the project through the shared risk and reward of the IPD contract.

Key Points and Questions

- Complete, thorough, and detailed planning, both at the beginning of a project and throughout the duration of the project, is key to successful completion.

Key Points

- Planning is required at many levels, from the project as a whole to each individual activity.
- Planning considers all resources that contribute to the job, including personnel, materials, construction equipment, the site, the environment, and anything else that might affect the job.
- Lean construction addresses all aspects of the construction process, including design, supply chain management, and field operations.
- BIM is beginning to change the way construction is managed and the construction project is organized. The changes will be ongoing and profound.

Review Questions and Problems

1. What are impediments to planning, both at the beginning of a project and throughout the project?
2. What critical elements are important to support the planning process?
3. Develop a checklist of items to consider in development of the site plan for either the bridge project or the building project.
4. Use the example building site plan provided as a basis to plan the layout of the site for the building example project.
5. Lay out a procurement log that could be used for the example building project. To demonstrate the log, include several representative material items for the building. This works well as a spreadsheet exercise.
6. Describe some of the ways in which BIM is changing how projects are managed.

7 Managing Time

7.1 Introduction

The schedule plays a central role in construction project management. In Chapters 4, "Project Planning," and 5, "Project Scheduling Concepts," the project plan was developed, and then, based upon an extensive set of calculations, the time characteristics were developed for the project as a whole as well as for each activity. In Chapter 6, "Production Planning," the production plan was developed.

Once the initial schedule is developed, it represents a powerful tool that can be used in managing various aspects of the project, including time, resources, production, and cost. As the project progresses and the project management team recognizes more of the details that affect the construction process, the schedule will be modified to evolve with the project.

This chapter concentrates on using the schedule to manage the time required to execute the construction processes. It begins by considering the project as a whole, determining how to shorten the overall project schedule, and looking at the cost trade-offs of expediting the project. It then focuses on current or upcoming parts of the project with the objective of managing the project components more effectively.

Learning objectives for this chapter include:

❏ Understand why the efficient use of time on a construction project is important.
❏ Learn procedures for shortening the project.

❏ Recognize how time affects both the cost of activities and the general and administrative costs of the project.
❏ Understand how to measure, report, and manage time invested in the project.

7.2 Time Schedule Adjustments

Frequently, project work schedules must be adjusted to accommodate adverse job circumstances. These revisions are often essential so that contract time requirements can be met. Many times, established time goals dictate that key stages of the work be achieved earlier than originally planned. The start or finish dates of major job elements often must be improved to satisfy established time constraints or commitments. Milestones, network interfaces, and final completion are common examples of key events that sometimes must be rescheduled to earlier dates. Such schedule advances are accomplished in practice by performing certain portions of the work in shorter times than had originally been allocated to them.

The following sections discuss why management action to reduce project time occasionally is needed and how the associated time reduction studies are conducted. The highway bridge is used for purposes of discussion and illustration.

7.3 Need for Time Reduction

There are many practical examples for which shortening the time of selected job elements can be desirable as a means of meeting important project target dates. For example, by terms of the construction contract, the owner may impose a job completion date that the current project plan will not meet. Failure to meet this contractual time requirement will put the contractor in breach of contract and make it liable for any damages suffered by the owner because of late project completion. On a job in progress, the owner may desire an earlier completion date than originally called for by the contract and may request that the contractor quote a price for expediting the work. It is entirely possible that the programmed project duration time may not suit the contractor's own needs. The contractor may wish to achieve job completion by a certain date to avoid adverse weather, to beat the annual spring runoff, to free workers and equipment for other work, or for other reasons. Financial arrangements may be such that it is necessary to finish certain work within a prescribed fiscal period. The prime contractor may wish to consummate the project ahead of time to receive an early completion bonus from the owner. A common motivation for time acceleration occurs when the work is well under way and delays have resulted in a substantial loss of time that must be recovered by the end of the project.

Although not involving the entire project, a similar situation can arise when attempting to meet an established milestone. It is not unusual for the computed early times of milestone events to occur later than desired. An analogous situation can arise with respect to network interface events.

These examples clearly disclose one of the great advantages of being able to establish advance construction schedules with reasonable accuracy. Such information makes it possible for the project manager to detect specific time problems well in advance and to initiate appropriate remedial action. Certainly this is preferable to having no forewarning of such problems until it is too late to do much, if anything, about them.

On first impression, it may appear incongruous to consider shortening the duration of a project when a contingency allowance has been added to the computed normal time. In the case of the highway bridge, a contingency of 6 days was added to the calculated time of 64 days to establish a probable completion time of 70 working days. This project time is the best advance estimate available of the actual work duration required. As a consequence of this fact, 70 working days is the most likely duration of the highway bridge, and any need to shorten the project should depend on how this time compares with the prescribed contract period or other established time limitation. Removing the contingency from the project is not a reliable means of accelerating the completion date.

7.4 General Time-Reduction Procedure

Perhaps it is appropriate now to mention that a variety of terms are applied to the process of shortening project time durations. *Least-cost expediting*, *project compression*, and *time-cost trade-off* are all used in reference to the procedures discussed in this chapter. The exact nomenclature is not especially important as long as the method and its application are thoroughly understood. The use of the term *expediting* in connection with shortening project time requirements is unfortunate because the same term is used with regard to actions taken to ensure timely resource support for construction operations. However, the double meaning of *expediting* is commonplace in the construction industry and is so applied herein.

To shorten the time period required to reach a milestone or interface event or to achieve project completion, one needs to be concerned with reducing the time durations of only a certain group of activities. As has already been shown, the time required to reach any future network event, terminal or otherwise, is determined by the longest time path from the current stage of project advancement to that event. Consequently, if the time required to reach a certain event is to be reduced, this can be accomplished only by shortening the longest path leading to that event. This observation is very revealing and important. In the absence of such management information, the usual reaction when a project is falling behind schedule is to haphazardly expedite all the ongoing activities in an attempt to make up the lost time. The inability to discriminate between those activities that truly control and those activities of little or no time consequence can make such expediting actions far more expensive than is necessary.

When the date of project completion is to be advanced, it is the network critical path that must be shortened. When reducing the time to achieve a milestone or interface event, the diagram critical path itself may not be involved in any way. The longest time path leading to the milestone or interface event must be shortened. This path may be entirely separate from the critical path that applies to the entire network. For purposes of discussing the reduction of project time, the longest time path leading to the event in question will be referred to as its critical path.

At this point, it must be recognized that when a longest path is shortened, the floats of other activity paths leading to the same event are reduced commensurately. It is inevitable, therefore, that continued shortening of the original critical path will lead, sooner or later, to the formation of new critical paths and new critical activities. When multiple critical paths are involved, all such paths must be shortened simultaneously if the desired time advancement of the event is to be achieved. Shortening one critical path but not another accomplishes nothing except to provide the shortened path with unneeded additional float.

When time reduction is done manually, the effect of each shortening action must be checked to ascertain whether it has produced new critical activities. The usual way to do so is to perform a network recalculation, following each step in the time-reduction process. Such recalculations can be done manually or by computer. If the network is not large, manual calculations can be fast and convenient. For a large network, many successive recalculations can become a substantial chore. Even though time-scaled diagrams are of limited value in making actual time-reduction studies, such plots are especially useful for explaining the total effect of a given time reduction. A network change can be visualized in terms of movements of rigid-frame portions of the diagram and the effect of these movements on its elastic (float) connections. Whenever any shortening action eliminates a dashed float line in a time-scaled network, a new critical path is formed automatically. Portions of Chart 5.3a (on the companion website) are used in this chapter to give the reader a better appreciation of what network time reduction actually involves.

7.5 Shortening the Longest Time Path

It has now been established that, if the date of a specified project event is to be advanced, the length of the longest time path leading to the event must be shortened. Basically, there are only two ways to accomplish this. One is to modify the job logic in some way, such that the longest route is diminished in length. Doing this involves a localized reworking of the original job plan, with time being gained by rearranging the order in which job activities will be accomplished or increasing lead and lag times to maximize activity concurrency. This time-reduction procedure does not reduce the durations of activities themselves; it gains time by the more favorable sequencing of selected job operations.

The other possible way to reduce the length of a critical path is to reduce the duration of one or more of its constituent activities. Each critical activity first must be examined to see if any shortening is possible. The compression of an activity can be achieved in a variety of ways, depending on its nature. Additional crews, overtime, or multiple shifts can be used. It may be possible to subcontract it. More equipment may be brought in temporarily and assigned to that activity. Earlier material deliveries may be achieved by authorizing the fabricator to work overtime, by using air freight or special handling, or by sending one of the contractor's own trucks to pick up and deliver the material. There usually are, of course, some activities whose durations cannot be reduced feasibly.

7.6 Project Direct Costs

Before the discussion of project time reduction can proceed further, it is necessary to discuss the nature of project direct costs and indirect costs. Costs are necessarily involved with time reduction because construction expense is a function of time. Although management discretion occasionally may dictate otherwise, an effort usually is made to achieve gains in time with the least possible increase in project cost. If project management is to make schedule adjustments at the least additional cost, it is necessary to understand how the costs of construction operations vary with time.

The direct cost of an activity is made up of the expense of labor, equipment, materials, and subcontracts. Each activity has its normal cost and normal duration. The "normal cost" is the least direct cost required to accomplish that activity and is the cost customarily ascribed to the work when the job is being estimated. "Normal duration" is the activity duration determined during the scheduling phase. Although there is nothing precise about a normal time, it still constitutes a reasonably distinct datum or reference point for accomplishing an activity at its least direct cost. It is obvious that the direct cost of the total project is equal to the sum of the direct costs of the individual activities and that the normal project duration is derived from the activities' normal durations. It follows, therefore, that the least total direct cost of the entire job is the cost associated with the normal project duration.

If the estimated activity times can be accepted as going hand-in-hand with minimum direct cost, then any variation in an activity time from that estimated, either more or less, must result in a commensurate increase in its direct cost. The degree to which this rule actually applies in practice is considerably more certain with respect to increased costs caused by reducing the activity duration than by extending it. However, contractors are seldom concerned with stretching out, or deliberately extending, the duration of a job activity.

The practical fact that shortening an activity time normally will increase its direct cost is easily demonstrated. The use of multiple shifts or overtime work obviously entails extra labor expense. Crowding in more work crews or pieces of equipment makes job supervision difficult, reduces operational efficiency, and increases costs of production. Early material delivery requires payment of premiums to the vendor or increased transportation and handling costs. All this leads to the conclusion that, if the project time requirement is to be reduced, the direct costs of the activities actually shortened usually will be increased.

7.7 Variation of Activity Direct Cost with Time

The direct costs of activities can vary with time in many ways, although here it is considered that these costs always vary in inverse proportion with time. A continuous linear, or straight-line, variation of direct cost with time is a common example. This is the result of an expediting action, such as overtime or multiple shifts, where the extra cost for each day gained is just about constant. Figure 7.1a is such a case, where the normal activity time of 15 days can be reduced by as much as 3 days. The increase in direct cost of this activity is a constant $100 per day, this being termed the *cost slope* of the activity. The contractor may elect to expedite the activity by one, two, or three days, for which the extra cost will be $100, $200, or $300, respectively.

The "normal points" in Figure 7.1 represent the activity normal times and normal direct costs discussed before. The expediting of an activity is often called *crashing*. As indicated in the figure, the minimum time to which an activity can be realistically reduced is called its *crash time,* and the corresponding direct cost is called its *crash cost*. The plotted intersection of these two values is called its *crash point*.

Figure 7.1b shows a continuous, piecewise linear, time-cost variation. The figure indicates that a time reduction of one, two, or three days is possible, but the cost slope increases for each additional day gained. An example of a piecewise linear variation is when crew overtime is involved and the use of a progressively larger crew is possible as the work advances. Figure 7.1c is a discontinuous or gap variation where the expediting action reduces the activity duration from 15 to 12 working days, with no time possibilities in between. The activity is reduced either by three 3 or not at all, and the extra cost of the reduction is a fixed sum of $300. This type of time-cost variation is common in construction. Paying a premium for an

7.7 Variation of Activity Direct Cost with Time

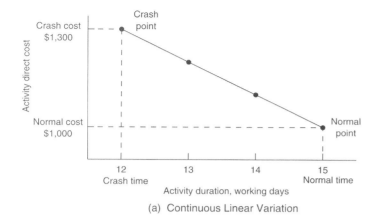

(a) Continuous Linear Variation

(b) Continuous Piecewise Linear Variation

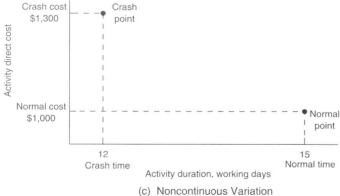

(c) Noncontinuous Variation

Figure 7.1
Activity time-cost variation

early material delivery and shipping by air rather than motor freight are examples.

Obviously, other forms of time-cost variation are possible. However, great accuracy in determining the extra costs resulting from expediting actions is seldom achievable, and the time-cost variations used, along with the expediting of construction activities, are generally limited to the three contained in Figure 7.1.

7.8 Project Indirect Costs

As described in Section 3.19, project overhead consists of indirect costs incurred in support of the fieldwork but that cannot be associated with any particular physical portion of the job. Figure 3.7 discloses that, on the highway bridge, the time-variable job overhead expense was estimated to be $43,690. Time-constant overhead expenses are not variable with project duration and need not be considered here. The probable duration of the highway bridge has been determined to be 70 working days. This means that the time-variable indirect expense on this job amounts to $43,690 ÷ 70 days = $624 per working day.

The preceding discussion discloses that crashing an activity, while increasing direct costs, will generally reduce indirect costs. If a specific activity is shortened, its direct expense increases, but if it leads to a corresponding decrease in overall project duration, the indirect cost is reduced.

7.9 Practical Aspects of Time Reduction

The computer is an invaluable tool to assist project managers in accomplishing many tasks. However, there are still many problems that are not suitable for computer solution. One of those tasks is project shortening, for which the computer does not normally serve as an adequate standalone device. Manual methods, relying on human insight and judgment, continue to play a commanding role in the process. The project time acceleration procedures discussed herein describe and emphasize such an approach. This does not mean that the computer plays a trivial supporting role, since management and computer can work together to achieve the best solution possible. The manager can originate and pass on matters of judgment, and the computer can process the decisions made by project management.

The process of least-cost shortening of actual construction networks can become enormously complex. Multiple critical paths can appear and make the shortening process a very complicated procedure. The number of possible expediting combinations to be tested, if an optimal solution is to be achieved, can become very large. It must be recognized, therefore, that the usual manual time reduction will certainly not always provide project management with truly optimal expediting combinations. However, mathematical precision with imprecise data is neither the only nor necessarily the most important consideration involved in such a process. Of necessity, the actual accomplishment of time reduction in practice must be concerned with a number of practical considerations beyond the matter of buying the most time for the least money.

Manual solutions for project time reductions, while perhaps not optimal, do provide invaluable guidance to the project manager in making decisions about whether expediting is practical and, if so, how to proceed. In most cases, guidance on how to make intelligent choices of time-reduction actions is as valuable as a theoretically optimal solution. Input data are uncertain, conditions change from day to day, and construction is simply not an exact or a completely predictable process. Even the critical path of a given network may change its routing occasionally as the work progresses. Project managers strive to find practical, reasonable answers rather than seek to achieve perfection. Expediting a project manually makes it possible to inject value judgments into the process and affords the project manager an intuitive feel for the effect of expediting actions on other aspects of the project. In addition, a project time-reduction study can easily include the critical evaluation of time gained by revisions in job logic as well as that gained by shortening individual activities.

The manual accomplishment of project time reduction is directed entirely toward reducing the length of the applicable critical path or paths. This is a step-by-step process using time-reduction measures that are considered feasible and best suited to the job context. These may be changes in the job plan, the shortening of individual activities, or both. The usual procedure is to gain each increment of time with the least possible increase in direct cost. Where other job factors are of greater importance than incremental cost, shortening steps are taken in whatever order project management believes is in the overall best interest of the work.

7.10 Reduction of the Highway Bridge Duration

For purposes of illustration, suppose the contractor on the highway bridge determines that the probable duration of 70 working days or 98 calendar days is unsatisfactory. Work on this job has not yet begun, and a study is to be conducted to investigate the feasibility and attendant cost of reducing the overall project duration by perhaps as much as 10 percent.

The essential question is, of course, how the project critical path can be reduced from its current time duration. Initially, common sense suggests that a second look be given to the present operational plan, the objective being possibly to gain time at no increase in direct cost. Although knowledgeable and experienced people carefully devised the job plan, the original planning effort cannot be expected to be perfect. It seems likely that a restudy of the operational sequence could sharpen the planning approach and perhaps indicate opportunities for greater time efficiency.

If reexamination of the original job logic does not produce the desired time gain at no additional cost, then the contractor has no option but to sacrifice project cost for time. Almost any construction operation can be performed in less time if someone is willing to pay for the additional expense. There are undoubtedly a number of opportunities for shortening the duration of the highway bridge by effecting changes in job logic or by reducing the times needed to accomplish individual critical activities. When extra cost is involved, project shortening is achieved by evaluating the feasible alternatives and, normally, adopting the least-cost combination of those that will produce the desired time adjustment.

It is the intent of the next five sections to present specific discussions of how the overall duration of the highway bridge might be reduced at little or no additional direct cost to the contractor. Such a reexamination of the programmed plan will not always result in a time gain, but the possibility is there and should be investigated. A word of caution is in order, however. When such restudies are being made to pick up some badly needed time, there is always a tendency to become optimistic. Those who make decisions concerning project time reduction must be sensible and pragmatic in their judgments.

A. Restudy of Critical Activity Durations

There is one obvious initial check to be made when reexamining the critical path of a project to be shortened: Review the time estimates of the individual critical activities. Errors can be made, and it is worthwhile to verify the reasonableness of the time durations originally estimated.

Another possibility also can be reviewed. When the time estimates were first made, it was not known which of the activities would prove to be critical. The original time estimates of some activities might have been made in contemplation of the limited future availability of labor crews or construction equipment. As a result, some of the activity duration estimates for what later turned out to be critical work items were based on smaller than optimum-size crews or equipment spreads. Now that the identities of critical activities have been established, it may be feasible to defer action on some noncritical activities, using their floats for this purpose, and to reassign resources temporarily to the critical activities, with the objective of accomplishing them in less time.

B. Restudy of Project Plan

A project restudy aimed at gaining time at no additional direct cost is essentially a critical second look at the established operational plan. The objective, of course, is to rework or refine the logic of a limited area of the network that will result in a shortened critical path. Here, innovative thinking and a fresh approach may result in important improvements of work methods. Too often, traditional and established field procedures are accepted as the only way of solving a problem. At times they should be challenged by inquiring minds seeking a better solution. A good old-fashioned brainstorming session occasionally will produce some ingenious ideas concerning new approaches.

In some cases, the contractor has the authority to make changes in project materials or design or perhaps is able to do so with owner approval. For example, in a design-construct contract, the project manager may decide to redesign a segment of the project to take advantage of a faster construction method, thereby reducing the overall duration of the project. In recent years, prefabrication with field assembly rather than "stick building" in the field has become much more common as a means not only to reduce the time and cost of field operations but also to increase safety and improve quality.

C. Critical Activities in Parallel

As an example of a change in project plan to shorten the critical path, it may be possible to perform certain critical activities concurrently or in parallel rather than in series. To show how this can work, reference is made to the time-scaled diagram of the highway bridge in Chart 5.3a on the companion website. Study of the critical activities will disclose the possibility that painting can be done concurrently with, rather than after, stripping the deck forms. Figure 7.2, which is excerpted from Chart 5.3a, illustrates this change in the project schedule, showing that activity 370, "Paint," can start at the same time as activity 350, "Strip deck," thus shortening the critical path by two days. If this were done, painting would no longer be critical; rather, the guardrail installation would take over the painting activity's position on the critical path, as indicated by Figure 7.2.

The possibility of accomplishing activities 350 and 370 in parallel with one another might generate an unsafe working condition. This would have to be resolved before this means of shortening the critical path could be approved for implementation in the field.

D. Subdivision of Critical Activities

In shortening a critical path, a check can be made to determine if each critical activity must necessarily be completed before the next one can start. The judicious subdividing of one critical activity may enable it to overlap another. In other words, it may be possible to subdivide a critical activity and perform a portion of it in parallel with another critical activity. There appear to be two or three such opportunities in Chart 5.3a, only one of which will be discussed here because of the similarity between them. Referring to activity 180, "Forms & rebar, abutment #1," one side of these forms must be erected before the steel can be tied. Consequently, the finish of activity 60, "Fabricate & deliver abutment & deck rebar," can overlap the start of activity 180 by one day, assuming that the time necessary to erect

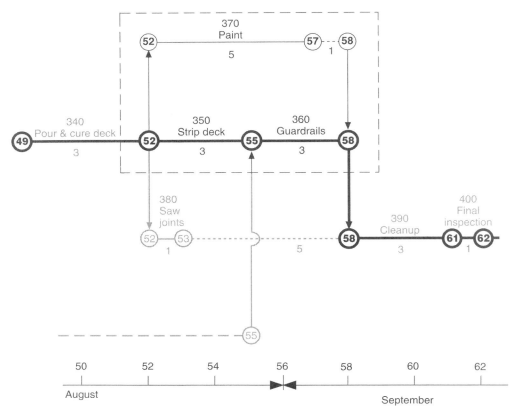

Figure 7.2
Highway bridge, critical activities in parallel

one side of these prefabricated forms is one day. Figure 7.3 shows how the original time-scaled diagram (Chart 5.3a) would have to be modified to reflect this change. As can be seen, this alteration would shorten the critical path, and hence the time to any succeeding critical activity, including the terminal activity, by one day.

As stated previously, the possibility of other critical paths being formed must be investigated each time the critical path is shortened. For the time reduction discussed in the previous paragraph, reference to Chart 5.3a shows that all activities in the network to the right of activity 180 have moved as a unit one day to the left. In so doing, the floats shown at the ends of activities 230 and 260 have been reduced by one day. In addition, the creation of the new activity 175, "Outside forms, abutment #1," reduces by one day the float following activity 80 and that following activity 170. This action does not, however, result in the formation of any new critical activities.

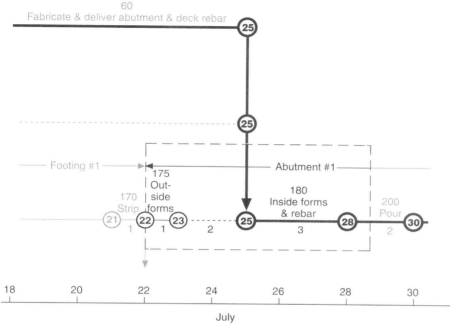

Figure 7.3
Highway bridge, subdivision of critical activity

Another possibility for shortening a project critical path may be to subcontract certain work that the general contractor originally intended to do with its own forces. The project plan may show certain critical activities to be in series with one another, not because of the physical order in which the work must be done, but because they require the same limited resource. Subcontracting all or a portion of the work involved to a specialty contractor that has adequate equipment and manpower may enable the activities to be performed concurrently rather than in series, thus saving considerable time.

It is not unusual for an equipment or labor restraint together with a dependent activity to be a part of the critical path. This situation usually represents a limitation on the availability of a general contractor resource. The prime contractor might consider subcontracting that part of the work to a firm whose labor crews or equipment would be available at an earlier date than its own. Determining whether the work could be subcontracted for the same cost that the prime contractor estimated would require study. If extra cost is involved, this becomes an expediting action that will have to be considered on a comparative cost basis along with the other additional expense possibilities.

E. Subcontracting

7.11 Time Reduction of Highway Bridge by Expediting

The preceding five sections discussed how a critical reexamination of the original job plan might result in shortening a project critical path at no increase in direct cost. If such a network study does not produce the desired time reduction, then project management must literally buy time by resorting to expediting actions that will increase direct costs. Expediting actions are thus distinguished herein from time gained at no extra expense.

The sections that follow consider the expediting process as it is applied to the highway bridge. For clarity, the discussion assumes that none of the possible time-reduction measures discussed previously has been adopted; therefore, the original plan and schedule of the highway bridge remain unaltered. Shortening this project by subcontracting is not considered either feasible or desirable, so the ensuing time-reduction measures are limited to those the prime contractor can achieve with its own forces.

The first step in expediting the highway bridge probably should be to determine if there are any changes in the job plan that would shorten the critical path. As has already been stated, modifications of network logic are often more fruitful than shortening individual activities. An obvious change that could be made in the logic of the highway bridge, and which involves the critical path, would be to prefabricate and use two sets of abutment forms. It will be recalled that the job plan for this project, as presented in Chart 5.1a on the companion website, calls for only one set of forms to be built, these forms being used first for abutment #1 and then reused for abutment #2. Consequently, according to the original plan, the start of the second abutment will have to wait the stripping of forms from the first abutment. A second set of abutment forms would eliminate the dependency line from activity 220, "Strip and cure, abutment #1," to activity 240, "Forms and rebar, abutment #2." This would enable the forming of abutment #2 to start just as soon as footing #2 was finished (activity 230). The effect of this change can be checked by a recalculation of the altered network or by reference to the time-scaled plot in Chart 5.3a on the companion website. In either case, if this change is made, the entire project will be shortened by 6 working days, from 70 to 64. The abutment #2 string of three activities will no longer be critical, as the critical path now is routed through activity 280.

A preliminary examination of the critical activities of the highway bridge discloses that there might be several possibilities for expediting individual activities. Suppose careful study, however, reveals that there are actually only four activities for which some expediting is considered to be practicable in the view of project management. Figure 7.4 lists the time reduction and additional direct cost for each of the ways in which the highway bridge may be expedited feasibly. Assuming that all of these alternatives are equally acceptable to project management, the process is now one of

7.11 Time Reduction of Highway Bridge by Expediting

Expediting Action	Time Reduction (Working days)	Direct Cost of Expediting
A Prefabricate two sets of abutment forms. Start placing Forms & rebar, abutment #2 (activity 240) directly after Strip footing #2 (activity 230)	6 (logic change)	$5,575
Computations One set of forms as estimated (Figure 3.6) Labor $5,144 Material $2,153 $7,297 Two sets of forms Material Plyform: 10% waste, 1 use, 70% salvage 3620 (1.10)($1.05)(.30) = $1,254 Lumber: 1.44 bf per sf, 1 use, 70% salvage 3620 (1.44)($0.85)(0.30) = $1,329 Labor 3620 ($1.96) + 45% indirect = $10,288 $12,872 Cost of expediting = $12,872 − $7,297 = $5,575		
B Steel fabricator agrees to work overtime and give abutment rebar special handling. Deliver abutment & deck rebar (activity 60) reduced from 15 to 11 days.	4	$2,376 (quoted by rebar vendor)
C Expedite Pour abutment #1 (activity 200) by bringing in an additional crane, hiring two more laborers, and working the concrete crew 2 hours overtime. Duration reduced from 2 days to 1 day.	1	$422
Computations Pour abutment #1 as estimated (Figure 3.6) Labor $12,269/2 = $6,134 Equipment $3,744/2 = $1,872 $8,006 Expedited Pour abutment #1 (10 hours) Labor Regular crew $284 x 10 x 1.35 = $3,834 Extra laborers and operator ($44 + $33) x 10 x 1.30 $1,001 Overtime premium ($284 + $77) x 2 x 0.5 x 1.35 $487 Equipment Cost of two cranes, etc. $2,500 Extra crane, in and out $605 $8,427 Cost of expediting = $8,427 − $8,006 = $422		
D Expedite Pour abutment #2 (activity 250) in the same way as abutment #1.	1	$422
E Expedite Strip deck forms (activity 350) by hiring two more laborers. Duration reduced from 3 days to 2 days.	1	$202 (not shown but similar to previous calculations

Figure 7.4
Highway bridge, direct costs of expediting actions

shortening the project, one step at a time, each increment of time being gained at minimum additional direct cost. Project management now can decide how much time reduction it is willing to purchase.

7.12 Least-Cost Expediting of the Highway Bridge

Using the information summarized in Figure 7.4, the highway bridge is now to be reduced in duration, with each successive increment of time compression being realized at a minimum increase in project direct cost. As each step is taken, a check must be made to determine whether that expediting action results in the formation of any new critical activities. This check is made by means of a network recalculation after each time-reduction step. Figure 7.5 summarizes the results of the successive expediting actions.

Examination of Figure 7.4 discloses that the least expensive first step in the shortening process is expediting by one day of critical activity 350, "Strip deck forms," at an additional direct cost of $202. In a network with a single critical path, as is the case with the highway bridge, the amount of any step decrease in the duration of a critical activity is subject to two limitations, one internal to the activity and the other external. The first of these is how much internal shortening of the activity is physically possible. In the case of activity 350, the physical limit has been established as one day. The second limitation is based on how much the activity can be shortened before a new parallel critical path is formed. Often, this limitation is referred to as the external or logical limit of an activity shortening. The logical limit of a given critical activity is equal to the total float of the shortest alternative path around that activity. Reference to Figure 5.1a shows that the path through activity 380 (TF = 7) is the shortest way around activity 350. Hence, the logical limit of activity 350 is 7, which just says that activity 350 could be shortened by as much as seven days before a new critical path is formed. Obviously, the first step in expediting the highway bridge using activity 350 is limited to the lesser of its physical limit (one day) or its logical limit (seven days), or a shortening of one working day. This information is summarized in step 1 of Figure 7.5a. Activities 200 and 250 are expedited in a similar manner and are shown as steps 2 and 3 in Figure 7.5a. In this figure, the creation of a new critical path as a result of crashing the duration is indicated by a "yes" or "no" entry in the "New Critical Path" column.

It is necessary to conduct three separate sequences of shortening actions in order to shorten the highway bridge to its full potential at the least additional cost. Figure 7.5a is the first of these sequences. The successive shortening by actions E, C, and D, as described in the figure, shortens the project by 3 days at a total extra direct cost of $1,248. If the project is to be shortened by only 3 days, this is the least expensive course of action. To

7.12 Least-Cost Expediting of the Highway Bridge

Step	Project Duration (Working days)	Expediting Action (from Figure 7.4)	Critical Activity Shortened	New Critical Path	Add'l. Direct Cost	Cumulative Direct Cost
(a)						
*1	69	E	350	Physical limit = 1 Logical limit = 7 No	$202	$202
*2	68	C	200	Physical limit = 1 Logical limit = 8 No	$422	$624
*3	67	D	250	Physical limit = 1 Logical limit = 6 No	$422	$1,248
(b)						
1	67	B	60	Physical limit = 4 Logical limit = 3 Yes	$2,376	$2,376
*2	66	E	350	Physical limit = 1 Logical limit = 7 No	$202	$2,578
*3	65	C	200	Physical limit = 1 Logical limit = 8 No	$422	$3,000
*4	64	D	250	Physical limit = 1 Logical limit = 6 No	$422	$3,422
(c)						
1	64	A	Logic change	Logical limit = 6 Yes	$5,575	$5,575
*2	63	E	350	Physical limit = 1 Logical limit = 7 No	$202	$5,777
*3	62	C	200	Physical limit = 1 Logical limit = 8 No	$422	$6,199

*Optimal time-cost relationship

Figure 7.5
Highway bridge, least-cost expediting

shorten the project additionally, an entirely new and different sequence of shortenings must be used. In other words, the time-reduction process must be started anew. Step 1 of Figure 7.5b shows that shortening activity 60 can reduce the project by 3 days, from 70 to 67 working days, at an additional cost of $2,376. Actually, the expenditure of $2,376 shortens activity

60 by 4 days, which is its internal or physical limit. However, when activity 60 is shortened by 3 days, a new branch on the critical path is formed, so the logical limit of this shortening action is 3 days and shortening activity 60 results in shortening the project only by 3 days. If the fourth day of the shortening of activity 60 is to become usable, some activity on the new critical branch also will have to be shortened by 1 day. Study of these newly critical activities discloses that it is not possible or feasible to shorten any of them. Therefore, the total effect of step 1 in Figure 7.5b is a reduction of 3 days. Steps 2, 3, and 4, shown in Figure 7.5b, reduce the highway bridge construction to a duration of 64 working days at a total additional cost of $3,422. The sum of $3,422 is the least additional cost for which the project can be shortened by 6 days.

To reduce the highway bridge duration below 64 days, a third new series of shortenings is needed. Step 1 of Figure 7.5c shows that prefabricating and using two sets of abutment forms (a logic change, not an activity shortening) reduces the highway bridge duration from 70 to 64 working days, a gain of 6 days. The time reduction achieved by a logic change is the difference between the lengths of the critical paths before and after the logic revision is made. A change in network logic, therefore, has no physical limit, only a logical limit. Steps 2 and 3 in Figure 7.5c reduce the duration of the highway bridge to 62 days at an additional cost of $6,199. It might be noted here that if activity 250 is now expedited, no further shortening of the project results. This is because, when the logic change in step 1 of Figure 7.5c is made, there is a change in critical path location, and activity 250 is no longer critical. Expediting it will only increase its float, not reduce the length of the critical path.

The project duration has now been reduced by a little more than 10 percent, which was the original objective. Hence, the time-reduction study is now complete and project management must decide how much expediting it wishes to pay for. Figure 7.6 summarizes the expediting costs involved in shortening the highway bridge. This information tells the contractor how to reduce the project duration by any given number of working days at the least cost, up to a maximum shortening of eight days.

7.13 Limitations on Time-Reduction Steps

As has been shown, a number of limitations apply to how much time reduction can be accomplished in any one step. These five limitations are summarized next.

1. *Physical limit of a critical activity.* This is the maximum shortening of a given activity considered to be practical. Although most activities can be shortened to some extent, some are considered to be intractable on practical grounds.

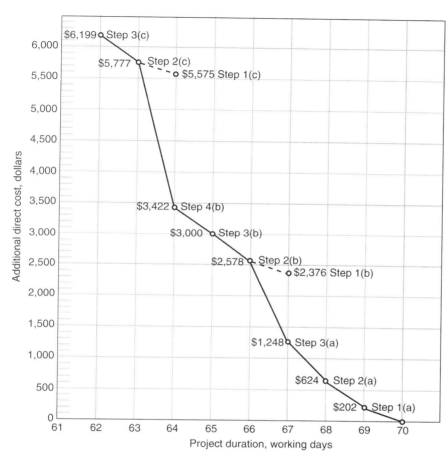

Figure 7.6
Highway bridge direct, cost of expediting

2. *Logical limit of a critical activity.* The reduction in duration of a critical activity reduces the total floats of other activities, which sometimes causes another chain of activities to become critical. This can, at times, prevent expediting an activity to its full potential. Step 1 in Figure 7.5b is an example of this.

3. *Logical limit of a network logic change.* A network logic change results in a set number of days being gained with no time change possibilities in between. Step 1 in Figure 7.5c is a logic change that reduces the highway bridge duration by six days. The six days is an irreducible time reduction—it is either that or nothing.

7 Managing Time

4. *Shortening limited by a parallel critical path.* Parallel critical paths, or subpaths, are common. If one branch of parallel critical paths is to be decreased in length, a commensurate decrease must also be made in the other branch. If only one branch is reduced, it simply becomes a floater, and neither the remaining critical path nor the project is reduced in duration. This situation will be encountered in conjunction with the inability to shorten activity 60 completely in Section 7.17.

5. *Shortening limited by an irreducible critical path.* When any given critical path has been shortened to its full capability, no further reduction in project duration is possible. A critical path that cannot be compressed makes fruitless the shortening of noncritical portions of the network.

7.14 Variation of Total Project Cost with Time

Prior discussions have shown that a general characteristic of construction projects is for their direct costs to increase and their field overhead costs to decrease as the construction period is reduced below the normal time. Figure 7.6 is a plot of the increase in direct cost required to expedite the highway bridge. It was determined in Section 7.8 that the time-variable field overhead expense for this project amounted to approximately $624 per working day. Figure 3.7 disclosed that the project constant overhead expenses totaled $26,178.40. Figure 7.7 shows how total project overhead, direct cost, and the sum of the two vary with time. The field overhead expense shown in Figure 7.7a is obtained by multiplying the number of working days by $624 and adding the sum of $26,178.40. The normal direct cost of the project is established from Figure 3.8 as $398,975. Adding the expediting costs from Figure 7.6 to this project normal cost gives the direct costs shown in Figure 7.7b. Combining the direct cost with the overhead expense for each project duration gives the costs shown in Figure 7.7c. Total project cost for any duration may be obtained by adding small tools ($4,649), tax ($14,205), markup ($73,155), and the cost of the contract bonds ($5,665) to the values shown in Figure 7.7c. The four values just cited are obtained from the original project estimate given in Figure 3.8.

To illustrate how the plot in Figure 7.7c can be used, suppose that the construction period prescribed by contract for the highway bridge is 62 working days and that liquidated damages in the amount of $400 per calendar day will be imposed for late completion. The best evidence now available indicates that 70 working days actually will be required unless the job is expedited. If a management decision is made not to expedite, the contractor is apt to be assessed $4,000 in liquidated damages (8 working days = 10 calendar days @ $400 per day). If the job is expedited to a duration of 62 working days, Figure 7.7c indicates that the total additional

7.14 Variation of Total Project Cost with Time

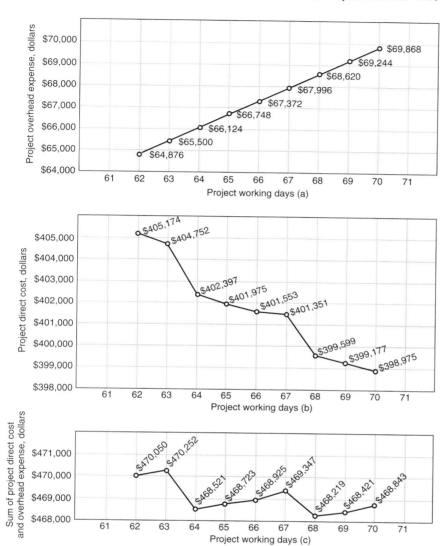

Figure 7.7
Highway bridge, variation of costs with duration

expense to the contractor will be $470,050 − $468,843 = $1,207. Very likely, the decision of project management will be to expedite the project.

Besides project shortening, there are other reasons the contractor must be able to associate costs with project durations. One such situation occurs when the contractor bids on a project in which the owner requires a range of price quotes for a variety of specified construction periods. Although this usually requires the contractor to submit a bid before a complete time study is available, the contractor still must make some judgments regarding the variation of direct and overhead costs with time.

7.15 Expedited Highway Bridge Schedule

Assume that the contractor has decided to expedite the highway bridge by 8 working days to an anticipated duration of 62 days. This expediting information must now be reflected in the construction schedule. The results of the network recalculation made after step 3 in Figure 7.5c will provide this information.

Reference to Figure 7.5c shows that four changes must be made to the original precedence diagram, Figure 5.1a, when the recalculations are made:

1. The *logic change* accomplished by building two sets of abutment forms and starting activity 240 immediately after activity 230 is accomplished by eliminating the dependency line from activity 220 to activity 240 (step 1 of Figure 7.5c).
2. Because two sets of abutment forms are now going to be used, the *duration* of activity 80, "Make abutment forms," is increased from three to six days (step 1 of Figure 7.5c).
3. The *duration* of activity 350 is reduced from three days to two days (step 2 of Figure 7.5c).
4. The *duration* of activity 200 is reduced from two days to one day (step 3 of Figure 7.5c).

These four network changes are made to the precedence diagram of the highway bridge, and the recalculations are performed. Figure 7.8 summarizes the activity times and float values that are obtained.

7.16 Milestone and Interface Events

The principal point of the preceding discussions has been the reduction of overall project duration. This reduction includes expediting an ongoing project that has suffered previous delays. It has already been pointed out, however, that sometimes it is necessary to advance the expected dates of milestone and interface events. If the scheduled time of such an event does not satisfy an established time requirement, then some action by the contractor is in order. The procedure is very much the same as that for shortening project duration, except that the project critical path may or may not be involved.

The longest path from project start to that event determines the early time for any event, and efforts to advance the event time must be concerned

7.16 Milestone and Interface Events

Activity (Bold type denotes critical activity) (1)	Activity Number (2)	Duration (Working Days) (3)	Earliest Start (ES) (4)	Earliest Finish (EF) (5)	Latest Start (LS) (6)	Latest Finish (LF) (7)	Total Float (TF) (8)	Free Float (FF) (9)
Start	0	0	0	0	0	0	0	0
Prepare & approve S/D abutment & deck rebar	**10**	**10**	**0**	**10**	**0**	**10**	**0**	**0**
Prepare & approve S/D footing rebar	20	5	0	5	7	12	7	0
Order & deliver piles	30	15	0	15	1	16	1	0
Move in	40	3	0	3	10	13	10	0
Prepare & approve S/D girders	50	10	0	10	1	11	1	0
Fabricate & deliver abutment & deck rebar	**60**	**15**	**10**	**25**	**10**	**25**	**0**	**0**
Fabricate & deliver footing rebar	70	7	5	12	12	19	7	6
Prefabricate abutment forms	80	6	3	9	19	25	16	16
Excavate abutment #1	90	3	3	6	13	16	10	0
Mobilize pile-driving rig	100	2	3	5	14	16	11	10
Drive piles abutment #1	110	3	15	18	16	19	1	0
Excavate abutment #2	120	2	6	8	18	20	12	10
Forms & rebar footing #1	130	2	18	20	19	21	1	0
Drive piles abutment #2	140	3	18	21	20	23	2	0
Pour footing #1	150	1	20	21	21	22	1	0
Demobilize pile-driving rig	160	1	21	22	24	25	3	3
Strip footing #1	170	1	21	22	22	23	1	0
Forms & rebar abutment #1	**180**	**4**	**25**	**29**	**25**	**29**	**0**	**0**
Forms & rebar footing #2	190	2	22	24	23	25	1	0
Pour abutment #1	**200**	**1**	**29**	**30**	**29**	**30**	**0**	**0**
Pour footing #2	210	1	24	25	25	26	1	0
Strip & cure abutment #1	**220**	**3**	**30**	**33**	**30**	**33**	**0**	**0**
Strip footing #2	230	1	25	26	26	27	1	0
Forms & rebar abutment #2	240	4	26	30	27	31	1	0
Pour abutment #2	250	2	30	32	31	33	1	0
Fabricate & deliver girders	260	25	10	35	11	36	1	1
Rub concrete abutment #1	270	3	33	36	44	47	11	11
Backfill abutment #1	**280**	**3**	**33**	**36**	**33**	**36**	**0**	**0**
Strip & cure abutment #2	290	3	32	35	33	36	1	0
Rub concrete abutment #2	300	3	35	38	44	47	9	9
Backfill abutment #2	310	3	35	38	39	42	4	4
Set girders	**320**	**2**	**36**	**38**	**36**	**38**	**0**	**0**
Deck forms & rebar	**330**	**4**	**38**	**42**	**38**	**42**	**0**	**0**
Pour & cure deck	**340**	**3**	**42**	**45**	**42**	**45**	**0**	**0**
Strip deck	**350**	**2**	**45**	**47**	**45**	**47**	**0**	**0**
Guardrails	360	3	47	50	49	52	2	2
Paint	**370**	**5**	**47**	**52**	**47**	**52**	**0**	**0**
Saw joints	380	1	45	46	51	52	6	6
Cleanup	**390**	**3**	**52**	**55**	**52**	**55**	**0**	**0**
Final inspection	**400**	**1**	**55**	**56**	**55**	**56**	**0**	**0**
Contingency	**410**	**6**	**56**	**62**	**56**	**62**	**0**	**0**
Finish	**420**	**0**	**62**	**62**	**62**	**62**	**0**	**0**

Figure 7.8
Highway bridge, expedited schedule

with shortening this longest path. The longest path would first be restudied to see whether the desired shortening could be achieved at no increase in direct cost. If not, then the contractor must resort to expediting at additional direct expense.

7.17 Project Extension

The entire emphasis of this chapter has been directed toward the shortening of a project. At times, though, it may be desirable to lengthen certain activities or even the project itself. An example of this would be a project whose costs were originally estimated assuming that expediting actions would be needed if the owner's time requirements were to be met. It is typical for a contractor to anticipate time difficulties and to build into the original cost estimate the extra expense of overtime, multiple shifts, and other means of expediting the work. Unfortunately, at the time the project is being bid, the contractor usually will not have made an accurate forecast of project duration nor have identified the critical activities. About the only ways in which the contractor can figure in the extra costs of expediting actions are either to expedite most or all of the job operations or to make some educated guesses about which work items may turn out to be critical.

As sometimes happens, a subsequent detailed network analysis reveals that all or some of the planned expediting procedures will be unnecessary. This case is the inverse of that previously treated in this chapter. The contractor now finds it desirable to relax the job plan and realize the attendant savings. Obviously, the contractor wants to do this in a manner that will maximize its gain.

Even if the overall project duration must remain as planned, the contractor usually can rescind the expediting actions for at least some of the noncritical activities. The maximum duration increase granted to any given activity would be limited to the length of its total float. Further relaxation would result in the formation of a new critical path or subpath. For this reason, a record must again be maintained concerning the effect of each activity's time change on the floats of other activities. If all expediting actions of noncritical activities cannot be rescinded, then the most expensive of these actions should be eliminated first.

If the overall project duration can be extended from the original plan, then certain critical activities also can be relaxed. The only effect of this relaxation is to lengthen the critical path and, hence, to increase the floats of all noncritical activities. In this case, the critical activities should be relaxed first, beginning with those most expensive to expedite. After this, the noncritical activities can be treated as before.

7.18 Time Management System

Until now, project time management has concentrated on work planning and scheduling. An operational plan and a detailed calendar schedule have been prepared to meet project objectives. This includes any necessary project shortening identifiable prior to the start of construction. The work can now proceed with assurance that the entire job has been thoroughly studied and analyzed. To the maximum extent possible, trouble spots have been identified and corrective action has been taken to eliminate them. A plan and schedule have been devised that will provide specific guidance for the efficient and expeditious accomplishment of the work. Project management now shifts its attention to implementing the plan in the field and establishing progress monitoring and information feedback procedures. The project has entered the execution phase. The remainder of this chapter focuses on a time management system that can provide effective control over the project as it evolves. Succeeding chapters will focus on putting this system to work.

It is axiomatic that no plan can ever be infallible; nor can the planner possibly anticipate every future job circumstance and contingency. Problems arise every day that could not have been foreseen. Adverse weather, material delivery delays, labor disputes, equipment breakdowns, job accidents, change orders, and numerous other conditions can and do disrupt the original plan and schedule. Thus, after construction operations commence, there must be continual evaluation of field performance as compared with the established schedule. Considerable time and effort are required to check and analyze the time progress of the job and to take whatever action may be required, either to bring the work back on schedule or to modify the schedule to reflect changed job conditions. These actions constitute the monitoring and rescheduling phases of the time management system and are the subjects of discussion in this and subsequent chapters.

7.19 Aspects of Time Management

In an environment of constant change, the established time goals of a construction project must be met. To that end, the time management sequence shown in Figure 7.9 is repeated regularly over the life of the project. The current operational plan and schedule, consonant with established project time constraints, underpin the time management system. Forward- and backward-pass calculations, based on the latest version of the job network, produce a work schedule with calendar dates given for the start and finish of each project activity. This schedule is used for the day-to-day time control of the project. Such a system constitutes an effective early-warning device for detecting when and where the project may be falling behind schedule.

The work plan, however, must respond to changing conditions if project objectives are to be accomplished successfully. If the overall time schedule of a construction operation is to be met, there must be a way to detect time slippages promptly through an established system of progress feedback from the field. The monitoring phase of time management involves the

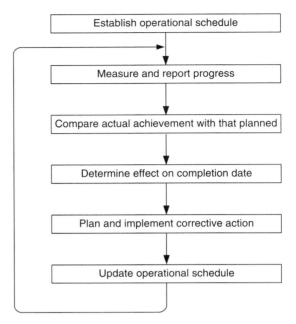

Figure 7.9
Time management cycle

periodic measurement of actual job progress in the field and its comparison with the planned objectives. Project monitoring, therefore, involves the determination of work quantities put into place and the reporting of this information in a format suitable for its comparison with the programmed job schedule. Network activities constitute a useful and convenient basis for progress measurement and reporting.

At regular intervals, the stage of project advancement is observed and reported. As of each cutoff date, note is made of those activities that have been completed and the degree of completion of those activities that are in progress. Review of this information by project management discloses where the project currently is ahead or behind schedule and by how much. Critical activities, and those with low float values, are monitored very closely because of their strategic importance in keeping the project on schedule. Corrective action to expedite lagging work items is taken after analysis of the reported progress data reveals what options are available.

No project plan or schedule can ever be perfect, and deviations will inevitably develop as the project progresses. As a result, the baseline version of the schedule will become increasingly inaccurate and unrealistic as changes, slippages, and other logic and schedule aberrations occur. Consequently, the network must be corrected as needed, and calculations must be updated occasionally so that the current job schedule reflects actual job experience to date. These updates often reveal shifts in critical paths and substantial changes in the floats of activities. The latest updated schedule reflects the actual current job conditions and the updated plan based upon those job conditions. It constitutes the current basis for project time control.

7.20 Key-Date Schedules

In formulating time schedules to be used for project control, consideration must be given to who will use the information. On a large job such as either of the example projects, a hierarchy of schedules may be needed. The detail used in connection with the preparation of a given job schedule can be highly variable with the level of management for which it is intended. Consequently, different schedules are prepared that are designed to meet the particular needs of the recipient. Craft supervisors are concerned primarily with those activities for which they are responsible, and their scheduling information needs to be specific and in substantial detail. Schedule information developed for owners, architect-engineers, and top-level field management generally will be in terms of milestones or key dates on which major segments of the project are programmed to start or finish.

One method of establishing a hierarchy of schedules is through the use of the project outline described in Section 4.8. The different levels of an outline provide a convenient work breakdown structure, consisting of various groups of network activities.

The project manager of the heavy civil example project likely will be provided with a listing of key dates for each of the several subprojects involved. In total, this will constitute a master schedule of time goals that will be monitored by top management during the construction process. Figure 7.10 presents an initial key-date schedule that has been prepared for the highway bridge. This key-date schedule is based on level two of the project outline. The original job plan and a 70-working-day duration constitute the basis for the presentation here. For reasons of clarity, the time-shortening actions treated earlier in this chapter are not carried forward to this discussion.

Early and late activity times discussed in Chapter 5, "Project Scheduling Concepts," constitute the basis for most field schedules. Contractors and scheduling software often use other names for these activity times. For example, early starts and finishes are commonly indicated as "scheduled" starts and finishes. Similarly, late starts and finishes are frequently shown as "required" starts and finishes. This new nomenclature is used with regard to the highway bridge schedules discussed in this chapter.

In Figure 7.10, the scheduled dates are the early-start and early-finish times for the construction operations listed. Operations 1 and 2 are

		Key-Date Schedule				Project No. 200808-05	
Project Highway Bridge							
		Scheduled		Required		Actual	
No.	Operation	Start	Finish	Start	Finish	Start	Finish
1	Procurement	Jun-14	Jun-19	Jun-14	Jun-19		
2	Field mobilization & site work	Jun-25	Jul-14	Jun-30	Jul-19		
3	Pile foundations	Jul-6	Jul-20	Jul-9	Jul-30		
4	Concrete abutments	Jul-20	Aug-17	Jul-20	Aug-30		
5	Deck	Aug-13	Aug-30	Aug-13	Sep-7		
6	Finishing operations	Aug-31	Sep-13	Aug-31	Sep-13		

Figure 7.10
Highway bridge, initial key-date schedule

exceptions to this because of a job circumstance that will be discussed in the next section. The dates listed in Figure 7.10 have been obtained using the activity times listed in Figure 5.3a. These were then converted to calendar dates using the calendar in Figure 5.4a. Figure 7.10 represents the first working version of a key-date schedule for the highway bridge and is prepared before field operations actually begin. Consequently, some of the dates contained in Figure 7.10 probably will change as the work progresses.

7.21 Adjustment of Move-in Date

In making up the key-date schedule, it is noted in Chart 5.3a on the companion website that activity 40, "Move in," has a total float of 12 working days. As depicted in Chart 5.3a, activity 80, "Prefabricate abutment forms," activity 90, "Excavate abutment #1," and activity 100, "Mobilize pile-driving rig," all follow activity 40. Chart 5.3a shows that these activities have total floats of 19, 12, and 13 days, respectively. What all this means is that, if these initial activities are scheduled and accomplished at their early times, field operations will be at a standstill for several days waiting for the delivery of steel piles and reinforcing steel. Consequently, there is nothing to be gained by starting up field operations immediately. From a time standpoint, activity 40, "Move in," and the subsequent three activities could be delayed by as much as 12 working days without affecting the completion date. This delay would result in a continuous chain of critical activities from "Move in" to "Cleanup." From the standpoint of resources, however, we will see in Chart 8.2 on the companion website that "Move in" could be delayed for six working days without any adverse effect on resource demand.

Based on both time and resource criteria, the project manager has decided to defer activity 40, "Move in," and the three following activities by nine working days, which will leave three days of float available for unanticipated problems. This postponement will in no way affect the project completion date and will eliminate an inefficient and unnecessary hiatus in field operations. Unfortunately, such a tardy move on the site by the contractor often can cause the owner considerable agitation unless the matter is explained beforehand. The ordering of reinforcing steel, girders, and steel pilings, however, still must be done at the earliest possible moment.

Key Points and Questions

Key Points

- There are many reasons why an activity or a project must be shortened.
- The schedule can be shortened in a number of ways. Some are without cost, but most require additional money.
- There are both practical and theoretical limits to the extent to which a project can be shortened.
- Since projects are dynamic and continually changing, the time management system must be able to recognize changes, make them apparent to the project manager, and then be modified to track the adjusted schedule.

Review Questions and Problems

1. What are some of the reasons schedules need to be shortened?
2. A common project manager responsibility is to estimate the cost of shortening activities. A project manager should have a checklist of those things that typically cost money when an activity must be shortened. Create such a checklist.
3. Explain the effect on project indirect costs when a job is lengthened and when it is shortened.
4. In the real world, a limited number of days can be taken out of the schedule before it becomes impractical to try to shorten the schedule further. Though it is impractical to take the exercise to the theoretical limit of shortening, there is a theoretical limit that prevents any further shortening of the project. What defines this theoretical limit, and what conditions exist in the schedule when this limit is reached?
5. Why is it important to retain successive schedule updates throughout the project?
6. For the following simple seven-activity project, develop a precedence network, calculate all relevant characteristics of each activity (ES, EF, LS, LF, TF), and the project duration. Then go through a repetitive process to expedite the project on a least-cost basis until the project is crashed. The solution should include the precedence diagram and calculations for each activity, and then the cyclical calculation to shorten the project activity by activity in a least-cost manner. What is

the total cost of the crashed project? Why is this less than the sum of the crashed costs for the seven activities?

Activity	Relationships	Normal Duration	Crash Duration	Normal Cost	Crash Cost
A		3	2	$3,000	$3,150
B	A: FS	4	3	$4,000	$4,250
C	B: FS	3	1	$6,500	$7,000
D	A: FS	4	2	$6,000	$6,500
E	D: FS	4	2	$3,500	$3,900
F	A: FS	5	2	$2,200	$2,725
G	C, E, F: FS	6	4	$3,900	$5,000

8 Resource Management

8.1 Introduction

Much of the job of a project manager, as well as the job of a field supervisor, focuses on the efficient investment of resources to achieve the project objectives. When we talk of resources in the context of construction, we typically think of manpower, equipment, and materials. These are fundamentally important resources, but in broader terms they are only some of the resources invested in the project.

In the broader context, a resource can be considered anything that adds value to the project. Certainly, manpower, equipment, and materials meet this definition of *resource*. However, the project manager must think outside the narrow confines of these big three to look for other resources that can be managed to the benefit of the project. Time is a resource that is so important that much of this book is dedicated to the efficient use of time. Money is another very important resource that will be discussed in some detail in the last two chapters of the book. In addition to what we normally understand *manpower* to mean—that is, craft workers who actually do the work on the project—there are many other people who add value to the project.

Within the company, the estimator, the purchasing agent, the financial unit, and the management information systems unit all support the project manager in achieving a successful project. Outside the company, subcontractors provide much of the labor and materials that go into the job. Meanwhile, material suppliers and distributors not only supply the materials and installed equipment that go into the job but can also serve as a resource to the project manager and field supervisor in designing more efficient ways to manage materials both offsite and onsite.

Thinking even more broadly, it is important to consider those that can add value to the project that are not under the control of the project manager. In light of managing up—that is, managing those resources not directly under our control—designers contribute in many ways to the ongoing project, not only in providing the initial design from which the project is built, but also in reviewing submittals, redesigning as required, in many instances to facilitate field operations, in approving payments and change orders, and in final signoff of the various components of the project. For a specialty contractor in a subcontracting position, the general contractor is another resource to be managed. The general contractor has control over project field operations. If the specialty contractor participates as part of the project management team, this can benefit not only the operations of the specialty contractor but also overall project operations.

Finally, the project owner should be considered a resource to the project manager. The owner is the prime mover that has provided the project for the project manager to manage. The owner makes decisions and timely decisions are essential to a smoothly running job. The owner also can guide and influence the design team to be more responsive to the needs of field operations. Finally, the owner controls the money, so the owner is a very valuable resource to the project manager.

It is the job of the project manager to manage all of these resources in support of efficient execution of the project. However, this chapter will focus on methods and procedures involved with the management of the three primary resources of manpower, equipment, and materials. Learning objectives for this chapter include:

❏ Understand the importance of resource management.
❏ Learn practical means to improve the allocation of labor resources.
❏ Understand the supply chain for construction materials.
❏ Recognize the importance of expediting to the successful completion of a project.

8.2 Objective of Resource Management

The completion of a construction project at maximum efficiency of time and cost requires the judicious scheduling and allocation of the primary resources of manpower, equipment, and materials. The supply and availability of these resources seldom can be taken for granted because of seasonal shortages, labor disputes, equipment breakdowns, competing demands, delayed deliveries, and a host of associated uncertainties. Nevertheless, if time schedules and cost budgets are to be met, the owner must be supplied with the necessary workers, equipment, and materials as they are needed on the job site.

The basic objective of resource management is to supply and support the field operations so that established time objectives can be met and costs can be kept within the construction budget. Field supervisors can achieve favorable production rates and get the most from their workers and equipment only when the requisite ways and means are optimally available. It is the responsibility of the project manager to identify and schedule future job needs so that the most efficient use is made of the resources available. The project manager must determine long-range resource requirements for general planning and short-term resources for detailed planning. He must establish which resources will be needed, when they must be onsite, and the quantities required. Arrangements must be made for their timely arrival, with regular follow-up actions taken to ensure that promised delivery dates are kept. Where shortages, conflicting demands, or delays occur, the project manager must devise appropriate remedial measures. The project plan and schedule may have to be modified to accommodate or work around supply problems. The scheduling and allocation of workers, equipment, and materials are interrelated. It is important to remember that an action affecting one often affects the others in some manner.

8.3 Project Resource Management

With respect to resource management on construction projects, a few general observations at this stage can serve as valuable guidelines for the practitioner. The long-term leveling of resources provides a good indicator of future resource needs, but only from a general planning point of view. Detailed resource planning months into the future is unnecessary and is usually a waste of time. Detailed resource leveling has its major advantage when applied to the near future (i.e., a maximum of 30 calendar days). Ample float makes efficient resource management possible, while low float values almost inevitably mean schedule delays or the need for large variations in applied resources. The concept of float providing resource efficiency becomes important in cases of dispute with the owner regarding the ownership of float. Disputes often occur when the owner orders extra work to be done but declines to grant additional time to the contractor because the added work is not on the critical path.

The highway bridge is used to illustrate resource management procedures throughout this chapter. The points illustrated with the highway bridge can be applied easily to the building project as well. In actuality, the resource planning and scheduling for a component of a project, like the highway bridge, would almost never be done independently of the several other parts of a project or even of other company projects. The reason for this is that there are usually conflicting demands for the same limited company resources from other job sites. There will surely be instances where workers, equipment, and materials will have to be traded back and forth among the various project segments, and even from job to job, to achieve maximum use of the resources available.

Resources often must be allocated on a project-wide or company-wide basis, with some system of priority being established among the separate points of demand. This is a complex and difficult problem for which only partially satisfactory solutions are practical. Management action in this regard cannot be stereotyped but must be based on judgment and economic factors intrinsic and unique to the particular situation. To develop the basic principles of resource management, the ensuing discussion treats the highway bridge job as if it were a totally separate and autonomous unit. These same principles will provide management guidance where different types of projects and multiple construction sites are involved.

8.4 Aspects of Resource Management

A prerequisite to manpower management is a detailed analysis of the labor requirements necessary to meet the project schedule. Previous chapters have provided detailed information regarding the specific manpower demands of each operation and the time frame within which these operations are to be completed. Using this information, the project manager may then compare these needs with an estimate of available manpower to determine whether the project schedule is generally achievable. Once this calculation has been performed and no adjustment to the overall job completion date appears necessary, the project manager may then turn his attention to leveling out the peaks and valleys in labor demand.

If the labor requirement take-off discloses that the demand will exceed the supply, the management situation can become considerably more complex. In such instances, time or cost overruns, or both, are almost inevitable. Remedial measures to combat an inadequate labor supply can include diverting labor from noncritical to critical activities, or resorting to some method of expediting the critical activities. The stretching out of noncritical activities, or the use of overtime or subcontracting on critical activities, may make it possible to maintain the originally established schedule, but often at an additional cost. Otherwise, the project manager faces the difficult task of allocating the available labor in a way that offers the greatest advantage while minimizing the duration of project time overruns.

With respect to equipment management, most of the major decisions concerning how the job is to be equipped were made at the time the job cost was estimated. Nevertheless, it is the responsibility of the project manager to see that the job is properly and adequately equipped. In a manner analogous to the checking of labor needs, a compilation is made of the total equipment demand on a daily basis. If there are conflicts among project activities for the same equipment items, rescheduling of noncritical activities often will solve this problem. If this is not possible, working overtime, on weekends, or multiple shifts might circumvent the difficulty. Another solution might be to arrange for additional items of equipment to be supplied from some outside source. If excessive equipment demands cannot be ameliorated in one of these ways, alternative equipment types or construction methods may need to be explored. Such changes from those used to estimate and price the operation may have substantial cost and time implications for the project and must be analyzed carefully. Should no other acceptable alternatives exist, the conflict must be removed by rescheduling activities with the least possible increase in project duration and cost.

Material management on a construction project is essentially a matter of logistical support. Job materials in the proper quantity and specified quality must be available at the right place and time. All aspects of material procurement, from ordering to delivery, are directed toward this objective, and a positive system of checks and controls must be established to assure its realization. A valuable resource in the management of materials both on and off the site can be the material supplier. Suppliers depend heavily on the ability to efficiently manage their materials, and they can help contractors develop better means to manage their materials on the site. In many cases, the contractor can work with the supplier to plan timely delivery of materials so that handling and storage onsite is minimized and suppliers and contractors working together might be able to prefabricate or prepackage materials to minimize loss and damage and eliminate onsite handling. It is often possible to gain savings in time and money by paying the supplier for prefabrication and prepackaging before materials get to the site.

Subcontractors can and often do play an important role in achieving project time and cost goals. Although there is much variation in actual practice, subcontractors typically provide their own workers, equipment, and materials. The project manager will seldom have any direct voice in the management of subcontractor resources. Rather, the manager's responsibility usually will be one of ensuring that each subcontractor commences work at the proper time and processes the work in accordance with the established time schedule.

8.5 Tabulation of Labor Requirements

The management of construction workers begins with a tabulation of labor needs, by craft, for each project activity. Figure 8.1 is a takeoff of the general contractor's labor requirements for the highway bridge. This figure does not include the manpower needed by subcontractors, such as those responsible for the reinforcing steel and painting. Much of the information contained in the figure is readily available from the original

Activity Number	Activity	Duration	Pile-driverman	Carpenter	Laborer	Equipment Operator	Oiler	Iron Worker	Cement Mason	Truck Driver
40	Move in	3		3	4	2				4
80	Prefabricate abutment forms	3		3	3					
90	Excavate abutment #1 days 1 & 2	3			3	1				
	day 3				7	1				
100	Mobilize pile-driving rig	2	3		2	2	1			2
110	Drive piles, abutment #1	3	4	1	2	1	1			
120	Excavate abutment #2 day 1	2			3	1				
	day 2				7	1				
130	Forms & rebar, footing #1	2		1	1					
140	Drive piles, abutment #2	3	4	1	2	1	1			
150	Pour footing #1	1		2	4	1	1		1	
160	Demobilize pile-driving rig	1	3		2	2	1			2
170	Strip footing #1	1			2					
180	**Forms & rebar, abutment #1**	**4**		3	3					
190	Forms & rebar, footing #2	2		1	1					
200	**Pour abutment #1**	**2**		2	6	**1**	**1**		**1**	
210	Pour footing #2	1		2	4	1	1		1	
220	**Strip & cure, abutment #1 day 1**	**3**		3	6					
230	Strip footing #2	1			2					
240	**Forms & rebar, abutment #2**	**4**		3	3					
250	**Pour abutment #2**	**2**		2	6	**1**	**1**		**1**	
270	Rub concrete, abutment #1	3			1				2	
280	Backfill abutment #1	3			3					
290	**Strip & cure, abutment #2 day 1**	**3**		3	6					
300	Rub concrete, abutment #2	3			1				2	
310	Backfill abutment #2	3			3					
320	**Set girders**	**2**			4	**1**	**1**	**4**		
330	Deck forms & rebar	4		3	2					
340	**Pour & cure deck day 1**	**3**		2	6	**1**	**1**		**4**	
350	**Strip deck**	**3**			3					
360	Guardrails	3			1			2		
370	Paint	5								
380	Saw joints	1			1					
390	**Cleanup days 1 & 2**	**3**			7					
	day 3				4	1	1			2
(Bold type denotes critical activities)										

Figure 8.1
Highway bridge, activity manpower needs

estimate in those instances where labor cost was estimated using crew size and production rate and the activity involved only one cost account. For instance, Figure 3.6 shows one foreman, one cement mason, six laborers, one equipment operator, one oiler, and one carpenter as the crew to pour the abutment concrete (activities 200 and 250). These labor requirements are entered directly into Figure 8.1. In doing so, it is assumed that the crew foreman will be a carpenter; consequently, a requirement for two carpenters is entered for activities 200 and 250. Similar assumptions regarding craft foremen are used throughout Figure 8.1. Where the labor cost was estimated initially by using a unit cost, crew sizes assumed when activity durations were being estimated can be used (see Section 5.5).

If activities involve more than one cost account and, correspondingly, more than one crew, the determination of their labor requirements is less direct. To illustrate this point, activities 90 and 120 both involve abutment excavation. Each of these activities involves two different cost accounts: excavation, unclassified; and excavation, structural. It was assumed in the planning stage of the highway bridge that the unclassified excavation would be performed with a tractor-dozer and would be done for both abutments in a single operation. Once this work is completed, the structural excavation will be done first for abutment #1 and then for abutment #2. Thus, activity 90 is really the unclassified excavation for both abutments as well as the structural excavation for abutment #1. This is reflected in the original plan, with a three-day duration for activity 90 and two days for activity 120. This is the rationale for the workers listed for these two activities in Figure 8.1.

8.6 Project Labor Summary

With the information given in Figure 8.1 and the time-scaled network shown in Chart 5.3a on the companion website, it is easy to determine the projected daily labor needs for the highway bridge based on an early start schedule of operations. Figure 8.2 is the labor summary for this project. Using the available floats of noncritical activities, it is possible to smooth or level the peak demands for manpower revealed by the figure. (Methods for doing this are discussed later in this chapter.) The early-start labor requirements usually serve as the best starting point for any adjustment that may be required of the daily labor demands. However, it must be noted at this point that an early-start schedule often turns out to be inefficient in terms of cost and resources.

An important characteristic of the highway bridge is revealed by the labor compilation in Figure 8.2. If the early start schedule is followed, there will be a period of time during which no work can proceed because of the lack of materials. After the abutment forms are built and the excavation is completed, the job will be at a standstill for seven working days awaiting delivery of piles and reinforcing steel. This is clearly disclosed in Figure 8.2 by the seven-day gap in the project labor requirement after the first few job operations have been finished. Often enough, the early start of field operations only results in a subsequent delay while awaiting the receipt of key resources. This matter was discussed in Section 7.21.

The labor summary of Figure 8.2 provides information concerning the local labor market and whether it can be expected to provide the numbers of craft workers required. However, contractors seldom make such a labor takeoff purely to ascertain the adequacy of the labor supply. They usually assume that they will be able to hire sufficient workers, but this matter often deserves more attention than it gets. The contractor's experience and knowledge of local conditions are valuable guides in this regard. Severe shortages of certain labor skills occur in many areas of the country

Craft	1	2	3	4	5	6	7	8	9	10	11	12	13	14	15	16	17	18	19	20	21	22	23	24	25	26	27	28	29	30
																										Working Days				
Piledrivermen				3	3											4	4	4	4	4	4	3								
Carpenters	3	3	3	3	3	3										1	1	1	2	2	3		1	1	2	3	3	3	3	2
Laborers	4	4	4	8	8	10	3	7								2	2	2	3	3	6	4	1	1	6	8	6	6	6	8
Equipment operators	2	2	2	3	3	1	1	1								1	1	1	1	1	2	2		1						1
Oilers							1	1								1	1	1	1	1	2	1		1						1
Iron workers																														
Cement masons																					1				1					1
Truck drivers	4	4	4	2	2																1									
	14	15	16	17	18	21	22	23	24	25	28	29	30	1	2	6	7	8	9	12	13	14	15	16	19	20	21	22	23	26
	June													July																

Figure 8.2
Highway bridge, daily manpower compilation

during peak construction periods. Arrangements to bring in workers with the requisite trade skills from outside areas must be made well in advance. Contractors will find that advance knowledge of the labor demands of their projects can be of considerable value in the overall planning and scheduling of their field operations.

8.7 Variation in Labor Demand

For purposes of discussion, we will assume that there will be an adequate labor supply for the highway bridge, or at least that the peak demands can be smoothed sufficiently for supply to meet demand. Figure 8.2 reveals that the requirements for different crafts vary widely and are at times discontinuous. Some variation in time demand for a given craft is normal because labor crews typically build up to strength at the start of the job and decline toward the end. However, the pronounced grouping of labor needs at different points during the construction period is decidedly undesirable and impractical. The recurrent hiring and laying off of personnel on a short-term basis is troublesome, inefficient, expensive, and scarcely conducive to attracting and keeping top workers. New craft workers on the job are not as efficient as they are after they become familiar with the work involved. Efficiency is a learning curve phenomenon, where a crew's production goes up and its unit labor costs go down with task repetition. New workers on the site are also in the highest risk category for safety infractions and incidents. Then, too, there is the practical consideration; when workers are laid off for a few days, they may no longer be available and they may be difficult to replace.

8.8 Manpower Leveling

Manpower leveling is the process of smoothing out daily labor demands. Perfection in this regard can never be attained, but often the worst of the inequities can be removed through a process of selective rescheduling of noncritical activities. On the highway bridge, crews are largely comprised

8.6 Project Labor Summary

31	32	33	34	35	36	37	38	39	40	41	42	43	44	45	46	47	48	49	50	51	52	53	54	55	56	57	58	59	60	61	62	63
2	3				3	3	3	3	2	2	3				3	3	3	3	2													
8	9				7	7	7	3	6	6	6		8	8	6	2	2	2	6		4	3	3	1	1	1				7	7	4
1									1	1			1	1					1													1
1									1	1			1	1					1													1
													4	4										2	2	2						
1					2	2	2		1	1			2	2	2				4													1
27	28	29	30	2	3	4	5	6	9	10	11	12	13	16	17	18	19	20	23	24	25	26	27	30	31	1	2	3	7	8	9	10
												August														September						

of carpenters and laborers. For purposes of discussion, the process of manpower leveling on the highway bridge will address only these two crafts. Because of the discontinuous nature of their work on this project, no amount of smoothing can level the daily job requirements for equipment operators, ironworkers, cement masons, and pile-driver operators. On the highway bridge, these workers will have to be provided when, and as, they are needed. It is for this reason that specialty crews often are shifted back and forth among company jobs. This kind of irregular labor demand is one of the drivers that prompt general contractors to subcontract portions of their work.

The peak demands and discontinuities for carpenters and laborers on the highway bridge can be leveled to some extent by using the floats of noncritical activities. To illustrate the basic mechanism by which resource smoothing is accomplished through rescheduling, a simple example will be discussed. Chart 8.1 on the companion website shows the daily requirements of the highway bridge for carpenters and laborers. These data are summarized in the form of two histograms to illustrate the peaks and valleys of labor demand required by an early start schedule.

Chart 8.1 has been prepared by plotting, for each activity, its daily demand for the designated crafts. Each activity is assumed to begin at its early start time. Opposite each activity, and under the working days during which it will proceed, is entered its daily labor demand. The symbol "C" is used for carpenters and "L" for laborers. The labor totals indicated by the histograms are obtained by adding vertically the labor demand for each day. The information contained in Chart 8.1 assumes that the same size crew will be used throughout the duration of any given activity. For activities that have long durations, this assumption is probably not justified. Generally, the crew for such an activity will start small, build up to full force, and taper off at the end. The assumption of constant crew size is reasonably accurate, however, for short-duration activities such as those used on the highway bridge.

Chart 8.1 discloses that there is a peak requirement for 10 laborers on project working day 6. This peak is caused by the fact that activity 80, requiring 3 laborers, and activity 90, requiring 7 laborers, are both scheduled to be under way on the same day. A usual way to remove or minimize such a conflict in an early-start schedule such as Chart 8.1 is to move one of the conflicting activities to a later date. When a noncritical activity with no free float is moved to a later time to level a resource, any succeeding activities must be adjusted forward by a like amount. If free float is available, the finish of the rescheduled activity can be postponed by an amount of time that is less than, or equal to, the amount of free float without affecting any of the following activities. Hence, when adequate free float is available, schedule changes to accomplish resource leveling are easily made. However, when a schedule adjustment bumps a whole chain of succeeding activities forward, all the resource needs of succeeding days can be affected substantially, thus complicating the calculation. These changes may serve to improve the overall situation or may only further complicate it. In the case of the 10 laborers needed on working day 6, Chart 5.3a on the companion website shows that activity 80 has 19 days of total float and 19 days of free float. Activity 90 has 12 days of total float and zero days of free float. This labor conflict can be remedied easily by moving either activity 80 or 90 to a later date; movement of activity 80 is preferable.

8.9 Heuristic Manpower Leveling

A number of operations research techniques are available for obtaining optimal solutions to manpower leveling problems. Numerous algorithms are available to accomplish such a time-critical analysis, but these require a computer to handle no more than a few resources. Computer-based methods can be time consuming and require a number of decisions up front regarding leveling priorities for which it is difficult to anticipate the full impact on the job. The resulting solution is often quite puzzling. For these reasons, heuristic methods often are used. Heuristic methods involve the application of approximations to solve very complex problems. As the resource information developed in these models is based on estimates, and subject to many external and unpredictable factors, a heuristic solution to manpower leveling offers sufficient accuracy for construction operations and provides a good basis for understanding manual methods of resource leveling.

In the absence of a resource-leveled schedule, field supervisors intuitively prioritize the commitment of resources to operations. On any given workday, prior to the start of the shift, the field supervisor reviews the resources available for the day and lists the critical activities to be accomplished. If after covering these critical operations, additional resources remain, the field supervisor will initiate other operations using any remaining resources. The decision as to which other operations receive resources is based on how close they are to becoming critical.

A simple manual heuristic method is used in Chart 8.2 on the companion website to level the demand for carpenters and laborers on the highway bridge. The method presented is based on priority rules that give reasonable results when used with a modest number of labor resources. (In the case presented here, the number of resources is two: carpenters and laborers.)

The priority rules mentioned pertain to the order in which project activities will be rescheduled from their early-start condition. In essence, the entire project is rescheduled. The critical activities are given highest priority and are therefore scheduled first. The noncritical activities are then scheduled with priority given to those activities with the earliest late-start dates. When more than one activity has the same late-start date, preference is given to the one with the least total float. The algorithm progresses through Chart 8.2 one day at a time, with activities being scheduled according to the priority rules just described. In this example, maximum limits on the daily labor demand have been set at three carpenters and seven laborers. These limits have been set equal to the maximum labor demands of individual critical activities, as shown in Figure 8.1.

8.10 Numerical Example

In Chart 8.2, the basic information given in the left-hand columns is a list of activities on the highway bridge, together with the resources needed, duration, and late-start value for each activity. The first step in the leveling process is to schedule the critical activities by listing the manpower required for each under the appropriate working days. Because critical activities 10 and 60 do not require labor, activity 180 is the first critical activity to be scheduled. It starts on day 26, finishes on day 29, and requires three carpenters and three laborers for each of its four days. Labor demand limitations are based on the maximum demands of the critical activities. Therefore, all critical activities can be scheduled for the same dates required in the original early-start schedule. If the labor demand of a critical activity were to exceed the supply, it will not be possible to keep the project within its originally planned duration unless special measures are taken. This case is discussed in Section 8.12.

After all the critical activities in Chart 8.2 have been scheduled, noncritical activities are scheduled in the order of their late-start dates. The first of these is activity 40, which must start no later than day 12. This activity, along with all the other noncritical activities, is scheduled to start as early as possible, subject to previously established resource limitations. Commensurately, it is scheduled to begin on day 1 and end on day 3. The activity with the next earliest late start is activity 90, with a late start of day 15. Because activity 90 follows activity 40 (see project network in Chart 5.1a on the companion website), activity 90 cannot be scheduled to start until day 4. When activity 90 is scheduled from day 4 through day 6, the project manpower limits are not exceeded, and its scheduling is acceptable. In a similar manner, activities 100, 110, 120, 130, and 140 are scheduled in the order of their late-start dates. Each activity is scheduled as close to its early start as possible while allowing the start date to slip as necessary to remain within the resource limits. It should be noted that activity 80, which can start as early as day 4 based on the network logic, has been allowed to slip to day 14 in order to maintain resource levels. Once all of the activities preceding critical activity 180 have been scheduled, it is noted that there is a six-day gap during which no resources are required. This is the same network characteristic that was noted in Section 8.6, where an early start of work on the site serves no purpose because of the delivery times required for pilings and reinforcing steel. As a result of this gap, all of the noncritical activities scheduled up to this point can be moved to the right in Chart 8.2 by at least six working days.

An important point must be made here. The logic of the network diagram shows that the start of field operations need not occur until day 12. The resource histograms in Chart 8.2 show that the start can be delayed six days without impacting resources. What this emphasizes is that float is necessary to allow for the shifting of activities to use resources efficiently. Both the time consideration and the resource considerations are important. A decision regarding "move in" was discussed in Section 7.21.

A comparison of the labor demand histograms in Charts 8.1 and 8.2 shows that the leveling efforts have resulted in considerable improvement. For example, the peak requirement for laborers has dropped from 10 to 7, and the number of days during which there is no need for laborers has been reduced from 15 to 12. However, if the opening activities were to be delayed by 6 working days, as described previously, the number of days when there will be no need for laborers will be further reduced from 12 to 6 working days. The leveled labor demands shown in Chart 8.2 are far from being uniform. In this regard, it must be mentioned that resource smoothing can be especially difficult on a project of limited extent with a large proportion of critical activities, as is the case with the highway bridge. The heavy civil project example would provide a much more flexible basis for making significant and meaningful leveling studies. The process just described is referred to as resource allocation, with time as the critical element. When the number of available resources is critical, a slightly different process is required and is described in Sections 8.12 and 8.13.

8.11 Labor Leveling in Practice

The smoothing of labor demands from those shown in Chart 8.1 to those in Chart 8.2 is performed every day by field supervisors all over the country. However, field supervisors usually perform this function in an intuitive and informal way. Through extensive experience, good site supervisors have developed a talent and instinct for manning a construction job efficiently, if not optimally. Manpower leveling, as a formal management procedure, is not standard practice in the construction industry. Rather, field supervisors generally are left to their own devices with regard to the field management of labor crews. While developing the ability to make these manpower decisions informally on a daily basis is sufficient for relatively simple projects with fairly constant resource demands, more complex projects will benefit substantially from formally completing the calculations. In addition to developing a better indication of overall project manpower requirements, these computations provide the project manager with a more realistic picture of the true criticality of project operations.

Charts 8.1 and 8.2 are attempts to resolve a complex matter with a simplified job model and heuristic rules. In reality, the starts and finishes of activities will never be defined as acutely as they are in the network diagram. There is almost certainly a degree of activity overlap that is not represented in the Critical Path Method (CPM) diagram. Nor is the job logic truly as rigid as it is made out to be. Craft workers are shifted about from one activity to another as they are needed. The daily fluctuations shown in Chart 8.2 for laborers will not really occur during the construction period. A relatively stable labor crew of five or six workers will be on hand throughout the job. A form of Parkinson's Law, which says that work expands so as to fill the time available for its completion, takes effect, and every worker is kept busy, even when more laborers are present on a given day than Chart 8.2 indicates actually would be needed.

Manpower leveling, using present-day algorithms, is a trial-and-error process and is made difficult by the fact that most activities use more than one labor classification as well as different types of equipment. A shift that improves one resource often complicates one or more other resources, either on that day or on succeeding days. Despite its limitations, manpower leveling does provide valuable information relative to the efficient use of job-site labor. Even if formal leveling studies are not attempted, the adopted rules of thumb pertaining to the priority of activity rescheduling can provide consistent and useful guidance. For example, it can be emphasized to field supervisors that critical and low-float activities have manpower priority and only those activities with large float values are to be used as fill-ins. Another guide for field supervisors could be to save the available float on activities for possible later use in resource leveling. Where people and equipment are not needed elsewhere, activities, even those with large float values, should be started as early as possible. On large and complex projects, or on jobs where the supply of labor is known to be limited, formal leveling efforts such as those previously discussed can provide valuable advance information for project management.

8.12 Restricted Labor Supply

The preceding example of labor leveling involved a case where only the rescheduling of noncritical activities was required to keep the job labor demands within established limits. Now assume that this condition is no longer true: There is a labor shortage situation that presents project management with a considerably more difficult rescheduling problem than that previously encountered. The basic implication of a labor shortage is that the durations of certain activities must be extended beyond their normal values if the manpower deficiency cannot be overcome by expediting actions such as subcontracting the work or working overtime. Subcontracting may well be the best answer, but such action depends on circumstances and will not be discussed further here. In any event, a restricted supply of manpower may or may not affect the overall project duration. Yet whatever the circumstance, if overtime or subcontracting cannot alleviate the situation, the contractor then has the problem of allocating, to best advantage, the available labor among the activities.

The first item to check when a labor shortage is expected is the requirement for this particular labor resource by the critical activities. Consider critical activity 330 on the highway bridge. This activity requires three carpenters per day for four days. Based on an 8-hour day, this would amount to 96 man-hours of carpentry work. If there are only two carpenters available, the contractor has a decision to make. If the activity is staffed by only two carpenters and if the usual 8-hour day is worked, the duration of the activity will be increased to six days. Because this is a critical activity, the project is commensurately delayed by two days. However, the project duration will be unaffected if the two carpenters work 12-hour days (4 hours of overtime) while engaged in this activity. Some carpenter tenders (laborers), and perhaps other trades, would probably also have to work overtime with the carpenters to keep the work in phase. Therefore, the project manager must analyze the situation and determine whether a two-day delay to the schedule would be more costly overall than the associated overtime needed to alleviate the labor shortage.

With regard to labor deficiencies on noncritical activities, the first action is to stretch out the duration of each activity concerned sufficiently to keep the labor requirement within the supply. However, when the extensions of noncritical activities equal or exceed the available floats, the contractor must again seriously consider the use of overtime. Otherwise, new critical paths and subpaths will materialize, possibly superseding the original critical path and delaying the entire project. These activities are often referred to as resource-critical activities.

8.13 Complex Labor Scheduling

When labor shortages of several crafts are involved and overtime is not a satisfactory solution, the matter of scheduling the available labor to minimize the project delay takes on an even greater degree of complexity. This situation mandates the use of a resource-critical analysis. This is a trial-and-error process that involves allocating the available labor to the activities while maintaining the necessary job logic to achieve project completion in the shortest possible time. In such cases, the complexity of the situation normally will preclude obtaining an optimal solution. The heuristic method described in Sections 8.9 and 8.10 also works well for resource-critical analysis. The durations of some critical activities probably will need to be extended or decisions made to expedite them. Critical activities requiring more of a limited resource than is available will have to be reexamined and probably will result in longer durations. The major difference in resource-critical scheduling is that both critical and noncritical activities are scheduled together, with priority given to those with the earliest late-start date. With a cap placed on specific resources, inevitably some activities will be delayed beyond their late-start date. The result is, of course, delay of the project completion date. At this point, a decision must be made: Expedite the delayed activities or delay the project. The best solution the project manager can hope for is to arrive at a practical compromise that appears feasible and reasonably efficient.

When deemed desirable and feasible, activities can be scheduled discontinuously at irregular time intervals. License to do this is, of course, dependent on the nature of the activity. Some job operations, such as a concrete pour, must be continuous (each day following the previous without intermittent pauses). Activities such as rubbing concrete or prefabricating concrete forms, however, can be performed intermittently and serve well as fillers. In fact, this is probably how activity 270, "Rub concrete, abutment #1," and activity 300, "Rub concrete, abutment #2," of the highway bridge will be accomplished. This would be especially true for activity 300 because, as shown in Chart 5.3a on the companion website, on the companion website, if it is started at its early-start time, it will parallel the setting of the steel girders. Performing both of these activities simultaneously can be unsafe, and activity 300 can afford to be delayed. Deck forming, pouring, and stripping are also possible safety conflicts with activity 300. Undoubtedly the job superintendent will put the cement masons to work rubbing abutment #2 whenever an opportunity presents itself.

Much attention has been given elsewhere to the matter of complex resource scheduling. At present, such resource scheduling gives a good approximation of the total resource quantities needed and their approximate timing. It illustrates graphically that network float is often an illusion, being in fact the leeway necessary to schedule resources efficiently. When complex labor shortages occur that cannot be avoided by overtime or subcontracting, skilled and experienced field supervisors are probably better able to work out an acceptable daily schedule as the work progresses. It is the project manager's duty to identify labor shortage problems well in advance, make the decisions necessary to minimize the problem, and allow sufficient time for the field superintendent to complete the job while allocating the available labor on a daily basis.

8.14 Equipment Management and Scheduling

On projects that require extensive spreads of construction equipment, the project schedule and the production costs are determined principally by the management of the equipment onsite. Since this is a highly specialized area, it will not be covered in detail here. It is important, however, to recognize certain characteristics of heavy-equipment management:

- To the maximum extent possible, equipment sent to the job should be of the type that will best perform the work under actual job conditions.
- Work should be planned and scheduled to achieve the fullest use of every equipment item, minimizing idle equipment time.
- Since equipment breakdowns not only idle the specific piece of equipment that is dysfunctional but also often disturb other operations, field maintenance must be a key part of the prejob planning.
- The production rate of an equipment unit depends not only on the physical characteristics of the piece of equipment but also on the operator and the field supervisors as well.
- The temptation to overload equipment in an effort to get more production is common but counterproductive.
- Actual production rates and costs need to be checked continually on the site for each major piece of equipment, as maintenance and repair vary greatly and affect the efficiency of operation.
- As with labor requirements, in order to achieve the greatest efficiency with an equipment spread, project managers regularly analyze operations to determine whether another unit of equipment is needed or should be removed.

Considerations in managing the equipment resource are analogous to those in managing the labor resource. Units required in each different category are assigned to schedule activities. By summing equipment demands for a class of equipment each day, a histogram for that class is developed showing the ongoing requirements for that equipment. Equipment requirements can then be leveled, using means similar to the way critical labor resources were leveled in earlier sections of this chapter.

8.15 The Construction Supply Chain

Supply chain management is a major area of study in business or management schools. It deals with managing the flow of goods. In construction, the focus is on managing the flow of materials from some point of origin, typically the point of manufacture or fabrication, to the point of installation on the job. This can be subdivided into offsite and onsite management of materials. Typically, the project manager is responsible for the offsite segment of the supply chain and the field supervisor, for the onsite segment of the supply chain. These concepts were introduced in Section 6.10, which focused on material ordering and expediting (the responsibility of the project manager), and Section 6.11, which focused on material handling, storage, and protection (the responsibility of the field supervisor).

Managing the supply chain is of critical importance to the project manager. If an item is not at the point of installation when needed, work stops. Time is lost. Cost goes up due to labors being idle. If this occurs regularly, morale problems begin to develop, resulting in lowered productivity and higher safety risk. The problem is not limited to missing material. It could be defective material. It could be the wrong item or an item that has not been approved.

Managing the supply chain affects all aspects of the project. As previously indicated, it directly impacts budget and schedule. Improper supply chain management can also affect cash flow, customer relations, and reputation. It behooves the project manager to understand general concepts of supply chain management and to be able to manage the supply chain for the specific materials on the project.

One way to understand the supply chain is to diagram the steps from the point of origin to the point of installation. The point of origin for construction would typically be a supplier or distributor, but the process actually begins with the specification by the designer and generally follows these steps:

1. Designer specifies item.
2. Supplier is selected and a purchase order is issued.
3. Supplier prepares submittal (cut sheet or shop drawing).
4. Submittal process is executed resulting in approval.
5. Item is fabricated or procured from manufacturer.
6. Item is shipped to the contractor.
7. Item is received at the site and goes to the workface for immediate installation or to a storage area.
8. If placed in storage, the item is withdrawn from the storage area and taken to the workface.
9. Item is installed.

There are many points at which this process can branch out or for which significant detail can be added. An example of a branching point would be where the item is either installed upon arrival at the job site or taken to a storage area. An example of a subprocess requiring more detail might be the approval process. Many contractors have a relatively small number of generic supply chain diagrams representing procurement of different classes of materials that they commonly use, such as concrete, structural steel, or hardware. At the beginning of a job, the project manager can develop diagrams for specific materials that then become the basis for other tools that are useful in tracking the procurement process.

8.16 Material Scheduling

Management control over materials is concerned with ensuring that the materials are on the job as needed, in the required quantities and qualities. Material management, both onsite and offsite, can have a significant impact on budget and schedule for any project. As stated earlier, the more the project manager and the field supervisor understand material management, the better they will be able to function in their management and supervisory responsibilities.

The contractor's purchase order customarily prescribes the quantity, quality, price, delivery date, and mode of transportation for the materials covered. Upon delivery, the quantity and quality of all material deliveries are verified by inspection, count, and test (if necessary) as they arrive. Apart from these standard procedures, management control of job materials is directed primarily toward achieving their timely delivery and safe storage. Considerable time and effort have been expended in developing a work schedule that will satisfy time and resource limitations. It should be obvious, however, that this schedule is meaningless unless it is supported by the favorable delivery of materials.

Lead times for material deliveries have already been included in the project schedule. This was accomplished by incorporating appropriate material constraints into the original project network. These constraints were based on the delivery terms included in the material quotations received from vendors when the job cost was being estimated. They represent the times required for shop drawing approval, material fabrication, and delivery to the job site. To the maximum extent possible, these lead times have been built into the operational schedule.

Immediately after the construction contract has been signed, it is necessary to fix the deadline dates by which purchase orders for the various project materials must be issued to the suppliers. In the case of the highway bridge, the critical path includes the preparation of shop drawings and the delivery of the abutment and deck reinforcing steel. Immediately issuing this purchase order, obtaining the necessary shop drawing approvals, and fabricating and delivering this material are each steps of major importance.

Noting when the work schedule requires the material to be on the job site, and allowing for the delivery interval, determines the latest possible order date for any particular item of material. The delivery period includes the times necessary for purchase order preparation and transmittal, shop drawing preparation and approval, fabrication of the material, shipping, and international customs clearing periods, if applicable. On the basis of this information, the project manager can prepare a purchase order schedule for use by the company's procurement department. This schedule will contain all of the necessary purchasing information as well as the deadline date by which the order must be processed and transmitted to the vendor.

Derived from this purchase order schedule is another valuable tool in material management, which is the material tracking log. It will contain a list of all materials required on the job. For each material, it will note critical points in the procurement process from the time it must be ordered through the approval chain, to shipping date, and ending with receipt of the material item by the contractor. This is an invaluable tool for project managers in smaller companies or for purchasing departments in larger companies to ensure that each material item arrives at the job site when scheduled. More will be said about this log in the next section, which focuses on expediting.

Establishment of the order lead time is an important aspect of material control. Ample provision must be made for the delivery interval, with a contingency present to allow for unforeseen delays, such as the required resubmittal of corrected shop drawings to the architect-engineer. Most field supervisors have a pronounced tendency to encourage the early ordering of materials because they regard this as insurance that the materials will be available when they are needed. Soon after the contract has been signed, the project manager or purchasing personnel are prone to buy out the job completely. This is often a good policy, since price escalations may be a problem. It is necessary, however, to coordinate material deliveries closely with the progress of the work. It is often undesirable to have excessively early delivery to the job site. Piles of unneeded materials can lead to serious problems of theft, damage, weather protection, interference with job operations, and rehandling. There is this practical factor, too: Early material delivery by the vendor can have a negative effect on the contractor's cash flow, requiring that the contractor expend funds early in the project for which reimbursement will not be made until later, when the material item is installed.

In the case of very restricted urban job sites, the careful scheduling of material deliveries is especially important. Early deliveries can be routed to temporary storage facilities owned or rented by the contractor when there is no suitable alternative. This does, however, involve the additional expenses of storage, handling, insurance, and drayage.

When off-site material storage is used, delivery to the job site must be anticipated in sufficient time to make the necessary arrangements. This applies particularly to congested urban areas where permits, police escorts, and off-hour deliveries may be involved. In recent years, the estimated cost associated with early purchase and storage of materials has been found to outweigh the benefits of early delivery on many construction projects and has led to the adoption of a just-in-time inventory management philosophy used in many parts of the manufacturing industry. Whether through this strategy or another, it is usually wise to arrange for the delivery of materials to the job site shortly before they are needed.

8.17 Resource Expediting

As mentioned earlier, in a construction context, the term *expediting* can have two different meanings. Expediting, as it applies to project shortening, was discussed in Chapter 7, "Managing Time." As used here, *expediting* refers to the actions taken to assure timely material delivery and subcontractor support for the project.

It is unfortunate that the stipulation of a required delivery date in a purchase order is no guarantee that the vendor will deliver on schedule. Neither does a letter of notification automatically ensure that a subcontractor will move in on the prescribed day. The project manager has an expediting responsibility to make sure that material and subcontract commitments are met. The project manager may carry out the expediting actions, or the company may have an expediting department. A full-time expeditor is sometimes required on a large project. On work where the owner is especially concerned with job completion or where material delivery is particularly critical, the owner sometimes will participate with the contractor in cooperative expediting efforts.

A necessary adjunct to the expediting function is maintenance of a check-off system, or log, where the many steps in the material delivery process are recorded. Starting with the issuance of the purchase order, a date record needs to be kept of the receipt of shop drawings, their submittal to the architect-engineer, the receipt of approved copies, the return of the approved drawings to the vendor, and the delivery of materials. Because shop drawings from subcontractors are submitted for approval through the general contractor, the check-off system will include materials being provided by the subcontractors. This is desirable because project delay occasioned by late material delivery is not influenced by who provides the material. This same documentation procedure is followed for samples, mill certificates, concrete-mix designs, and other submittal information required. General contractors sometimes find it necessary, on critical material items, to determine the manufacturer's production calendar, testing schedule if required, method of transportation to the site, and data concerning the carrier and shipment routing. This kind of information is especially helpful in routing production and transportation around strikes and other delays.

Each step in the approval, manufacture, and delivery process is recorded, and the status of all materials is checked daily. At frequent intervals, a material status report is forwarded to the project manager for his information. This system enables job management to stay current on material supply and serves as an early-warning device when slippages in delivery dates seem likely to occur.

The intensity with which the delivery status of materials and the progress of subcontractors are expedited depends on the float of the activities concerned. Critical activities, for obvious reasons, must be the most closely watched. If delays appear forthcoming for such activities, strong appropriate action is necessary. Electronic messages, letters, telephone calls, and personal visits may be required to keep progressing on schedule. Low-float activities can be similarly categorized because only minor slippage in these can be tolerated before they, too, become critical. If priority among activities becomes necessary, it obviously would be established on the basis of ascending float values.

Weekly job meetings and reports that include all major suppliers and subcontractors are very helpful in expediting, as well as in coordination of all activities, as will be discussed in Chapter 10, "Project Coordination." The project network, updated schedule, and latest material status report provide the basis for an agenda for these meetings. The advance recognition of impending trouble spots enables early corrective action to be taken.

Key Points and Questions

Key Points

- ❑ The objective of resource management is judicious allocation of the physical resources: manpower, equipment, and materials.
- ❑ The goal of manpower leveling is practical improvement of the allocation of labor resources, rather than a theoretical optimal allocation.
- ❑ Allocating equipment resources on a large civil job is similar to allocating labor resources on a building job.
- ❑ Supply chain management focuses on getting the correct materials to the workface when needed.
- ❑ Expediting is required because the stipulation of a required delivery date in a purchase order and letters of notification provide no guarantee that the material supplier or subcontractor will deliver on schedule.

Review Questions and Exercises

1. What is the basic objective of resource management?
2. Project managers have many resources that are not directly under their control, such as the designer and the owner. Provide some bulleted

points describing how project managers can manage (influence) resources outside of their control or above them in the contractual hierarchy.
3. What is the primary concern of material management on a project?
4. Provide some reasons why seeking the optimum manpower allocation is impractical.
5. Select a material used in the example building and diagram the supply chain for that material.
6. Why is expediting required on construction jobs?
7. For the following simple seven-activity project, develop a precedence network, calculate all relevant characteristics of each activity (ES, EF, LS, LF, TF), and calculate the project duration. Then draw a time-scaled network and under it a labor resource histogram for the ES. Do the same for the LS. In both cases, use the maximum allowable number of workers. Finally, make adjustments to the noncritical activities, within the constraints of the ES and LF and the maximum and minimum workers allowed to provide what you feel is the best labor distribution.

What do you feel is the biggest problem with the ES and LS configurations, and how has this been alleviated with your solution?

Activity	Predecessor Relationships	Duration	Workers Maximum	Allowed Minimum
A		3	3	3
B	A: FS	4	3	3
C	A: FS	3	4	2
D	A: FS	4	8	4
E	B: FS	4	5	5
F	B,C: FS	5	3	2
G	E, F: FS	6	2	2

8. A material tracking log is a critical tool for managing the material resource and is an excellent application for a spreadsheet. Lay out a spreadsheet that could be used to track materials for the concrete subcontractor on the example building.

9 Project Scheduling Applications

9.1 Introduction

Whereas previous chapters focused more on tactical use of the schedule to manage specific components of the project, such as production, time, resources, and costs, this chapter will consider strategic scheduling applications as they relate to the overall project, including consideration of the legal aspects of the schedule.

This chapter begins by considering the role of the schedule and the variety of operational schedules available to the project manager, which leads to a discussion of the ways in which scheduling information can be organized and can be presented.

The first application considered is the line-of-balance schedule used as a tool for managing projects that are repetitive, either horizontally or vertically. Disturbances and delays to the project are a normal part of construction. The schedule can be used to identify delays, analyze their impact, and devise means to mitigate and remediate the delays. Changes and delays often lead into the legal realm. One dominant cause of delay is weather, which affects almost all construction.

Learning objectives for this chapter include:

- Recognize the key role the schedule plays in construction.
- Understand the most useful ways to sort and present scheduling data.
- Be able to develop line-of-balance schedules to represent repetitive projects.

❑ Understand legal ramifications of the schedule for all parties.
❑ Anticipate and mitigate the effects of weather on the project.

9.2 Role of the Schedule

All contracts must have an object, usually relating to the value of, and the expectations of, goods or services. Most contracts relate to an object that is palpable, physical, already exists, and is therefore easily defined. For example, typically when a buyer and seller enter into a contract for the sale of an automobile, the automobile already exists. It can be seen, felt, examined, tested, and driven. To assure that the automobile will meet the buyer's needs, the buyer is expected to inspect the automobile diligently prior to entering into the contract.

A construction contract is quite different in this regard. At the time both parties enter into a contract, the object of the agreement is only a vision. It does not exist and hence cannot be examined by either the seller or the buyer. In fact, the object of the contract is given definition only through explanation.

The buyer, or owner, in the case of a construction contract, must define the vision through design drawings and technical specifications. As exemplified by the extensive set of drawings on the companion website, for the relatively simple example building project, construction drawings are extensive and detailed so that they can accurately communicate everything that goes into the project. By their very nature, these documents are imperfect in their ability to precisely and fully define the owner's vision. Often vagaries, omissions, or lack of detail lead to differences in interpretation. Definitions provided by the owner concentrate on the final product and exclude the construction processes needed to achieve the final product, which are the responsibility of the contractor.

The seller, who in this case is the contractor, must be equally diligent in defining the object of the contract. Typically, this is accomplished through a detailed cost estimate and a baseline schedule. Although the estimate is not generally given to the owner, the price, and often a breakdown of the price, is provided. The baseline schedule is developed before or very soon after the start of construction and sent to the owner for approval. These two documents provide the contractor's definition of their contribution to the project, and thus, they concentrate specifically on the construction processes that will produce the product that the owner has specified.

Both process and final product definitions are needed for a proper and comprehensive understanding of the object of the construction contract. Both parties must make every effort to clearly communicate their concept of the project at the time the contract is drafted and signed. Failure by one will give the other the opportunity to unilaterally add detail to ambiguous

9.3 Operational Schedules

areas when conflicts arise. Even in the simpler case of the automobile, conflict often arises from an incomplete understanding.

9.3 Operational Schedules

As the project progresses, scheduling data take on different forms depending on the state of the project and the intended use. Four such forms of the schedule are presented.

A. Baseline Schedule

The baseline schedule, sometimes referred to as the as-planned schedule, reflects the construction firm's original understanding of the project and the contractor's original intention for completing the work. For the highway bridge example project, this schedule was generated in Chapter 5, "Project Scheduling Concepts," and is shown in Chart 5.2 on the companion website.

The baseline schedule encompasses the entire project, including all subcontracted work, and provides an overall snapshot of the project's scope. Because of its breadth, a baseline schedule often generalizes operations at a summary level. As each baseline activity moves closer to commencement, the project team often expands and communicates current activities in the form of a detailed short-term schedule. For the baseline schedule, priority is placed on capturing the project in its entirety and planning it from a macro perspective. A high level of detail at this stage is often counterproductive in that the larger concepts and interactions may be lost in detail.

In general, both the contractor and the owner need to agree on the validity of the baseline schedule early in the project, preferably prior to starting work. Early approval is important because it is often difficult for an owner to review a baseline schedule impartially once project changes have started to take place. Once submitted and approved, the baseline schedule serves as the basis for the evaluation of all subsequent progress reporting and project changes.

B. Updated Schedule

Once construction is under way, the baseline schedule becomes the project schedule and is updated periodically so as to reflect the status of work completed to date. The update also provides an opportunity to replan the remainder of the project based on the current status and the experience gained thus far. For the example bridge project, a monthly update will be generated in Chapter 10, "Project Coordination," and is illustrated in Chart 10.1 on the companion website.

The objective of this periodically updated schedule is to reflect the dynamics of a living and changing project plan. An updated schedule is an as-built schedule in progress. In case of dispute, the existence, timeliness, and continuity of a regularly updated schedule is vital, and the ability to clearly link these updates with the original baseline schedule is indispensable. For this reason, anytime a schedule is updated, the old schedule is retained to build the as-built documentation of the construction process.

C. Short-Term Look-Ahead Schedule

Just as the baseline schedule is a general plan for the entire project, and the updated schedule reflects the status of work completed to date, a short-term schedule (sometimes referred to as a short-interval schedule) is needed to focus detailed attention on the specific activities scheduled over the coming weeks.

Frequently, this type of schedule concentrates on the next four to six weeks of the project. It often forms the basis of the weekly project coordination meetings at either the supervisory or management levels, which will be addressed in some detail in Chapter 10, "Project Coordination." Here again, short-term schedules are closely linked to the baseline and updated schedules. In this way, the project team is able to add detail to the project plan while assuring that the macro planning contained in the baseline schedule is fully supported.

Short-term schedules are employed in a lean construction environment as part of the Last Planner Process© (see Section 10.6) to track the reliability of planning by evaluating in each planning cycle the accuracy of the previous cycle's plan. At the beginning of the coordination meeting, the previous plan is reviewed to determine how many of the planned activities were actually completed. The percentage of activities actually completed is a measure of the reliability of the planning process, and with additional information, chronic problems in planning and execution can be revealed. This is accomplished by asking, for each activity not completed, what deterred it from being completed. When a chronic problem continues to reoccur, the project manager knows where to concentrate to improve the planning and production processes. Improvements in the planning process are revealed by seeing an increase in percent of activities actually completed on the previous week's plan.

D. As-Built Schedule

The as-built schedule is the final result of the periodic project schedule updates. This schedule documents the actual start and finish dates for each activity. It also includes any delays, change orders, extra work, weather, and other factors that impacted the construction. While it records all the changes, it does not apportion responsibility or assign liability for delays.

From a purely practical standpoint, the as-built schedule should be updated on a weekly basis to ensure that actual start and finish dates are recorded accurately and any changes or delays are documented properly. The as-built schedule and the periodic updates provide an important project record, as they are the only documents that capture and model the project history for later analysis and discussion. For this reason, project managers must capture and record actual project logic. This information is essential if later analysis is to be undertaken for the purpose of claims support or process improvement studies.

9.4 Organizing Schedule Data

To serve a variety of different purposes, network activities can be listed in several different manners. The ways that activities are grouped together, or the order in which they are listed, are called sorts or filters. Sorting provides emphasis on different criteria and makes different forms of information easier and quicker to find. Computers have the capability of sorting large numbers of activities quickly and accurately. By sorting activities on different bases, different forms of useful information become available to project management personnel. Various types of sorts are described next.

- *Activity number sort.* In this case, the activities are listed in ascending order of their assigned numbers, as was the case in Figures 5.3a and b. This sort eases going back and forth from the network diagram to the activity data. With the activity number from the diagram, the schedule dates and float values are quickly found from the corresponding activity number in this sort listing.

- *Early-start sort.* Listing activities in the calendar sequence of their earliest possible start times is an optimistic schedule and is commonly used by project field personnel. Figure 5.5a and b are examples of this type of sort. Although it is unlikely that all activities will be started by their early-start dates, such a listing does serve as a daily reminder. It focuses attention on the necessity of meeting these dates in the case of critical activities and on the fact that little time slippage can be tolerated with low-float activities. It also keeps field supervisors aware that float is being consumed by those activities whose beginnings are delayed beyond their early-start dates. Such schedules are also desirable for vendors and subcontractors that are responsible for providing shop drawings, samples, and other submittal information. Design professionals should also work from an early-start schedule so that such submittals will receive their timely attention.

- *Late-start sort.* The late start of an activity is the time by which it must be started if the project is not to be delayed. Failure of an activity to start by its late-start date is a strong indication that the project may be in time trouble.

- *Late-finish sort.* A late-finish sort—one in which the activities are listed in the order of their latest allowable finish dates—is a convenient monitoring device. If an activity has not been finished by its late-finish date, the project will be behind schedule, according to the established action plan.

- *Total float sort.* A listing of activities in the order of their criticality—that is, in ascending order of total float values—can be very helpful to job management in pinpointing those areas where timely completion is of top priority. Figures 9.1a and 9.1b, which provide a partial listing of the activities of the highway bridge and the building example projects, bring into sharp focus the identity of those activities that require the closest attention insofar as timely completion is concerned. Those activities appearing at the top of such a list can be given the special attention they deserve. If the timing of an activity begins to slip as the project progresses, its float will decrease and it will move up the list on this type of report.

- ❏ *Project responsibility.* Each network activity can be assigned the name of the organization or person responsible for its timely completion. Some activities will be assigned to the general contractor, while others are the responsibility of a named specialty contractor, the owner, or the design professional.
- ❏ *Combined sorts.* A valuable aspect of sorting lies in combining sorting criteria. For example, sorts can be produced that will provide a specific job supervisor of a named contractor with a listing of those activities for which that contractor is responsible. These activities can be ordered by early start and be limited to those activities that are scheduled to begin within the next 30 days. With this kind of sorting capability, it is possible to provide specific information to each of the people and organizations that share the responsibility for timely project completion.

Activity (Bold type denotes critical activities)	Activity Number	Duration (Working Days)	Expected Start Date A.M.	Total Float
Prepare & approve S/D, abutment & deck rebar	10	10	June-14	0
Fabricate & deliver abutment & deck rebar	60	15	June-28	0
Forms & rebar, abutment #1	180	4	July-20	0
Pour abutment #1	200	2	July-26	0
Strip & cure, abutment #1	220	3	July-28	0
Forms & rebar, abutment #2	240	4	August-2	0
Pour abutment #2	250	2	August-6	0
Strip & cure, abutment #2	290	3	August-10	0
Set girders	320	2	August-13	0
Deck forms & rebar	330	4	August-17	0
Pour & cure deck	340	3	August-23	0
Strip deck	350	3	August-26	0
Painting	370	5	August-31	0
Cleanup	390	3	September-8	0
Final inspection	400	1	September-13	0
Guardrails	360	3	August-31	2
Order & deliver piles	30	15	June-14	3
Drive piles, abutment #1	110	3	July-6	3
Forms & rebar, footing #1	130	2	July-9	3
Drive piles, abutment #2	140	3	July-14	3
Pour footing #1	150	1	July-13	3
Demobilize pile driving rig	160	1	July-14	3
Strip footing #1	170	1	July-14	3
Backfill abutment #2	310	3	August-13	3

Figure 9.1a
Highway bridge, total float sort

9.4 Organizing Schedule Data

ID	Task Name	Start	Finish	Late Start	Late Finish	Free Slack	Total Slack
1	Start	Mon 12/16/13	Mon 12/16/13	Mon 12/16/13	Mon 12/16/13	0 days	0 days
4	Site Work	Mon 12/16/13	Tue 3/18/14	Mon 12/16/13	Tue 3/18/14	0 days	0 days
10	Drilled Piers	Wed 1/15/14	Tue 1/28/14	Wed 1/15/14	Tue 1/28/14	0 days	0 days
11	Grade Beams & Slabs	Wed 1/29/14	Tue 3/11/14	Wed 1/29/14	Tue 3/11/14	0 days	0 days
12	Structural Erection	Wed 3/12/14	Tue 4/1/14	Wed 3/12/14	Tue 4/1/14	0 days	0 days
15	Framing	Wed 4/2/14	Tue 5/13/14	Wed 4/2/14	Tue 5/13/14	0 days	0 days
18	Insulation	Wed 5/14/14	Wed 6/25/14	Wed 5/14/14	Wed 6/25/14	0 days	0 days
19	Gypsum Drywall	Thu 6/26/14	Thu 8/7/14	Thu 6/26/14	Thu 8/7/14	0 days	0 days
27	Paint	Fri 8/8/14	Fri 9/26/14	Fri 8/8/14	Fri 9/26/14	0 days	0 days
28	Millwork	Mon 9/29/14	Fri 10/24/14	Mon 9/29/14	Fri 10/24/14	0 days	0 days
40	Electrical, RI Partitions	Wed 5/14/14	Wed 6/25/14	Wed 5/14/14	Wed 6/25/14	0 days	0 days
44	Closeout	Mon 10/27/14	Fri 10/31/14	Mon 10/27/14	Fri 10/31/14	0 days	0 days
33	HVAC Finish	Mon 9/29/14	Fri 10/17/14	Mon 10/6/14	Fri 10/24/14	5 days	5 days
35	Plumbing in Partitions	Wed 5/14/14	Wed 6/18/14	Wed 5/21/14	Wed 6/25/14	5 days	5 days
36	Plumbing Fixtures	Mon 9/29/14	Fri 10/17/14	Mon 10/6/14	Fri 10/24/14	5 days	5 days
38	Fire Protection, Finish	Mon 9/29/14	Fri 10/17/14	Mon 10/6/14	Fri 10/24/14	5 days	5 days
43	Fire Alarm, Finish	Mon 10/6/14	Fri 10/17/14	Mon 10/13/14	Fri 10/24/14	5 days	5 days
7	Site Paving, Finish	Fri 8/1/14	Thu 8/28/14	Fri 8/15/14	Fri 9/12/14	0 days	10 days
8	Irrigation	Fri 8/29/14	Fri 9/19/14	Mon 9/15/14	Fri 10/3/14	0 days	10 days
9	Landscape	Mon 9/22/14	Fri 10/10/14	Mon 10/6/14	Fri 10/24/14	10 days	10 days
14	Roofing	Wed 4/2/14	Tue 4/29/14	Wed 4/16/14	Tue 5/13/14	0 days	10 days
16	Dampproof	Wed 5/14/14	Wed 6/18/14	Thu 5/29/14	Wed 7/2/14	0 days	10 days
17	Masonry	Thu 6/19/14	Thu 7/31/14	Thu 7/3/14	Thu 8/14/14	0 days	10 days
29	Signage	Mon 9/29/14	Fri 10/17/14	Mon 10/13/14	Fri 10/31/14	0 days	10 days
20	Aluminum Frames	Wed 5/14/14	Wed 6/25/14	Thu 6/5/14	Thu 7/17/14	0 days	15 days
21	Wood Doors	Thu 6/26/14	Thu 7/17/14	Fri 7/18/14	Thu 8/7/14	15 days	15 days
24	Ceilings	Fri 8/8/14	Fri 9/5/14	Mon 9/8/14	Fri 10/3/14	0 days	20 days
25	Ceramic Tile	Fri 8/8/14	Fri 9/5/14	Mon 9/8/14	Fri 10/3/14	0 days	20 days
26	Flooring	Mon 9/8/14	Fri 9/26/14	Mon 10/6/14	Fri 10/24/14	20 days	20 days
30	Toilet Partitions	Mon 9/8/14	Fri 9/26/14	Mon 10/6/14	Fri 10/24/14	20 days	20 days
31	Toilet Accesseries	Mon 9/8/14	Fri 9/26/14	Mon 10/6/14	Fri 10/24/14	20 days	20 days
41	Electrical Finish	Fri 7/18/14	Fri 9/26/14	Fri 8/15/14	Fri 10/24/14	20 days	20 days
22	Storefront, Glazing	Thu 6/26/14	Thu 7/17/14	Fri 8/15/14	Fri 9/5/14	15 days	35 days
23	Plaster	Wed 5/14/14	Wed 6/11/14	Fri 7/11/14	Thu 8/7/14	40 days	40 days
34	Plumbing, under Slab	Wed 2/5/14	Tue 3/25/14	Wed 4/2/14	Tue 5/20/14	35 days	40 days
39	Electrical, under Slab	Wed 2/5/14	Tue 3/18/14	Wed 4/2/14	Tue 5/13/14	40 days	40 days
13	Miscellaneous Steel	Wed 4/2/14	Tue 4/29/14	Thu 6/12/14	Thu 7/10/14	10 days	50 days
6	Site Paving, Start	Wed 3/19/14	Tue 5/20/14	Thu 6/12/14	Thu 8/14/14	50 days	60 days
3	Utilities	Mon 12/16/13	Tue 2/18/14	Wed 4/9/14	Wed 6/11/14	20 days	80 days
32	HVAV Rough in	Wed 4/30/14	Wed 6/11/14	Fri 8/22/14	Fri 10/3/14	75 days	80 days
37	Fire Protection Rough in	Wed 5/14/14	Wed 6/4/14	Mon 9/15/14	Fri 10/3/14	80 days	85 days
42	Fire Alarm Rough in	Wed 4/30/14	Wed 5/28/14	Mon 9/22/14	Fri 10/17/14	90 days	100 days
5	Site Electrical	Wed 3/19/14	Tue 4/8/14	Mon 10/6/14	Fri 10/24/14	140 days	140 days
2	Mobilize	Mon 12/16/13	Fri 12/20/13	Mon 12/8/14	Fri 12/12/14	60 days	255 days

Figure 9.1b
Commercial building, total float sort

The preceding discussion is intended to give the reader insight into some of the types of sorts commonly used and the project management uses to which they are put. It is possible to list many other types of project information whose use can be very valuable to different members of the management team.

9.5 Schedule Presentation Formats

Once a schedule has been prepared, it may be presented in a number of different visual formats depending on its purpose and audience. Presenting a schedule in the proper visual format highlights the information the project manager is trying to convey and makes the schedule a better communication tool. This section explores several industry standard presentation formats and the information and audiences they target.

A. Precedence Networks

Precedence networks have been used throughout this book to depict the plan and schedule of the example projects (see Charts 5.1a and b on the companion website). The authors have selected this presentation format because of its ability to illustrate the logical relationships between project activities. Since activities are depicted as boxes interconnected by logic arrows, tracing logical paths through the network is straightforward. Precedence networks are especially advantageous during planning, as they allow project teams to quickly list the activities needed to complete the project and link them together with logic. A precedence network has the additional advantage of incorporating a large and varied amount of information about each activity within the activity box without cluttering the schedule.

A precedence network is an analytical tool as well as a presentation format. When job-site problems occur, the network helps determine their impact on the job. Alternative methods for solving the problem can be evaluated quickly based on their effect on the critical path and project duration.

B. Gantt Charts

Gantt charts, or bar charts, are used often because they are easy to read. Activities are illustrated as bars on a horizontal time line. The beginning and end of a bar coincides with the activity's starting and ending dates (see Charts 5.3a and b on the companion website). This format highlights the time and operational concurrency aspects of the schedule and clearly illustrates how activities are to be distributed over the project time line. Bar charts are good for illustrating resource utilization and sequencing. Once activities requiring a particular resource have been identified, the daily resource demand can be illustrated as a histogram, as illustrated in Charts 8.1 and 8.2 on the companion website. This allows the project manager to quickly assess resource requirements and allocate accordingly.

The time-scaled logic diagram is a blend of a precedence network and a Gantt chart (see Charts 5.3a and b on the companion website). By combining logic and duration in one representation, the time-phased interplay between activities is emphasized. However, complex network logic, such as start-or-lead or lead/lag information, is difficult to represent. The necessary overlapping of logical relationships in time-scaled networks often diminishes their usefulness in communicating the construction plan.

C. Time-Scaled Logic

Line-of-balance schedules represent both a presentation format and a scheduling technique. Also known as linear schedules, they are a very visual method of representing a particular type of project. Their use typically is limited to linear projects such as railroads, highways, transmission lines, and pipelines, although they have been used successfully to represent high-rise building construction. The common element of these projects is that they are repetitive, either strung out along long distances or up through numerous stories in a building.

D. Line-of-Balance Schedules

Using a combination of sloping lines, bars, and blocks, operations can be sequenced so that space and resource conflicts are quickly identified and addressed. In general, potential conflict exists anytime one schedule element intersects or crosses another. This method of scheduling allows the planning team to quickly explore a variety of sequencing options and select the most desirable. Line-of-balance scheduling techniques are described next.

9.6 Schedules for Repetitive Operations

For repetitive operations, such as the pipeline relocation described in Section 5.22, a form of bar chart introduced in the previous section, often referred to as a line-of-balance schedule, better represents the field schedule information than a precedence diagram or Gantt chart. Figure 9.2 illustrates a line-of-balance schedule for the pipeline project. In this figure, the horizontal axis represents time, as is typical in schedules, and the vertical axis represents distance along the right-of-way. Figure 9.2 has been prepared on the basis that the pipeline construction will be initiated on Monday, July 12. Each of the sloping lines represents the work category indicated and is plotted on an early-start basis. Figure 5.9 provides the necessary early-start (ES) and early-finish (EF) values. A conversion calendar such as that in Figure 5.4a, with the numbering of working days starting on July 12, serves for the translation of these early times to calendar dates. The horizontal line at mile 3 in Figure 9.2 represents the construction of the pipeline crossing structure whose beginning will also be on Monday, July 12. The crossing structure is scheduled to be ready for pipe by the afternoon of August 3 and to be completed on August 11. The scheduled interface between the pipeline and the crossing structure is shown as August 6.

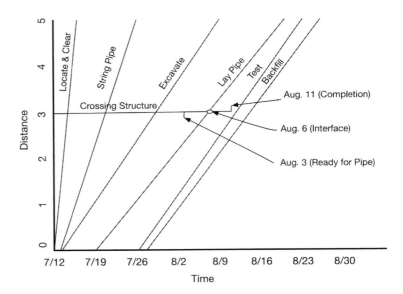

Figure 9.2
Pipeline relocation line-of-balance schedule

Further study of Figure 9.2 yields additional information about the pipeline relocation project that is not readily discernable in either a precedence diagram or Gantt chart. The slope of the line represents the production rate for that activity. It is clear that for each of the first four activities, the production rate of the successor activity is slower than that for the predecessor. What this means in practice is that if each of these activities proceeds as scheduled, there will be no conflicts with the successor overtaking the predecessor. However, when we get to the "Test" activity, its production rate is faster than the predecessor so that if there were not a delay of several days in starting the "Test" activity, there is the risk that testing would overtake "Lay Pipe," which could not happen in the real world. This would be clearly indicated by the "Test" activity's intersecting the "Lay Pipe" activity.

Figure 9.3 provides a series of illustrations showing ways in which conflicting activities might be rescheduled, other than by simply delaying the start of the successor activity. Figure 9.3a shows a situation in which pipe laying moves at a faster pace than excavation. It is clear that soon after passing mile 2, pipe laying catches up to "Excavate." Figure 9.3b provides a solution in the form of simply suspending pipe laying at mile 2, giving the excavation crew the opportunity to get ahead again. The pipe-laying crew could be reassigned to another project for several weeks and then brought back in to complete laying the pipe. The same effect could be achieved

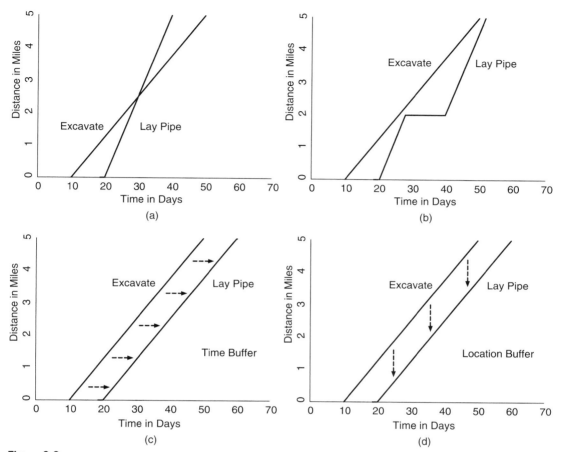

Figure 9.3
Resolving line-of-balance conflicting activities

with several shorter delays for the pipe-laying crew rather than a single longer delay; however, each time the pipe-laying crew is reassigned, costs of demobilization and remobilization on this job would increase.

Figures 9.3c and 9.3d illustrate the use of buffers to separate potentially conflicting activities. The time buffer illustrated in Figure 9.3c imposes the condition that the successor activity (in this case, laying pipe) will arrive at a point on the route no less than a designated number of days (in this case, 10 days) after the predecessor has passed that point. The location buffer illustrated in Figure 9.3d imposes the condition that the successor activity (in this case, laying pipe) be no closer than a set distance (in this case, a mile and a quarter) from the predecessor activity.

There is another solution not illustrated here. If, after beginning the pipe laying, it becomes apparent to the project manager that the pipe-laying crew is moving faster than the excavation crew, steps could be taken to increase the rate at which excavation is accomplished. This might mean bringing in a second excavation crew or providing different equipment for the existing crew that would enable excavation to go more quickly. The excavation crew might also work 10-hour days instead of 8-hour days, or on a 6-day schedule instead of a 5-day schedule. If these steps were taken to increase the rate at which the pipe was laid, a discontinuity in slope would show up on the chart at the point at which the rate was increased with the new slope steeper than the original slope. The key here is that the line-of-balance representation makes it quickly and clearly obvious to the project manager on a timely basis that a problem is developing so that the project manager is able to work out the best solution prior to the potential problem turning into a real problem.

There also may be other information that comes to light, telling the project manager that even though a problem is anticipated if measures are not taken to accelerate the excavation, it might actually resolve itself before it does become a problem. If soil conditions or terrain are such that excavation is difficult early in the job, but after the first mile the soil or the terrain changes so that the excavation will naturally move more quickly, the problem may well resolve itself. Since the slope of the activity line indicates the rate at which that activity is being executed, such a change in conditions would reveal itself in the plan by indicating a change in the slope of the excavate activity line.

Illustrations to this point consider the activities as moving continuously across the terrain without any delays. However, in the real world, progress is often slowed or even stalled through part of the route. This can be illustrated on the diagram by giving width to the activity line such that at a point in the route where work takes excessive time to perform, the activity line is thickened through that portion where execution is slow.

Figure 9.4 represents a transmission line that illustrates a number of the concepts discussed earlier. It also illustrates that relationships between activities can be represented by any of those provided in precedence scheduling, including start-start and finish-finish relationships with lag time between activities.

9.7 Impacted Baseline Schedule

There are times when even the contractor's best scheduling effort is inadequate to keep pace with changes on the job. Problems are encountered that are so severe that the entire scope of work is placed in flux. Some changes alter the entire direction of the work, while others create repercussions that ripple through activities for months into the future. For example, the owner issues a change in the paint specified for the guardrails on the highway bridge. The contractor agrees to the change

9.7 Impacted Baseline Schedule

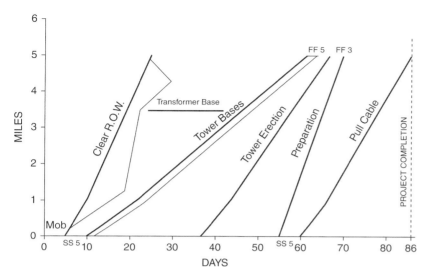

Figure 9.4
Five-mile transmission line

with a marginal cost adjustment for the paint. The contractor applies the new paint, but within weeks it begins to chip and peel. Over the next month, the contractor reapplies the paint several times using a variety of methods, all without success. Finally, the owner amends the change and instructs the contractor to readopt the original paint specification. By this time, however, cold weather has suspended painting operations for the winter, and project completion is delayed until spring. Under such circumstances, the contractor's original change order price was wholly inadequate to cover the full impact of the change. Additionally, due to the consequential effect of this change, the contractor is unable to identify and quantify the full magnitude of its impact until the project is completed. The contractor must notify the owner immediately of its intent to claim for the ongoing impact and carefully document the claim. Once the project is complete or the full impact of the change is realized, an analytical schedule is prepared to determine the cause, liability, impact, and cost of the change.

Over the years, many analytical techniques have been developed for quantifying and assigning these impacts, but, in general, these techniques fall into two basic categories: adding the delays to the baseline schedule to illustrate their effect and quantify them, or removing the disruptions from the as-built schedule to illustrate what would have happened had the delays not occurred. This section is concerned with the first of these two categories, generally referred to as the impacted baseline schedule.

When generating an impacted baseline schedule, the original baseline schedule is used as the foundation. To determine the time impact of a series of delays, remove the logical linkages for all activities impacted by the first delay from their successors. Move the successors to the right on the network diagram and add the new activities caused by the first delay. Then link the impacted activities to the delay activities. Finally, link the delay activities to the original successors. The next example is for a delay involving a single delay activity. If an activity A experiences a delay that affects its successor, activity B, then the linkage between A and B is removed. A delay activity, D, is added as a successor to A and given a duration equal to the delay. Activity B is then made a successor to D. This process continues until all delays, from whatever source, have been accounted for in the model.

Once all of the project delays have been introduced with actual activity durations and logic, the displayed early-start and early-finish dates for all completed work in the impacted schedule should be identical to the actual start and finish dates in the as-built schedule.

For this modeling technique to be effective, any constraints or fixed dates within the as-planned schedule must be removed and replaced by activities with adjusted durations or lead/lag logic. This will allow the schedule to accommodate logical and durational changes made during forward- and backward-pass calculations.

The analysis model now contains all of the project delays. A forward and backward pass on the network establishes their combined effect. Because some of these delay activities were caused by the contractor while others were caused by the owner, their relative impact must be apportioned properly. Liability for each of the added delay activities is assigned to the responsible party. If the period of time spanned by one party's delay does not overlap the delay caused by the other party, then the delay is considered *simple*. Reducing all of one party's delay activities to zero duration will quantify the other party's impact on the project. Should both parties have delays on independent branches of the network, but overlapping the same period of time, then the delays are said to be *concurrent*. In this case, the overlapping portion of the delay is attributable to both owner and contractor.

With concurrent delay, it is necessary to determine whether the contractor's delay was independent of the owner's delay or occurred as a result of the owner's delay. When the contractor's delay is independent, extra time frequently is added to the schedule but without extra compensation. If it can be shown that the contractor's delay was in some way a result of the owner's delay, both extra time and compensation are claimed.

Because both contractor and owner delays are addressed, the impacted baseline analysis provides a sufficiently level playing field for both parties to expose the causes of delays, assign responsibility, and determine the time impact and its associated cost to the project. Delay analysis practitioners, while pleased with the generally manageable cost and level of effort required to support an impacted baseline analysis, often disagree regarding the validity of the baseline schedule as a point of departure for project analysis. Owners may argue that the purpose of their review and acceptance of the project baseline schedule was based on its conformance to the schedule specification requirements and level of detail. The means and methods, and therefore the schedule activities, their chronology, duration, and interlinking logic, were proposed by the contractor and were the contractor's responsibility. They therefore were given only a cursory review. Owners may claim that the baseline schedule was always unachievable, pointing to repeated failures of contractor performance as supporting evidence. In such circumstances, both parties may become preoccupied with the question of the appropriateness of using the baseline schedule, preventing advancement of the larger delay issue.

Consequently, impacted baseline techniques are appropriate for the analysis of projects that have a well-developed and supported baseline schedule and have largely proceeded according to this schedule with few, easily defensible deviations. The delays that have occurred should be finite in number, with clear schedule impact boundaries. In such cases, the impacted baseline model can focus attention on the changes that have taken place and their time impact on the project completion date.

9.8 But-for or Collapsed As-Built Schedule

The but-for or collapsed as-built analysis uses the as-built schedule as a foundation and works backward to quantify and assign project delays. To begin this process, a working copy of the as-built schedule is made. The actual start and actual finish dates of the network activities are removed from the copy. If actual durations, actual logic, and appropriate lags are properly in place, the calculated early-start and early-finish dates will correspond precisely to the actual dates in the as-built schedule. This enables the working copy of the as-built schedule to accommodate the removal of impacted activities for each delay and the recalculation of the project duration. Any reduction in project duration reflects the project outcome minus this specific delay.

The person preparing the claim should list all known delays that occurred on the project, whether caused by the contractor, owner, or others. The delays should be listed in reverse order of occurrence. Then the delays are analyzed one at a time. The latest delay is analyzed first by removing it from the network. If the delay is in the form of an increased duration to an original activity, that duration is reduced to its unimpacted duration. If the delay resulted from added activities, those activities are removed. The network is then recalculated, and the effect on project duration is noted and attributed to the responsible party. Clearly, only delays on the critical path need to be analyzed, although it should be noted that the critical path will change as activities are shortened or removed.

This process is repeated for each of the delays on the list. The reduction in the network time calculation is cumulative, as the first delay is not reinserted into the network before removing the second. Once all of the delays have been removed, the network should reflect the project's as-built condition except for the delays. Because the contractor was not awarded additional time for each of these delays as they occurred, their effect on the project was cumulative. In order to analyze this accumulation of delays, the as-built schedule must be reduced by the last delay first; thus, the removal of delays starts at the end of the project.

As a direct result of a delay, contractors frequently suspend parallel operations and redirect resources to unaffected portions of the project. This creates multiple critical paths around the delayed activity. In order to measure the true effect of the delay, these artificially suspended activities must also be reduced to their original durations.

Each time a delay or change is made to the as-built schedule, it must be carefully recorded and substantiated in a log. This log will later become the audit trail for the schedule delay analysis and is often the source of criticism and controversy. Therefore, particular attention should be paid to making this document as unbiased and unassailable as possible.

This type of delay analysis is best applied to projects where activity logic has undergone a significant adjustment and the baseline schedule is no longer an adequate reflection of the project. The addition, deletion, extension, or reduction of activities and their associated logic may have changed for a number of reasons unrelated to the impact analysis at hand. Rather than attempting to account for each of these nonissues in the adaptation of the baseline schedule, they may simply be assumed in the as-built schedule analysis.

But-for analysis can account for the combined effect of a large number of interdependent and autocorrelated project delays. At the same time, it can focus analysis on only those delays that directly impact the critical path and hence the project completion date.

Changes made in one group of operations often adversely affect other seemingly unrelated operations by forcing them into unfavorable weather seasons. Using the weather calendar methods discussed later in Section 9.10, impacted baseline and but-for techniques are able to analyze and quantify the extent of these consequential effects.

Despite being touted as one of the most accepted methods for proving delay claims, but-for analysis presents several problems. In practice, but-for analysis can be extremely complex and is often plagued with procedural disputes. In each instance where the delayed portion of an as-built schedule activity is collapsed, proper justification for this reduction must be researched and presented as support. In most instances, these delays and their net effect have not been properly documented, and a large degree of recollection and judgment must be exercised. In such circumstances, analysis is reduced to negotiation and the real issues of the claim may become lost in a debate over nonissues.

9.9 Legal Schedules

This section considers some of the legal aspects of a project schedule. The discussion here is general and should not be considered a substitute for legal advice. The topics that follow are important considerations in planning, scheduling, and submitting schedules for review and approval. The schedule is a double-edged sword with the potential to damage the position of either party in a dispute.

A. Right to Finish Early

Most construction contracts have a specified completion date. This date is determined by the owner and based on the owner's criteria. The contract construction time may be based on a perceived reasonable period or on the date the facility is needed.

Once the contractor has been selected, it may choose to finish the project earlier than the contract completion date. The contractor's reason for finishing early is generally self-interest. The firm may be able to reduce project overhead or free up equipment and resources so that they may be used on another project, or finish before winter or the rainy season begins.

Owners may resist early contract completion and refuse to approve an accelerated schedule. Generally, the owner has spent a significant amount of time planning for and coordinating the broader aspects of construction and commissioning prior to procuring the contractor. As the contractor usually has not been a member of the project planning team, the owner has had to produce a time schedule for planning and orchestrating issues such as project finance and logistical support.

Submitting an accelerated project construction schedule forces the owner to free up assets at an accelerated rate for progress payments. The increased pace of construction may force the owner to employ more construction inspectors and provide increased design professional services. The owner may have multiple contracts with complex coordination responsibilities. One contractor's desire to accelerate may impact carefully formed plans. Owners may not be able to utilize a facility prior to the specified completion date regardless of early construction completion. In this case, early completion transfers the ownership costs, operational responsibility, and insurance risks to the owner while offering few or no benefits. In short, the contractor's notification of its intention to complete early forces the owner to reconsider and replan many of the project's fundamental logistics.

Some owners are pleased to accept and accommodate a contractor's accelerated schedule. They receive the benefit of reduced bid prices from the contractor and a shorter project financing period. They may also be able to minimize the length of time that inspection staff, project supervision, and design professional support are needed on the job site. An earlier completion date often enables the owner to begin operating the facility earlier and begin recouping construction costs through early revenues.

Owner reaction to an early-completion schedule has been mixed and is largely dependent on the owner's experience and particular situation. Yet from a legal standpoint, the contractor's right to finish early is well documented. Court decisions have largely been in favor of the contractor's right to complete the project in less than the specified contract period. Owners have been held liable for actions that prevent the contractor from finishing in accordance with an early-completion schedule. This means that the owner may not delay the start of the project or hinder the contractor's progress once construction begins.

In some cases, owners have circumvented the issue of early completion by adding preclusive clauses to the contract, forbidding the submission of schedules with accelerated completions. Generally, this has had adverse effects, as such clauses normally do not preclude the contractor from recovering delay damages if it is proven that early completion would have occurred but for owner disruptions. Such a clause only serves to prevent the contractor from giving the owner prior notice of its intentions. Other contract clauses state that the contractor may submit an early-completion schedule, but the submitted completion date becomes the contract completion date, and all damages and penalties are assessed against this revised target. Such clauses enable the owner to take advantage of early completion while still maintaining enforceable milestones. This type of clause puts the contractor's resolve to the test but allows it the flexibility to change the contract completion dates to better suit conditions.

Should a contractor decide that early completion is advantageous, six points should be considered before proceeding.

1. *The contractor should clearly inform the owner of its intention to complete the contract early.* This is best accomplished with a project schedule clearly indicating an early completion date.

2. *The early-completion schedule should be submitted prior to commencing construction operations.* Schedules generated during construction or following contract completion for the explicit purpose of substantiating a delay claim are at great risk of being criticized and disregarded.

3. *The early-completion schedule must be complete, showing all aspects of the work, including subcontracted work, and must be realistic and achievable.* Schedules that do not indicate some aspects of the work may be deemed deficient and prejudicially optimistic. Time savings should not include the reduction of contractually provisioned activity durations, such as submittal review periods or requests for information response times. In general, the owner's duty is limited to the requirements of the contract.

4. *Once submitted, the project schedule must be updated on a regular basis and must consistently demonstrate both the feasibility of the plan and the intent of the contractor.* The updates should also clearly indicate the nature and effect of the delays occurring to date. Frequent updates are necessary, as contemporaneous data are vital to the defense of any later claims for delay.
5. *To improve the reliability of an early-completion schedule, both the contractor and the owner should use and depend on it during construction.* It is difficult to refute the acceptability of a schedule that has been used by both parties to plan, control, and guide decision making during construction.
6. *In the event of a delay claim, the contractor must prove that progress up to the point of disruption was in accordance with its early-completion plans.*

Timely notice and documentation of project delays is of utmost importance, should any party later present a claim in order to recover from the effects of change or delay. The simple fact that a delay has been documented does not necessarily indicate an intention to claim on that issue. Rather, it provides for that possibility at a later time should the need arise. A delay that is identified and documented immediately can be discarded later if no serious impact results. Yet a delay that is not documented may be lost entirely.

B. Owner Approval of the Schedule

Confusion often surrounds contract clauses requiring the owner's approval of a project schedule. Typically, construction contracts do not specify the means and methods to be used by the contractor. Design drawings and specifications generally are limited to providing information about the final constructed product rather than dictating the specific manner in which the work is to be done. Owners frequently specify certain project milestones that must be achieved but rarely dictate the contractor's work schedule as a contractual obligation. Yet many contracts require that the contractor prepare and present the construction schedule and receive the owner's approval before any contract payments are made. This clause, however, does not limit the contractor's choice of construction means and methods. Nor does it remove any of the contractor's responsibility and liability associated with implementing the chosen means and methods. The owner's approval of the contractor's construction schedule covers three elements:

1. *The contractor's schedule and work plan must meet all requirements as set forth in the contract documents.* In reviewing the contractor's schedule, the owner should ensure that contract milestone requirements are met, that all permitting and environmental constraints are addressed, and that all other contract requirements are satisfied. The contractor should be given free rein over the methods used to meet these contract requirements.

2. *The contractor's schedule must provide the owner with sufficient detail to develop other project-related schedules.* It must provide owners with sufficient detail to allow them to monitor the progress of construction and evaluate the impact of a delay. Owners must be provided with the tools to monitor progress and explore alternatives for mitigating delays.

3. *The contractor's schedule must provide the owners with sufficient detail to analyze the effects of a proposed project change.* Owners must have enough information regarding the contractor's intended means and methods to be able to conduct project "what-if" scenarios and arrive at reasonably accurate answers without involving the contractor directly. The same argument applies to justification of a claim. The contractor's schedule must provide sufficient information to enable owners to analyze and justify any change or delay claim. It was stated earlier in this chapter that the contractor's schedule is a double-edged sword. The schedule not only communicates the contractor's intentions, and thereby protects the contractor from the negative effects of a project change, but also provides owners with a measuring stick that may prevent a contractor from recovering the requested amount for a claim. Consequently, it is the owner's duty to determine whether the submitted schedule communicates the contractor's plan in sufficient detail to protect the owner's interests.

C. Resource-Loaded Schedules

Success on construction projects depends on the efficient utilization of limited and costly resources. Materials, labor, and equipment are the costliest parts of a project. Contractors that reduce waste and rehandling of materials, operate construction equipment efficiently, and maximize the efficiency of onsite labor have a significant competitive advantage. Consequently, success in the construction business often is determined by the contractor's ability to manage and utilize resources efficiently.

Submitting a construction schedule without specifying resources implies that the contractor has unlimited resources available. A resource-loaded schedule, however, clearly indicates to all parties the fundamental interdependencies between activities and resources under which the contractor will be performing construction. Failure to provide this information often is misconstrued to indicate that the contractor has the flexibility to apply all necessary resources to a project change without incurring additional costs. When it comes to job-site scheduling, contractors tend to schedule resources rather than activities. Their published schedules must reflect this approach and constraint.

D. Legal Need for a Schedule

The construction schedule is the only project document that fully communicates the contractor's intentions for delivering the contracted scope of services over the full course of the project duration. It spells out the planned timing and sequencing of key construction activities; it anticipates constraints, forecasts resource utilization, and estimates unknown external factors such as weather. It is vital for the contractor to communicate this information to the owner. The construction schedule does this; it also offers a primary avenue for avoiding costly misunderstandings. The schedule allows the owner to completely consider contractor issues when project-related decisions are being made, thereby protecting both the contractor and the owner from misalignment. It defines where contractor flexibility is possible with little cost and time impact as well as changes that may be devastating to the contractor's cost and schedule constraints.

Once a change or delay has occurred, the baseline schedule may be the only document with the capacity to convey the contractor's original intentions and subsequently quantify the impact of such a change. Failure to provide this information to the owner early in the project may limit or prevent the contractor from fully recovering the cost of a change or delay. The legal necessity for a project schedule is clear. Even when a schedule is not required by the terms of the contract, or when such a requirement is later waived, or when obtaining the owner's approval appears impossible, the project schedule is still a necessary legal document and should not be neglected.

9.10 Handling Weather Effects and Other Unknowns

Contractors estimate project costs and project time with limited information. Lump-sum competitive bidding demands that contractors set their prices weeks before construction begins. They may endure risks of bad weather, unknown subsurface conditions, and other uncertainties that threaten to change the project dramatically. Handling such unknowns and their associated risks requires experience, patience, luck, and a certain amount of audacity. Although assuming these risks is part of the business, successful contractors minimize their exposure through solid planning, documentation, and research.

Construction contracts often require the contractor to assume the risk of normally occurring or commonly anticipated conditions. The definition of these conditions is subjective and difficult to isolate in arguing a claim. Therefore, it is vital that the construction firm provide, unambiguously, its interpretation of a normal occurrence or commonly anticipated condition. This definition must be based on research; it must fit the location and conditions of the project and be incontrovertible. Timing is essential. This information must be provided to the owner at the beginning of the project so that the contractor's intentions are indisputable and both parties have time to compromise should a disagreement arise.

A good example of the contractor's assumption of risk associated with normally occurring conditions is weather. Coping with weather in the schedule may be accomplished in a number of ways. The project manager may choose to add duration to affected activities, add a weather contingency activity to the end of the schedule, or schedule activities on an artificially abbreviated workweek. While all of these methods add time for weather, they do not account for the randomness of weather, the effect on activities pushed into harsher seasons by other unrelated delays, or the establishment of a baseline definition for "normal" weather. Accounting for weather through the use of weather calendars based on local meteorological data is preferred. Airports, universities, laboratories, agricultural associations, nautical authorities, and oceanic and atmospheric organizations are all good sources of local meteorological data.

The discussion that follows relates only to rainfall and its effect on construction. Snow, extremes of temperature, visibility, or other weather factors can be handled in the same manner. For construction purposes, 10 years of rainfall records provide a statistically significant period. These records are averaged for each month corresponding to the project construction period. For the highway bridge, this is from June through September. Next, the project activities are examined for their sensitivity to rainfall. Highly sensitive activities may have to be suspended due to 0.1 inches or more of rain in a 24-hour period. Similarly, moderately sensitive activities are affected by 0.2 inches or more of rainfall, and less sensitive activities can withstand as much as 0.5 inches. Figure 9.5 shows the corresponding data summary for these criteria.

Weather calendars now can be established to reflect differing amounts of rainfall. The first calendar, covering highly sensitive activities, randomly reflects the number of days in each month in which 0.1 inches or more of

	Average Number of Days		
	> 0.1 inches Rainfall	> 0.2 inches Rainfall	> 0.5 inches Rainfall
January	8	3	1
February	10	3	1
March	9	2	0
April	7	1	0
May	5	1	0
June	10	5	2
July	12	6	2
August	8	4	1
September	4	1	0
October	4	1	0
November	5	2	1
December	7	3	1

Figure 9.5
Average rainfall summarized by month and volume

rain are expected. A second calendar shows the days when 0.2 inches or more are expected, and the third calendar shows the days with more than 0.5 inches. The second and third calendars are for moderately sensitive activities and less sensitive activities, respectively.

Applying this technique to the highway bridge, weather records show that, on the average, there are 10 days in June with an expected rainfall in excess of 0.1 inches. There are 5 days with rainfall in excess of 0.2 inches and 2 days with an excess of 0.5 inches. The calendar reflecting 0.1 inches of rainfall will therefore have 10 days randomly assigned as rain days, with no work scheduled. The other two calendars have similarly assigned rain days. The months of July, August, and September are handled in a similar fashion.

Referring to the highway bridge plan, activities are assigned to one of the three weather calendars, according to their sensitivity to rain. "Move in" is moderately sensitive. "Excavate abutment #1" and "Pour & cure deck" are examples of highly sensitive activities. "Mobilize pile-driving rig" is less sensitive and assigned to the third calendar.

This process of using historically relevant data to account for project uncertainties applies modeling and simulation techniques to better define construction risks. Therefore, it is imperative to follow standard modeling and simulation guidelines. First, the model should resemble the historical data. In the case of rain, the selection of individual rain days cannot be reasonably predicted, so a random distribution offers the best approximation. Hence, in our example, 10 days are selected at random from the month of June to represent 0.1 inches of rainfall. Second, historical data should be applied to the simulation under statistically identical conditions. If the historical data set includes weekends, then the selected rain days should include both weekdays and weekends. Using these guidelines, project weather can be accounted for realistically in the project schedule. This method also clearly defines normal weather and demonstrates that its impact has been included in the contractor's plan. The final project weather calendar is shown in Figure 9.6.

An important benefit of the weather calendar method is its ability to distribute the delay effects of weather realistically over the entirety of the schedule. When a weather contingency is added to the end for a project, regular updating generally will indicate that the project is behind schedule if weather delays occur early in construction or on schedule, should they occur at the end of the project. In either case, material deliveries and other interfaces with outside organizations will occur too early or too late if they follow the published schedule. The weather calendar method provides a truer status of the project at each update, allowing for better forward planning. As the schedule is updated, it is important to remove predicted rain days and add actual rain days. This information is fundamental to any later schedule analysis.

Now that normal rainfall has been defined and accounted for in the project schedule, the project manager must address all of the contractual

Days with Greater than 0.1 Inches of Rain

	No. of Days	\multicolumn{13}{c}{Randomly Selected Date}

	No. of Days	1	2	3	4	5	6	7	8	9	10	11	12	13
January	8	14	24	2	27	15	3	20	5					
February	10	8	1	10	21	22	17	24	18	19	23			
March	9	20	4	2	27	30	8	11	10	13				
April	7	1	8	25	6	20	10	17						
May	5	17	21	3	21	30								
June	10	11	12	19	3	22	30	7	4	21	1			
July	12	30	26	2	21	17	14	4	9	15	5	18	25	
August	8	10	12	31	7	3	22	19	4					
September	4	8	5	14	17									
October	4	30	1	10	14									
November	5	18	23	3	12	29								
December	7	9	28	26	24	14	3	5						

Days with Greater than 0.2 Inches of Rain

	No. of Days	1	2	3	4	5	6	7	8	9	10	11	12	13
January	3	14	24	2	27									
February	3	8	1	10										
March	2	20	4											
April	1	1												
May	1	17												
June	5	11	12	19	3	22								
July	6	30	26	2	21	17	14							
August	4	10	12	31	7									
September	1	8												
October	1	30												
November	2	18	23											
December	3	9	28	26										

Days with Greater than 0.5 Inches of Rain

	No. of Days	1	2	3	4	5	6	7	8	9	10	11	12	13
January	1	14												
February	1	8												
March	0													
April	0													
May	0													
June	2	11	12											
July	2	30	26											
August	1													
September	0													
October	0													
November	1	18												
December	1	9												

Figure 9.6
Selection of weather days based on historical summary

responsibilities associated with weather. Some construction contracts clearly state that weather and other project unknowns are the responsibility of the contractor. These contracts are generally more expensive, as the contractor must include this additional risk in the price. Other contracts reduce the contractor's responsibility to normally occurring weather in order to attain a more favorable bid. In this case, the contractor must develop a competent system for defining normal weather, logging actual project weather conditions, contrasting these conditions to the plan, and managing any variances. The first step is to define normal weather using the processes presented earlier in this section. Next, the project manager must record actual weather conditions on the project. Whether actual weather conditions are recorded on the daily time cards or in the project manager's daily log, they should be documented formally and maintained consistently throughout the life of the project. This information often becomes the foundation of valuable claims for both additional money and time.

Comparisons between actual and planned weather should be made frequently; monthly is standard. Generally, these comparisons should coincide with project progress payments and other project reconciliations so that all aspects of the project may be brought up to date on a regular basis. It is imprudent to postpone decisions on requested time adjustments until the end of the project. Just as lump-sum construction contracts must have a distinct and commonly understood contract value, the contractor and the owner need a legally defined contract completion date so that critical project milestones can be set. If worse-than-expected weather causes the project to fall behind in any one month, a contract extension should be requested immediately. If an extension is not granted, the contractor must immediately begin constructively accelerating the work at additional cost in order to recover the lost time and complete the project by the contract completion date. In such circumstances, the contractor may then claim for an equitable adjustment resulting from the acceleration costs incurred. Consequently, resolving any discrepancies between planned and actual weather conditions on a monthly basis, and making the necessary changes in the contract completion date immediately, is in both the contractor's and the owner's best interests.

The ability to acknowledge and incorporate the unknown into a construction plan is a skill that requires extensive experience, effort, and planning. Many project managers are good at accounting for the predictable aspects of construction projects but have difficulty when several outcomes are equally likely. In such cases, planning and scheduling frequently are abandoned for a more existential approach. Yet most successful contractors will agree that the unknown does not preclude project planning but rather necessitates better planning with more contingencies. When project managers arrive at such junctures, they should focus on each of the possible outcomes, try to understand their likelihood and their anticipated effect on the project, and then develop a contingency plan for each outcome. A methodical approach, with the assistance of good research, often makes resolution of project unknowns more attainable. The objective should be to isolate the uncertainties, better understand them, and then adopt a set of plans to work around them. No plan means no goals, no benchmarking, no risk management, no hope of countermeasures, and little opportunity for recovery.

9.11 Presenting the Schedule

Frequently, the manner in which a schedule is presented significantly affects its acceptance and utilization on the project. Schedules, by their very nature, are complex documents that strive to communicate complex information in a limited amount of space. Therefore, an essential element of good schedule preparation is composition of a guidebook or narrative that explains the assumptions, concepts, and thought that went into the schedule's development. Without such documentation, the full value of the schedule may not be realized.

A schedule narrative should exhibit the contractor's understanding of the work that is to be performed. This is frequently the contractor's own interpretation of the contract documents. During the estimating stages of the project, the contractor read the contract documents and made a series of interpretations regarding the scope, risks, assertions, and definitions. This information became the foundation for the contractor's bid. These interpretations formed the basis for the contractor's understanding of the project and for all its subsequent planning and preparation. In most cases, project disputes originate with divergent interpretations of the contract documents at this initial stage. Therefore, a key focus of the narrative is to communicate the representations, assertions, and requirements that were relied on during the planning and bidding stages of the project. Supplying this information at the onset of the project aligns interpretations, protects both the contractor and the owner from future claims, and facilitates better communication and collaboration.

The narrative report is a user's guide to the schedule. It may include information about the structure of the schedule, such as how the project was divided into schedule activities, or an explanation of specific activity coding systems that were utilized. This explanation is particularly useful when scheduling software is used and automated sorting and filtering functions are available. The guide should explain the schedule calendar system and the process used to create the weather calendars. An improved understanding of the schedule and the level of effort dedicated to its development facilitates its acceptance, approval, and use.

Typically, projects have key features or equipment whose proper and efficient use is credited with providing the bidder with a winning advantage. These advantages generally are given high merit in the contractor's plan, and changes that lessen their effectiveness frequently result in costly claims. The narrative report should call special attention to these situations and clearly define the conditions under which their successful use is anticipated. This information acts as a warning to the owner and clearly distinguishes between those changes that can be absorbed by the contractor and those that are likely to have cost implications. If this competitive advantage comes from the efficient and cost-effective use of a unique resource, the narrative should spell out the contractor's plan for its use, detailing operationally how the resource is to be utilized and sequenced throughout the project. If the contractor has developed a special process for handling a particular aspect of the work, the narrative should detail this process. It should highlight the conditions under which this process is applicable and the representations in the contract documents that led the contractor to believe that these conditions would exist onsite.

Fundamental to the contractor's construction plan and schedule is the productivity of key construction crews and equipment. Their productivity forms the basis of the schedule. Achieving this productivity is key to staying on schedule. One may argue that the selection of specialized equipment is a trade secret, but it is difficult to conceal big, yellow equipment on the job once things get going. As for production rates, knowing them is not the same as knowing how to achieve them. Most owners, should they spend the time, can quickly determine crew and equipment production rates by watching the work. The fact is that hiding these things is counterproductive and damaging in the long run. Freely providing equipment and production information to the owner has several advantages. Knowledge of the contractor's planned crews and productions will allow the owner to establish more quickly the validity of change orders and claims. When work quantities are increased, requests for additional time due to resource unavailability may be substantiated more easily. In instances where the project has been delayed or halted, the contractor's standby costs may be given greater merit. But, arguably, the greatest advantage is derived from the owner's ability to approximate the effects of a proposed change prior to its being issued. With this in mind, the contractor may elect to provide specific crew makeup, availability, and expected production rates to the owner as part of the narrative report.

The objective of the narrative is to facilitate an improved understanding and utilization of the contractor's project schedule. Consequently, each issue should be introduced and presented factually and fairly. When the document is presented, the project manager should explain its objective to the owner so that the intent is not misunderstood. Construction projects are largely about people and relationships, and most successful projects are built on trust and good communication. Every effort made at improving mutual understanding and coordinating goals will pay large dividends with the successful conclusion of the project.

Key Points and Questions

Key Points

- ❑ The schedule defines the contractor's contractual obligations by showing the means, methods, and sequencing that will produce the owner's product.
- ❑ As the project progresses, scheduling data take on different form, depending on the state of the project and the intended use.

- ❑ A useful schedule representation for projects that are strung out horizontally in a line or stacked vertically is the line-of-balance schedule.
- ❑ Schedules have definite legal ramifications. The schedule is a double-edged sword, with the potential to support or damage the position of either party in a dispute.
- ❑ The manner in which a schedule is presented significantly affects its acceptance and utilization on the project.

Review Questions and Problems

1. A construction contract differs from contracts that we generally deal with that focus on an easily defined physical object because the physical object can be examined. How is the object of a construction contract defined?
2. What does the owner contribute to the project and what does the contractor contribute to the project?
3. For what type of project is a line-of-balance schedule particularly suited? Identify specific examples.
4. Describe the type of project for which impacted baseline techniques are most appropriate.
5. Describe the situation for which but-for or collapsed as-built analysis is most appropriate.
6. Many contracts require that the contractor prepare and present the construction schedule and receive the owner's approval before any contract payments are made. What are the three elements of the contractor's schedule that the owner will typically be interested in?
7. The construction schedule is the only project document that fully communicates the contractor's intentions for delivering the contracted scope of services over the full course of the project duration. What, specifically, does it accomplish?

10 Project Coordination

10.1 Introduction

In this chapter, we focus on managing the ongoing project. Chapter 6, "Production Planning," considered production planning with primary focus on preplanning prior to starting field operations. Chapter 7, "Managing Time," provided a general discussion of managing time for the project as a whole. Chapter 8, "Resource Management," addressed the management of resources, primarily workers, materials, and equipment. Chapter 9, "Project Scheduling Applications," broadened the focus to strategic scheduling topics, including consideration of claims and the legal ramifications of the schedule.

This chapter begins by looking at detailed schedules used by the field supervisor to plan crew work on specific activities in the near term. It then moves on to measurement and reporting of progress. Progress reporting provides the opportunity to analyze the current status of the project. Often, this will lead to rescheduling and corrective action to bring the project back within specified time parameters. This cycle of planning and executing activities, measuring and reporting progress, revising the plan based on current status, and updating the schedule is continued repetitively throughout the project.

Learning objectives for this chapter include:

- ❑ Recognize the role of the field supervisor in planning and executing day-to-day activities.

❏ Gain an introductory understanding of the application of lean principles to improve production.
❏ Learn about progress measurements and progress reporting.
❏ Understand the importance of continually updating the plan and schedule to reflect current job status.

10.2 Schedule Information on the Job

Although the project manager is responsible for the overall application and direction of the project time management system, field supervisors also play key roles in keeping the project on schedule. It is the field supervisors who put into action the plans and schedules devised by project management. Consequently, if project time management is to work, there must be some established means of communication between office and field. The project manager must keep field supervisors currently and accurately informed concerning the schedule of operations and the time status of the work.*

The information transfer cannot be accomplished merely by relaying piles of computer data to the supervisors, who have neither the time nor the inclination to search out the information relevant to their individual responsibilities. The project manager must provide concise, short-range schedule information that will meet the specific needs of each recipient. Limited to pertinent subject matter, and in a level of detail appropriate to the user of the information, the data provided should quickly and clearly communicate schedules, current time status, scheduling leeway, and trouble spots.

10.3 Detailed Schedules

Considerably more detail is needed by the immediate field supervisors than was included in Figure 7.10. The site superintendent on the highway bridge will require a substantially expanded time schedule, which will provide a day-to-day forecast of field operations. Detailed project schedules are customarily prepared using activities as a basis. Different time information can be provided with reference to the activities in such schedules, but early-start and early-finish dates are usual. Such a schedule is an optimistic one, and there are usually many instances where this schedule will not be met because of limited resources, inevitable delays, and time slippages.

---

*A much more detailed description of the responsibilities of the construction supervisor and the supervisor's relationship to the project manager is found in Jerald L. Rounds and Robert O. Segner, Jr., *Construction Supervision* (Hoboken, NJ: John Wiley & Sons, 2011).

10.3 Detailed Schedules

Nevertheless, a project schedule based on activity early starts is the one generally used to establish project time objectives. The detailed job schedule should also note which activities are critical and indicate values of free float. This information is of considerable significance to those who direct the work in the field. Labeling an activity as critical stresses its importance to lower-level managers. Knowledge of free float is also valuable in that it offers field supervisors the possibility of using such extra time to meet unexpected job conditions.

The release of total float values to field supervisors is not always considered to be good practice. Total float, unlike free float, usually is shared by a series of successive activities. The use of total float in conjunction with one activity normally has an effect on the float values of other activities. Total float data can be misleading to those who are involved with only restricted portions of a project. Only those responsible for the overall scheduling are in a position to evaluate and control the usage and allocation of total float. Free float is a readily usable commodity, while total float is a shared resource and must be carefully apportioned by knowledgeable planners through the give-and-take of negotiation.

On large projects, detailed schedules are prepared for each craft supervisor and subcontractor. Each of these schedules is the projected time program of the work for which that particular supervisor is responsible. The time spans of these schedules are limited, typically covering only the next two weeks to 30 days. The amount of time depends considerably on the nature of the work involved. It is seldom worthwhile to produce detailed job schedules for more than a month in advance. There is no need for longer periods if much of such a schedule may be rendered obsolete by subsequent changes and updates. Revised and updated schedules are issued as the work goes along. Tabular listings and computer-printed bar charts are the most common forms of short-term work schedules. The network diagram is certainly a necessary adjunct to the job schedules. Copies of the network can be provided to the various supervisors or may be kept in the field office.

The highway bridge is sufficiently small that only one detailed work schedule is required, and this is for the job superintendent. The time schedule for the first month of construction operations on the highway bridge is shown in Figure 10.1. "Move in" and the three following activities have been delayed by nine working days, as was explained in Section 7.21. As the work proceeds in the field, the job superintendent will enter the dates on which the activities actually start and finish into the schedule shown in Figure 10.1. Work schedules, like the one in the figure, are prepared for field supervisors and may or may not include information concerning the delivery of job materials. Practice varies in this regard, which does not mean that such information is unimportant to the supervisor, because this person is vitally concerned with what materials are ordered and when they will arrive. However, the scheduling of material deliveries is normally handled separately from the field operations. The project expeditor provides the project manager and site superintendent with periodic reports concerning the status of job materials, and this information is incorporated into the project weekly progress reports (see Section 10.15). Consequently, work schedules prepared for field supervisors often include only those activities that are physical parts of the work and omit material delivery information that is provided separately. As this is the procedure followed herein, no material delivery information is included in Figure 10.1.

10.4 Subcontractor Scheduling

Management control of subcontractors depends upon having them on the job when they are needed and ensuring that they accomplish their work in accordance with the established job schedule. There are three main considerations involved in carrying out this responsibility. First, the project manager should involve the major subcontractors during the planning and scheduling of the project. If a subcontractor participates in preparing

Detailed Schedule						
Project Highway Bridge				Prepared by G.A.S.		
Project No. 200808-05				Date June 14		

Activity	Activity Number	Scheduled Duration (Working Days)	Scheduled Start Date	Scheduled Completion Date	Free Float (Working Days)	Actual Start Date	Actual Completion Date
Move in	40	3	25-Jun	29-Jun	0		
Prefabricate abutment forms	80	3	30-Jun	2-Jul	10		
Excavate abutment #1	90	3	30-Jun	2-Jul	0		
Mobilize pile-driving rig	100	2	30-Jun	1-Jul	1		
Excavate abutment #2	120	2	6-Jul	7-Jul	1		
Drive piles, abutment #1	110	3	6-Jul	8-Jul	0		
Forms & rebar, footing #1	130	2	9-Jul	12-Jul	0		
Drive piles, abutment #2	140	3	9-Jul	13-Jul	0		
Pour footing #1	150	1	13-Jul	13-Jul	0		
Demobilize pile-driving rig	160	1	13-Jul	13-Jul	3		
Strip footing #1	170	1	14-Jul	14-Jul	0		
Forms & rebar, footing #2	190	2	15-Jul	16-Jul	0		
Pour footing #2	210	1	19-Jul	19-Jul	0		
Strip footing #2	230	1	20-Jul	20-Jul	8		
Forms & rebar, abutment #1	**240**	**4**	**20-Jul**	**23-Jul**	**0**		
Critical activities in bold							

Figure 10.1
Highway bridge, initial detailed schedule

the job schedule, that subcontractor may well have a better appreciation for the role played during construction and have a better understanding of how the work fits into the project plan as a whole. Often, this small matter is the difference between a subcontractor that facilitates project performance and one that hinders it.

The second consideration is the form and content of the subcontract agreement. A carefully written document with specific requirements in terms of submittals, approvals, and schedule often can strengthen the project manager's hand in obtaining subcontractor compliance. The timing of the issuance of subcontracts is not at issue here because a prime contractor normally will proceed with subcontract preparation immediately after the construction contract has been signed.

The third consideration is assuring that the subcontractors order their major materials in ample time to meet the construction schedule. Some general contractors find it advisable to monitor their subcontractors' material purchases. This is sometimes accomplished by including a subcontract requirement that the subcontractor submit unpriced copies of its purchase orders to the general contractor within 10 days after receipt of the subcontract. In this way, the general contractor can oversee the expediting of the subcontractors' materials along with its own.

In a manner similar to that for material ordering, the project manager must establish a lead-time schedule for notifying subcontractors when they must report to the project. These notification dates are established on the basis of the project work schedule and a lead time that may vary from one week to a month or more. Subcontractors must be given adequate time to plan their work and move onto the site. Notification dates can be listed in chronological order of their appearance. As each notice date arrives, the project manager advises the subcontractor in writing of its report date and follows this up with a telephone call. It is equally important for the project manager to assure that the job is completely ready for the subcontractor to move in and proceed with its work efficiently.

10.5 Activity Planning

In order to assure timely completion of scheduled project activities, a foreman is assigned the responsibility of planning for the activities' success. Approximately a week before an activity is scheduled, a number of details have to be checked. Are the drawings related to this work complete? Are any changes expected? Are the shop drawings approved? If the shop drawings are being prepared sequentially, are they sequenced to the order in which the work must be done in the field?

If this activity requires any special tools or equipment, now is the time to see that they will be available. If tools and equipment are to be shared with parallel activities, an understanding is required regarding who will get them and when. In the case of equipment failure, available options should be identified.

Are the required materials onsite? Do you know where they are stored and whether they are accessible? If they are not onsite, how reliable is a timely delivery of them? Are the materials subject to being of a wrong quality, quantity, or dimension? Are all the support materials on hand? Remember, "… for want of a nail, the shoe was lost. …"

Inspect the work space. Check on access for equipment and materials. Check that the area is uncluttered. If other work will be occurring concurrently, work out a plan to share the space. Check to see that the layout is correct and that support utilities, such as compressed air, electric power, and ventilation, are available. Make sure there is sufficient light to assure quality work. Check for safety hazards and provide for waste disposal.

10.6 The Last Planner Process©

In Section 6.4, planning was discussed in the context of lean construction principles. A key component of the implementation of lean construction in the field is found in the Last Planner Process©, which focuses on the reliability of foreman planning.

Reliable foreman planning, which forms the basis of assigning work to crews, is essential to the efficient execution of a construction operation. Activity planning is much more likely to be reliable if the plan exhibits four quality characteristics:

- *Definition:* The assignment is clearly defined, with a definite start and finish.
- *Soundness:* All required resources will be available when needed.
- *Sequence:* The activity fits into the sequence required by the project, or if several options in sequence are available, the best sequence is chosen to move the project forward.
- *Size:* The amount of work assigned to the crew is correct so that the work can be completed within the allotted time without overworking the crew or leaving slack.

Work assignments that are made taking into account these considerations will be much more likely to be completed as planned than if such considerations are not adhered to. This can be verified by measuring the effectiveness of planning, which is accomplished by comparing, at the end of the planning cycle, the number of activities that were indeed completed to the number of activities planned. On a traditional project, about 50 percent of the activities planned are actually completed. Research has shown that planning based on the four quality characteristics will generally increase the reliability by about 30 percent. This increase in reliability of planning results in many positive outcomes, such as lowering cost, increasing safety, improving the schedule, and, not least of all, improving trust on the job. Craft workers begin to trust the supervisor's planning. Other planners who depend on the plans of the reliable supervisor have more trust that the predecessor's plan will be completed, which helps them in making their plans. Project executives are impressed by the positive outcomes of reliable planning.

Another aspect of this process is to collect data on planning failures. When there is a planning failure (i.e., a planned activity is not finished), if a root cause analysis is carried out to find the reason(s) for the failure, chronic problems can be identified and steps can then be taken to eliminate these chronic problems. For example, if material supply is identified as a chronic problem, the project manager can work with the supplier to correct the problem. If timely and correct information is the problem, improvements can be made to the information dissemination process.

The project manager should take care in collecting data on planning failures, as this can create other problems. If planning failures are the result of a project executive's impeding information flow or an owner's representative making delayed or poor decisions, dealing with such problems can be very tricky and require exceptional diplomacy.

10.7 Production Checklists

Each construction operation involves a large number of important details. With the attendant confusion, noise, and interruptions of a typical project, it is easy to overlook important details. These omissions lead to rework for which there is neither time nor money in the budget. Well-thought-out checklists can provide an excellent way to keep track of the myriad of details that accompany each construction operation.

Checklists are valuable for many phases of construction work. Examples include:

- Permitting process
- Preconstruction site inspection
- Layout and excavation
- Utility location
- Concrete formwork
- Concrete flat work
- Piping

Figure 10.2 is an example of a concrete wall and column checklist.

Concrete Wall and Column Checklist

- ☐ Shop drawings are approved and onsite
- ☐ Concrete ordered is correct quantity and type
- ☐ Forms are clean and oiled
- ☐ Forms have proper chamfers and architectural features
- ☐ Previous pour is clean and prepared
- ☐ Electrical conduits, boxes, etc., are in place
- ☐ Plumbing cans and chases are in proper location
- ☐ Weld plates and other imbedded items are in proper location
- ☐ Check all measurements and tolerances
- ☐ Top of pour clearly marked
- ☐ Bulkheads are secure
- ☐ Columns are properly oriented
- ☐ Forms are plumb and braced
- ☐ Blockouts are properly located
- ☐ Work surfaces are secure and have handrails
- ☐ Verify special finishes
- ☐ Finish and cure exposed surfaces
- ☐ Provisions made for rain, hot or cold weather protection
- ☐ Clear access for delivery of concrete to site
- ☐ Inspectors notified of pour
- ☐ Testing equipment is available

Figure 10.2
Concrete wall and column checklist

Good checklists are the result of years of experience and are developed by contractors over a long period of time. Mistakes and omissions occur far too often, but good construction practice builds on these mistakes by adding them one by one to the relevant checklist, thus decreasing the chance of a mistake being repeated.

It is easy to rationalize that construction supervisors do the same jobs over and over and therefore have no need for a written checklist. It is fair to point out, however, that airline pilots also do repetitive operations but never start an engine without referring to a written checklist. Mistakes on a construction project can endanger many lives. The concrete wall checklist shown in Figure 10.2 provides the experienced concrete foreman with a comprehensive list of items to be checked off as work progresses. For the inexperienced foreman, the checklist provides the insight, skill, knowledge, judgment, and background that would otherwise be provided by more experienced people.

Well-thought-out checklists help prevent mistakes and omissions. They are also a valuable and time-saving aid to the quality control person checking the work.

10.8 Look-Ahead Schedules

There are two types of look-ahead schedules. First, from a project scheduling perspective, the schedule is detailed in a four- to six-week rolling schedule. This type of short-term schedule was briefly discussed in Section 9.3.

Second, production short-term or look-ahead schedules are created by foremen and show exactly how the work will be accomplished. Each foreman is assigned a specific task from the project schedule. Typically, a look-ahead schedule is a rolling schedule and will cover the next 7 to 10 days. The next example from the highway bridge illustrates a typical look-ahead schedule.

A carpenter foreman is assigned the task of forming abutment #1. The updated project schedule shows this work to begin August 2, which is a Monday. This is a critical activity, and it appears that the forms will be ready as scheduled. This foreman is also responsible for stripping the abutment forms from abutment #1, so the start of work will not have to be coordinated with any other foreman. The first step is to check the budget for the man-hours available and the schedule to see the time allowed.

The budget, Figure 3.10a, shows that there are 1,810 square feet of forms to place and 3,656 labor dollars to do the work. Since the average hourly labor cost for each member of the crew is $29, the foreman calculates that he has about 126 man-hours to get the work done. According to the schedule, the work must be completed in three days.

Planning this operation can be done with a simple precedence diagram or with a bar chart. Figure 10.3 is a bar chart showing each step of the process and assigning crew members to each task. The first item in the bar chart—"Clean, repair, and oil forms"—is charged to the previous account,

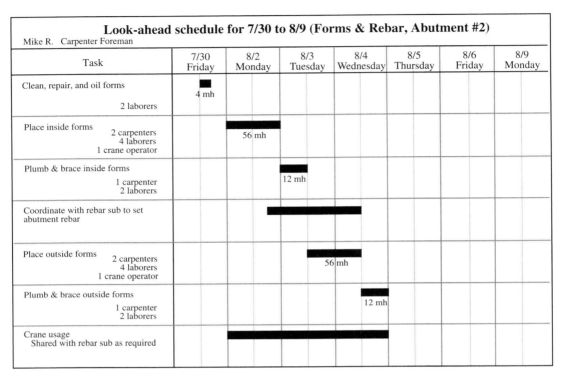

Figure 10.3
Highway bridge, seven-day look-ahead schedule

stripping abutment forms. The balance of the items in the bar chart shows a total of 136 man-hours, just slightly above the budget. The work is scheduled to take three days, keeping the project on schedule.

10.9 Planning the Paperwork

There is a staggering amount of paperwork associated with a construction project. Building a custom home generates one or two file boxes of documents. When the highway bridge is complete, the documents will fill a four-drawer filing cabinet. The paperwork for a high-rise building will fill an office. Keeping track of all these documents requires planning and diligence. The Construction Specifications Institute MasterFormat© provides a numbering system to assist in organizing correspondence, requests for information, change orders, project meeting minutes, progress payments, and claims for the commercial building segment of the industry. Several computer programs are available that will track submittals, change orders, and other time-sensitive documents. Whatever system is adopted to bring order to the documents, serious planning at the beginning of the project will save countless hours later.

The beginning of the project is the time to plan for as-built drawings and warranty considerations. Making sure that each and every deviation from the original plans is documented on the as-built drawings is difficult. Leaving these changes to the end of the job is time consuming and leads to inaccuracies. Warranties often require that an agent of the manufacturer be present during installation. Each of these requirements needs to be identified and a reminder placed in the project schedule or checklist.

A successful technique used by some contractors to handle weekly project meetings is based on consecutively numbering each meeting. In the first project meeting, all the items discussed are classified as new business and numbered 1.1, 1.2, and so on. At the next project meeting, old business is covered first and involves resolution of items 1.1, 1.2, and so on. As items are resolved, they are dropped from the list. New business items in the second project meeting are numbered 2.1, 2.2, and so on. In this way, every project meeting problem is numbered. Each number remains as old business until resolved. Not only does this system keep track of each problem, but the length of time the problem remains unresolved is indicated by its number.

10.10 Putting the Plans on Paper

Just as project planning results in a documented project network, complete with resources and job costs, production planning has to be reduced to paper as well. Sketches, lists, and narratives record the production plans. This information needs to be organized in such a way that the plans can be found at the proper time. When a delivery of transformers is due, the sketch showing where they are to be stored must be readily available. If you cannot find the record of a carefully formulated plan when it is needed, all of the planning effort spent in developing it is wasted. Putting the planning to paper and communicating it to the owner and the project team is further discussed later in this chapter.

10.11 Progress Measurement

To make periodic measurements of progress in the field, network activities serve as exceptionally convenient packages of work. The advancement of an activity in progress can be expressed in different ways. Three commonly used methods are:

1. *Estimated number* of working days remaining to complete the activity.
2. *Estimated percentage* completion of the activity in terms of time.
3. *Quantities* of work units put into place.

How a contractor chooses to express activity completion depends on the type of work involved and whether these same data are also used to check field costs. However, the number of working days remaining to finish an activity is fundamental to the workings of project time management. Progress data in the other forms are readily converted into days to complete by using these relationships:

$$\text{Working days to complete} = d(1 - P/100)$$

$$\text{Working days to complete} = d(1 - W/T)$$

where
d = total activity duration in working days
P = estimated percentage of completion
W = number of work units put into place
T = total number of work units associated with the activity

Most scheduling software allows the user to input progress in any of these formats and makes the conversion automatically. Inherent in the relationships is the assumption of a straight-line variation between time and work accomplishment. If an activity requires a total of four days for its performance, it is assumed that one-quarter of the work quantities to be installed will be completed each day. This is normally an acceptable assumption if an activity is limited in scope. A more realistic relationship between time and work accomplishment may have to be used where activities are of substantial extent.

As with all computer analysis, a time management system is no better than the quality of the input information. If progress reports from the field are inaccurate, then management decisions will be made on the basis of fictitious situations. It is very important that progress measurements be done conscientiously and with reasonable accuracy. Management action must be based on what actually happened rather than on what should have happened. The person responsible for progress reporting must recognize the importance of factual and correct determinations. Project progress records often are important in settling later disputes regarding project delays. In fact, progress reports can form the basis for claims and litigation.

The conclusion of a given activity must be viewed in terms of its substantial completion rather than its absolute finalization. As work progresses in the field, there are many items that, at least temporarily, are not completely finished, as, for example, small deficiencies that are remedied subsequently when the opportunity presents itself. Progress reporting must make allowances for such minor shortcomings. Therefore, finish dates generally are recorded for substantial completion rather than the technical completion of activities.

10.12 Progress Reporting

How often field progress should be measured and evaluated depends on the degree of time control perceived to be desirable and feasible for the particular work involved. Within limits, the greater the frequency of feedback and response, the more likely it is that the project time objectives will be met. However, this rule must be tempered with other considerations. For example, some kind of balance must be struck between the cost and effort involved and the management benefit gained. Another consideration is that the same field progress report often serves for field cost management (Chapter 11, "Project Cost System") as well as time management. Consequently, the cycle times for both management applications frequently are matched.

Fast-paced projects, using multiple shifts, may demand daily progress reports. Large-scale jobs such as earth dams, which involve a limited range of work items, may use a reporting frequency of a monthly basis or even longer. Weekly progress reporting is typical on commercial building projects where a weekly payroll cycle is also the norm. It is difficult to generalize because management control must be consonant with project characteristics and peculiarities. On projects of the size, duration, and type of the highway bridge, progress reporting probably would be done on a weekly basis. This is the premise here. It should be noted that although formal progress reporting may be made on a weekly or monthly basis, critical operations may need to be tracked with a quicker and less formal form of cost and production control and on a more frequent basis.

The project manager must see that progress measurement and reporting are done properly and that the progress information receives prompt management review and analysis. A standard procedure for collecting and transmitting the weekly progress data must be established. Progress measurement requires direct visual observation in the field by someone familiar with the type of work involved. This may include a physical count of work units in place or may be reduced to evaluating quantities of work accomplished from the project drawings. At times, suitable measurement can come from delivery tickets for materials like concrete or load slips for earth moving. On many projects, the project manager personally carries out the measurement and reporting functions. Otherwise, a staff member, such as the field engineer, will perform these duties. In any event, an independent review of work accomplishment is preferred. Field supervisors are not usually best suited for progress reporting because they are very busy people who are not normally inclined toward handling paperwork of this kind. In addition, a field supervisor may fail to report unfavorable progress in the hopes of working problems out later or may withhold quantities for future use on a less fortunate day.

10.13 Bar Charts

On receipt of progress measurements from the field, management must compare the information with the latest project schedule. This can be done in different ways, depending on management preferences and procedures. For example, a tabular listing can be prepared that shows the scheduled start and finish dates and the actual start and finish dates for each activity. Although progress data in this format may be useful for some purposes, it is not usually the best medium for making a comprehensive evaluation of the current time status of construction operations.

For the day-to-day time management of a project, some form of graphic display is effective and convenient. A widely used method for recording job progress is the bar chart, the general characteristics of which were discussed in Section 5.29. Several different styles and conventions are used in drafting project bar charts. Two of the most common are described in subsequent sections in this chapter, Sections 10.14 and 10.17, and both are widely used by the construction industry. One procedure is used to depict progress on the highway bridge as of July 14, and the other method is the basis for the July 21 progress report. For obvious reasons, the contractor would use the same bar-charting procedure throughout the construction period of a given project, though the two different approaches are illustrated here on the same example project so the workings of each can be observed.

The bar chart is an excellent medium for recording progress information and portraying the current time status of individual activities or other project segments. However, it is not a proper tool for evaluating the overall time status of the project or for planning corrective measures when the work falls behind schedule. Only the project network can perform this function adequately.

The manual preparation and updating of project bar charts can involve considerable time and expense. Software is widely used to print out updated bar chart schedules. When project outlines are used for a work breakdown structure, bar charts of differing levels of detail can be produced with ease.

10.14 Highway Bridge as of July 14

Figure 10.4 has been prepared to show progress on the highway bridge up through the week ending Wednesday, July 14 (working day 22). This bar chart shows the scheduled and actual beginning and completion dates for each activity up through July 14. Plotted progress data have come from past field measurements that were made and reported at weekly intervals. Thus, this bar chart is updated once a week and shows the current status of each activity and how its accomplishment compares with the schedule. Practice varies concerning the entry of material deliveries and other resource information on bar charts. In this book, the bar charts portray the

Construction Progress Chart

Project: Highway Bridge
Date: July 14
Job No.: 200808-05

Activity	Activity number		Working Days — June / July
Move in	40	Scheduled / Actual	
Prefabricate abutment forms	80	Scheduled / Actual	
Excavate abutment #1	90	Scheduled / Actual	
Mobilize pile-driving rig	100	Scheduled / Actual	
Excavate abutment #2	120	Scheduled / Actual	
Drive piles, abutment #1	110	Scheduled / Actual	
Forms & rebar, footing #1	130	Scheduled / Actual	
Drive piles, abutment #2	140	Scheduled / Actual	
Demobilize pile-driving rig	160	Scheduled / Actual	
Pour footing #1	150	Scheduled / Actual	
Strip footing #1	170	Scheduled / Actual	
Forms & rebar, footing #2	190	Scheduled / Actual	
Pour footing #2	210	Scheduled / Actual	
Forms & rebar, abutment #1	**180**	**Scheduled / Actual**	
Strip footing #2	230	Scheduled / Actual	
Critical activities in bold			

Figure 10.4
Highway bridge, bar chart as of July 14

255

advancement of the physical aspects of the work only and do not include dates for material deliveries or the availability of other resources.

In Figure 10.4, the upper row of shaded ovals for each activity extends between its scheduled start and finish dates. As has been discussed, bar charts customarily are made on an early-start basis; however, exceptions to this general rule on the highway bridge are the first four activities listed in the figure. As previously explained, the scheduled beginnings of these four activities have been delayed by nine working days beyond their early-start times. The white ovals that extend to the right of the noncritical activities represent the total float.

The lower row of black ovals opposite certain activities shows the time period during which work actually progressed on these activities. The start of each of these rows is plotted at the date on which work commenced. The line is then plotted to the right at weekly intervals, either to the current date or terminating at the day of completion. For each activity, the cumulative percentage completion is entered at the right end of the lower line until it is completed. The numbers 0 and 100 on the lower lines show the actual time periods required to progress from zero to 100 percent completion. If an activity is in process at the time of a progress measurement, but not yet completed, the actual percentage of completion is marked. An example of this is activity 140, which was only 30 percent completed on July 14, the date of the last progress report. Comparison of the actual and planned percentage completions as of a given date reveals the time status of an activity then in progress. One advantage of the form of progress recording used in Figure 10.4 is that it provides an exact historical record of calendar times when job activities were actually in process.

An examination of the figure will provide a quick and informative review of how the job progressed as of July 14. Activity 40, "Move in," was accomplished exactly as it had been scheduled. Activity 80, "Prefabricate abutment forms," took one working day longer than planned and finished six days late. However, it was completed within its free float period and creates no problem. The experience with activity 120, "Excavate abutment #2," was much the same. Progress on activity 140, "Drive piles, abutment #2," is far behind schedule. This activity should have been completed by July 13. Instead, it was only 30 percent finished as of July 14. Figure 10.4 shows that activities 160 and 170 should have been completed by the July 14 cutoff date, but they were not even started.

10.15 Weekly Progress Reports

The reporting of work advancement in the field is accomplished by listing those activities that started, finished, or were in progress during the week just ended and indicating their stages of completion at the cutoff date. Also noted are the dates on which activities either started or fin-

10.15 Weekly Progress Reports

	Weekly Progress Report				
Project	Highway Bridge		Week ending	Wed., 21-Jul (Working day 27)	
Job No.	200808-05		Prepared by	GAS	
Activity	Activity Number	Date Started	Date Completed	Percent Complete	Working Days to Complete
Fabricate & deliver abutment rebar	60	–	15-Jul	100	0
Fabricate & deliver deck rebar	65	–	–	–	13
Drive piles, abutment #2	140	12-Jul	–	80	2
Strip footing #1	170	15-Jul	15-Jul	100	0
Forms & rebar, abutment #1	180	16-Jul	21-Jul	100	0
Fabricate & deliver girders	260	–	–	–	10

Figure 10.5
Highway bridge, weekly progress report

ished during the reporting period. The field superintendent enters these actual start and finish dates into the detailed field schedule, Figure 10.1, as the work progresses. There is no need to list activities previously completed or not yet started. Figure 10.5 is the weekly progress report as of Wednesday afternoon, July 21 (working day 27) on the highway bridge. Because there normally are a limited number of activities in process during any given week, progress reporting is seldom a burdensome task, although some time is required if the measurements are to be properly determined. A file of the weekly progress reports can be maintained as a historical job record.

To report the overall project standing, the weekly progress reports must include procurement and delivery information as well as the measurements of physical progress. Each time field progress is determined, a check is made with the material expeditor (see Section 8.17) to ascertain the delivery or availability status of materials, subcontractors, and construction equipment. Data pertaining to these resources are entered into the weekly progress report, as shown in Figure 10.5.

Some explanation is required concerning the first two entries in the figure. These are the result of a change in delivery of the reinforcing steel for the abutments and bridge deck. The steel fabricator proposed an earlier rebar delivery for the abutments if the deck rebar could be delivered somewhat later. The abutment rebar (activity 60) was received on July 15 (originally promised July 19). A new activity 65, "Fabricate and deliver deck rebar," has been established with delivery promised August 9 (working day 40), or 13 working days from the date of the weekly progress report in Figure 10.5.

Progress reporting on a weekly basis is a common procedure in the construction industry and is being used herein for discussion purposes. This means that some day of the week must be chosen as the cutoff date. The day selected often is picked so that the same weekly progress measurement can serve for both time management and labor cost accounting purposes (Chapter 11, "Project Cost System"). This is why Wednesdays are used here as progress measurement days. Payday in the construction industry is often on Friday, and the daily time cards serve both payroll and labor cost accounting purposes. When construction craft workers receive their weekly wages on Friday afternoon, they are commonly paid up through the previous Wednesday. This gives contractors time to prepare and distribute their job payrolls. As a result, contractors frequently use Wednesdays as the cutoff day for weekly payrolls and for their labor cost accounting as well. Measuring field progress on Wednesday afternoons enables a single weekly measurement of field progress to serve for both time management and cost control. It must be understood that there are many exceptions to this procedure and that there is nothing sacred about using Wednesdays for progress measurement purposes. The management principles discussed herein do not in any way depend on the day of the week used for cutoff purposes.

10.16 Field Progress Narrative

The weekly field progress report usually is accompanied by a brief narrative discussion of salient project features. For example, it could include a general statement about the time status of the job, a discussion of critical activities or low-float activities (also referred to as subcritical activities) now in difficulty or behind schedule, a description of potential trouble spots, and a notation of project areas that are going along exceptionally well. As a specific illustration of this point, Figure 10.4 shows that the pile driving for abutment #2 is in trouble and behind schedule. The weekly progress report that conveyed this information would explain that the difficulty was caused by the unsuspected presence of boulders and tightly packed gravel. An assessment of the problem would be given, including a forecast of how much delay ultimately would be involved.

10.17 July 21 Status of Highway Bridge

The progress information contained in Figure 10.5 is now plotted on the project bar chart in Figure 10.6. This bar chart is updated through July 21. Recall that bar charts can differ with regard to how actual progress is recorded. One procedure has been discussed in conjunction with Figure 10.4. Figure 10.6 depicts progress data in a completely different fashion.

In Figure 10.6, the early-start schedule for each activity is plotted to a horizontal time scale as before. The schedule bar for an activity appears as a narrow horizontal white box extending from its scheduled start-to-finish date. The dashed line at the end of each activity represents that activity's

10.17 July 21 Status of Highway Bridge

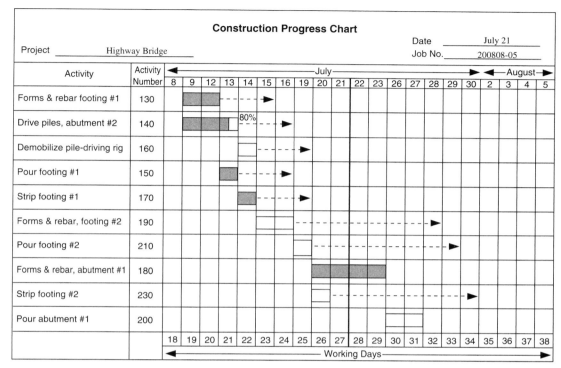

Figure 10.6
Highway bridge, bar chart as of July 21

total float. When using this second convention of recording actual work progress on a bar chart, it is assumed that the physical progress of an activity varies linearly with time. Actual progress, however, is not plotted to the established time scale as it was previously. Rather, the progress of an activity is indicated as a shaded portion of its schedule bar. The usual procedure is to shade in a length of the schedule bar in direct proportion to the physical advancement of the work. For example, activity 140 was reported in Figure 10.5 to be 80 percent complete as of the afternoon of July 21. In Figure 10.6, 80 percent of the length of the schedule bar for activity 140 has been shaded. The percentage of completion also can be entered above the bar if desired. If, at the time a progress report is made, an activity has been completed, the entire length of the bar is shaded.

The use of this mode of progress recording makes it possible to determine at a glance which activities were ahead of schedule and which were behind as of the cutoff date, and by approximately how much. Figure 10.6 has been updated as of July 21, and a heavy vertical line has been drawn at this date. Any activity that has an unshaded bar to the left of the July 21 line was behind schedule as of that time. Activities 140, 160, 190, 210, and

230 are all in this category. Any activity that has a shaded bar to the right of the July 21 line was ahead of schedule. Activity 180 is an instance of this. The form of progress recording used in Figure 10.6 will not provide any historical record of true activity periods, unless actual start and finish dates are jotted onto the bar chart as the work progresses.

Reference to Figure 10.6 shows that activity 140, "Drive piles, abutment #2," was only 80 percent completed as of July 21 and is now more than six working days behind schedule. This has prevented work from starting on footing #2. Figure 5.3a shows that activities 190, 210, and 230 have eight days of float associated with them. Whether this will be sufficient to absorb the delay in pile driving will be indicated by the next network update (see Section 10.21). As indicated by Figure 10.6, activity 180, "Forms and rebar, abutment #1," was finished two days ahead of schedule. This was made possible by the delivery of abutment rebar (activity 60) two working days earlier than had been originally expected.

10.18 Progress Analysis

The analysis of job progress is concerned primarily with determining the effect of this latest information on the project completion date and any intermediate time goals that have been established. The successful attainment of set time objectives is, after all, the essential purpose of the time management system. Considerable attention has been given to the fact that the length of the critical path determines the time required to reach a given network event. Practically, then, the analysis of progress data is concerned essentially with determining, as closely as possible, the present length and location of the applicable critical path.

When a progress report is received from the field, the status of the currently identified critical activities is probably the first item noted. This is a quick and simple check. If these critical activities have been accomplished by their scheduled finish times, there is no problem with the project time goal insofar as the current critical path is concerned. If a critical activity has been started late, a setback in project completion is likely, unless the delay can be somehow made up by the time the activity is completed. If a critical activity has been finished late, the completion date is delayed commensurately.

The next step in the analysis is to check the possibility that a new critical path has been formed. This can be done by subjecting the noncritical activities to either of two different checks. One of these checks involves using either a late-start activity sort or a late-finish activity sort. A late-start sort is an activity listing ordered according to the latest allowable starting times (LS). A late-finish sort is a similar listing in terms of the latest allowable finish dates (LF). These sorts can be obtained manually or by computer. If an activity has missed its LS date, the project is behind, according to the current action plan. If an activity did not finish by its LF date, there is now a new critical path, and the job is delayed by the amount that this activity completion trailed the LF time.

Another way to make much the same check involves the use of total float (TF) values. If the completion of an activity is delayed beyond its early-finish (EF) time, its TF is reduced by the same amount. If the delay is equal to the TF, a new critical path is formed. If more than the TF is consumed, there is a new critical path longer than that of the original critical path, and project completion is automatically set back by the amount of the overrun. The discussion that follows illustrates this point.

On the highway bridge as of July 21, activity 140, "Drive piles, abutment #2," is obviously behind schedule, and it is now necessary to determine what effect this is having on the critical path location and job duration. Reference to Figure 5.3a shows that activity 140 has an LF value of 24. Figure 10.5, prepared as of working day 27, discloses that this activity is now expected to finish in 2 working days, or as of day 29. Consequently, activity 140 is now critical, and there is a new critical path through the network. Noting in Figure 5.3a that activity 140 originally had a total float of 3 and learning that it will now finish 8 days behind its EF value of 21 provides the same information. The new critical path is 5 days longer than the previous one (64 working days), so project completion is now delayed by 5 days, making a construction period (not including contingency) of 69 working days. Since the original project completion time was determined to be 70 days, including 6 days of contingency, there still remains 1 day of contingency available.

At this point, it must be noted that the preceding discussion has been limited to the effect of only activity 140 on the duration of the highway bridge. The overall time status of the project cannot be known until a complete updating calculation has been made as of the date of the last progress report (July 21). This will be done in Section 10.22.

10.19 Schedule Analysis to Determine Project Delays

It was stated previously that the project schedule is updated weekly. When a delay occurs on a project, the delay is reflected in the updated schedule by adding more time to the original activities or through the addition of new activities. When the contractor is responsible for the delay, the lost time must be made up at the contractor's expense. When the owner causes the delay, the contractor must request additional contract time. It is the contractor's responsibility to notify the owner, in writing, of the delay. The request for additional time must be supported with evidence satisfactory to the owner. Although simple in concept, this process can become very complicated and burdensome.

A substantiated request for extra time is made using a network schedule updated to a point in time immediately prior to the commencement of the delay. A forward- and backward-pass calculation shows the unimpacted completion date and the network critical path. Then the delaying event is added to the schedule. Often, new activities are added or the durations of existing activities are modified. Again, a forward- and backward-pass calculation is made and the project duration is determined. In this way, any resulting change in the critical path or project duration is the sole consequence of the delaying event. When done properly, this impacted schedule provides compelling evidence of the delay and its consequence on the project. The next example illustrates this procedure.

Assume that on the afternoon of July 14, day 22, the status of the highway bridge project is as indicated in Figure 10.4. However, the trouble with the pile driving on abutment #2, activity 140, has become much more serious than described in Section 10.16, and the pile driving is making little progress. During the afternoon of July 14, the owner's engineer directed the contractor to stop work on the pilings for abutment #2 until the problem could be studied. All other work can proceed as planned. As shown in Figure 10.7, the contractor now inserts a milestone entitled "Stop order on pile driving, abutment #2" and adds two new activities. Activity 155 shows that the engineer required five days to devise a suitable solution to the pile-driving problem. Upon receipt of this solution in the form of an engineer's directive, the contractor removed the troublesome material and replaced it with compacted backfill. The required pilings were then driven, and the entire process required a total of eight working days to complete as shown by activity 165. This information is shown in Figure 10.7, and the forward-pass computations are completed to project completion on this basis. Figure 10.7 shows that the project now has an anticipated finish time of 77 working days, six days later than indicated by Chart 10.1 on the companion website. It is to be noted that Figure 10.7 clearly shows the cause of the delay and its effect on the time for project completion.

Should the owner authorize additional project time based on the evidence in the figure, a contract change order is issued and the matter of the delay is concluded. If the owner rejects the delay and refuses to issue a change order, the contractor must give notice of a formal claim. Most contracts require the contractor to continue working, leaving the claim to be adjudicated at the end of the job.

The contractor is now faced with a time dilemma. The contractor can add the unauthorized extra time to the schedule, continue working, and hope to prevail with the claim for extra time at the end of the project. Alternatively, the contractor can accelerate the project at its own expense and recover the lost time. The second alternative is the safest and the one chosen by many contractors in this situation. At this point, the working relationship between the contractor and owner may become taut.

As noted in Chapter 7, "Managing Time," acceleration of project activities requires additional resources or overtime work to get the project back on schedule. The cost of acceleration must be documented carefully, added

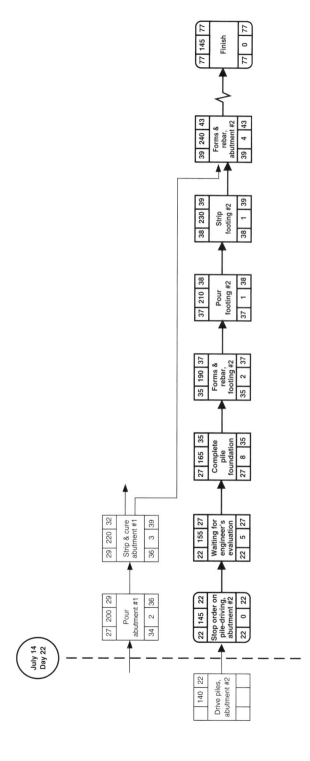

Figure 10.7
Revised network problem with activity 140

264 10 Project Coordination

to the cost of the original change, and claimed at the end of the project. Figure 10.8 is a decision flowchart showing the process just described.

Regardless of whether the owner grants extra time for a delay or not, the updated schedule showing the effect of the delay, and prepared

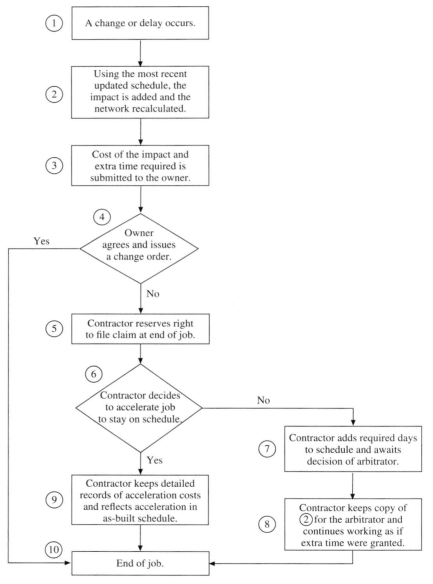

Figure 10.8
Change order process flowchart

contemporaneously with the delay, is the best method for proving a delay claim that will prevail.

At times, delays are difficult to identify and quantify. For instance, the pile-driving problem described previously occurs and the work stops while the engineer decides on a solution. Several days go by and nothing happens. At this point, the schedule needs to be updated to show that the information is not forthcoming. A new activity entitled "Waiting for engineer's decision on …" is appropriate in this case, with a duration that continues to increase with each update. A recalculation of the project duration with each update shows the impact of the engineer's delay to the project.

Whenever possible, the effects of a change should be identified, quantified, and resolved concurrently by the people currently involved with the project. When delay decisions are deferred, memories fade, project personnel change, and the dispute becomes personalized. If resolution is delayed, eventually the dispute is adjudicated, attorneys are involved, the cost of the claim increases dramatically, and a fair and equitable resolution is more remote.

When no possibility of a timely analysis exists, the contractor must, after the fact, utilize an analytical scheduling technique to re-create the project on paper in order to prove entitlement. Impacted baseline and collapsed as-built schedules, described in Chapter 9, "Project Scheduling Applications," are two of the most common techniques for such analysis.

10.20 Corrective Action

After each progress report has been analyzed, a decision must be made concerning what corrective action, if any, is required. Small delays, actual or potential, usually will not require any particular corrective action, provided that the usual contingency allowance has been included in the project duration. With a contingency of only six working days included in the highway bridge schedule, the trouble with the driving of piles for abutment #2 will cause considerable concern. Where time slippages of the approximate length of the network time contingency become involved, project management must give serious consideration to taking some form of corrective action. In general, remedial steps are indicated when situations like these arise:

- Activities begin to fall appreciably behind their late start or late finish dates.
- Substantial delays of resource availability are indicated.
- The time durations of future activities appear to have been materially underestimated.
- Logic changes in work yet to be performed have become necessary.

Where some form of corrective action is necessary, it must be done promptly so that the desired objective is achieved.

As with project time reduction discussed in Chapter 7, "Managing Time," corrective action here is focused on reducing the duration of the critical path. The critical path, or paths, can be shortened by compressing individual activities or by localized reworking of the network logic. As before, this often involves testing possible alternative courses of action. If there has been no recent updating of network calculations, the analysis of the weekly progress reports gives management only a general idea of the time status of the job. Consequently, the efficacy of a given corrective measure can be known only approximately unless it is simulated by incorporating it into a network update.

The entire process of project time management can be enhanced considerably by scheduling periodic job progress meetings. The objective of such meetings is to provide the opportunity for all stakeholders to plan and coordinate the next period. Held weekly, biweekly, or monthly, all entities that could be affected by the work of the upcoming period should be represented, including project management, field supervision, major subcontractors, material suppliers, owner representatives, designers, and other parties, as appropriate.

Although such meetings can and do serve as a forum for a variety of job topics, the time status of construction operations is always a major consideration at such sessions. The meeting begins with a brief review of progress since the previous meeting, where the data referred to in Section 10.6 on planning reliability and root causes of planning failures can be developed. Then the meeting turns to detailed planning for the upcoming period, focusing on coordination to minimize interferences between trades and identification and removal of impediments that might delay progress. Face-to-face discussions can be very productive in eliciting ideas and obtaining cooperation from those who play major roles in keeping the work on schedule.

Preparation for the meeting should include development of an agenda and a reminder to all regular participants to attend. An invitation to join the meeting should be sent to participants not normally attending but whose involvement is important to the planning for this period. After the meeting, minutes should be prepared that summarize discussion and commitments made during the meeting.

At such a job progress meeting on the highway bridge, it has been decided that some form of corrective action must be devised to remedy the job delay caused by the difficulty with activity 140. Discussion reveals that considerable time recovery could be possible by making a network logic change involving construction equipment. This change would involve removing activity 160, "Demobilize pile-driving rig," as a predecessor to activity 180, "Forms & rebar, abutment #1." Discussion in Chapter 5, "Project Scheduling Concepts," required this sequence so that a crane would be available for the placing of forms and rebar for abutment #1. The decision is now to eliminate this equipment dependency by providing another, somewhat smaller crane for the forming and placing of rebar at abutment #1. This would remove activity 160 as a necessary prerequisite for activity 180. For now, activity 160 will be made a prerequisite for activity 240, "Forms & rebar abutment #2." This logic change can be effected with little additional expense. The effect of this job logic change will be determined in Section 10.22.

10.21 Network Updating

As construction proceeds, diversions from the established plan and schedule inevitably occur. Unforeseen job circumstances result in changes in activity durations, activity delays, and changes in project logic. Remedial actions such as resource leveling and time expediting can produce similar effects. As such deviations occur and accumulate, the true job status diverges further and further from that indicated by the programmed plan and schedule. At intervals, therefore, it becomes necessary to incorporate the changes and deviations into the working operational program if it is to continue to provide realistic management guidance. This is accomplished by a procedure called *network updating*.

The basic objective of an update is to reschedule the work yet to be done using the current project status as a starting point. Updating reveals the current time posture of the job, indicates whether expediting actions are in order, and provides guidance concerning how best to keep the job on schedule. An update is also very valuable in testing the effectiveness of proposed time-recovery measures.

Updating involves making necessary network corrections and recalculating activity and float times. It is concerned entirely with determining the effect of schedule deviations and plan changes on the portion of the project yet to be completed. Included here are both unexpected departures from the program and those corrective actions initiated to remedy specific time progress and resource availability problems.

From the viewpoint of time management effectiveness only, an update as of each weekly progress report could be advantageous. It would keep job management continuously up-to-date on the time status of the work and would assure prompt and informed remedial action when needed. However, such updates could involve considerable effort and expense. No definite rule exists concerning the timing of network updates. The need for recalculation depends more on the seriousness of the plan and schedule deviations than on their number. There is a point of diminishing returns in retaining an outdated plan and schedule. Attempting to make the project fit an obsolete program is an exercise in futility and literally does more harm than good. Time control, to be optimally effective, must be based on a correct and current job model. What is important is not how often the network is recalculated but how well the plan and schedule continue to fit the actual conduct of the work.

Updating calculations normally are accomplished with scheduling software, although manual calculations are discussed in the next section (Section 10.22) to provide a complete understanding of the process involved. Before an update can be made, the network must be corrected to reflect the latest information concerning the logic and durations of all work yet to be performed. This is an important point. The activities that have been completed are now history, and they are immutable. To determine the time condition of the remaining activities, the new calculations must start from the current project status as of the designated cutoff or data date.

To make a network update, information as of the cutoff date is required concerning the work that has yet to be completed for:

❏ New activities that must be added to the network
❏ Existing activities that are to be deleted
❏ Changes in job logic
❏ Changes in original material delivery or other resource availability dates
❏ Estimated times to complete all activities presently under way but not yet completed
❏ Changes in estimated activity durations
❏ Changes in the scope of the work

With regard to information needed for a project update, subjective input from the field supervisors and job expeditor can be especially valuable. For instance, the project supervisor can furnish revised information about future progress based on recent job experience. This information may relate to expected activity durations or to future logic changes. The expeditor may be able to input information regarding labor disputes, business disruptions, and other pertinent factors pertaining to supplier plants and transportation facilities. It may be possible to combine material shipments for one job with those of another to expedite deliveries—conditions that were unforeseeable when the original plan was devised. All this information needs to be reflected in the updated schedule.

Figure 10.9 summarizes all the information needed for updating the highway bridge as of the afternoon of July 21. The first four items have been discussed previously in conjunction with the weekly progress report in Figure 10.5. Item 5 is a revision upward in the estimated duration of a future activity. Item 6 was discussed in Section 10.20.

10.22 Updating Calculations Manually

Item	Network Update Information as of July 21 (working day 27)
1	Add activity 65, "Fabricate & deliver deck rebar."
2	Anticipated delivery date of deck rebar is August 9 (working day 40). Activity 65 requires 40 − 27 = 13 working days to complete.
3	The estimated time to complete activity 140 is 2 working days. Driving rate is very slow due to boulders and tightly packed gravel.
4	Anticipated delivery date for the girders is August 4 (working day 37). Activity 260 requires 37 − 27 = 10 working days to complete.
5	The estimated time to accomplish activity 330, "Deck forms & rebar," has been revised upward from 4 to 6 working days.
6	A 25-ton crane was brought in and used to place the forms and rebar in activity 180. This removed the dependency between activity 160 and 180. A dependency was established between activity 160 and 240.

Figure 10.9
Highway bridge, network update information

10.22 Updating Calculations Manually

To perform a manual update on the highway bridge as of July 21 (working day 27), the information contained in Figure 10.9 must first be incorporated into the project network. Chart 10.1 on the companion website is the corrected precedence diagram used for the recalculation of activity times and floats. In the chart, those activities that have been completed as of day 27 are to the left of the heavy dashed line. All of the completed work items show only the activity description and number. Activities in progress, but not yet completed, are indicated by the dashed line labeled with the notation "July 21, Day 27." For these activities, the time durations shown have been changed to the number of working days estimated to reach completion. The dashed line shows the current stage of advancement of the work as of day 27 and is often referred to as the data date or time contour.

A number of diagrammatic conventions other than that shown in Chart 10.1 can be used with regard to making manual updating calculations. For example, the durations of all completed activities can be set equal to zero, and the durations of activities in progress can be set equal to the days to complete them. The start activity is replaced with an "Elapsed time" activity whose duration is set equal to the data date (27 in this case). Computations are started at the beginning of the network by setting the early start of the elapsed time box equal to zero. Another scheme is to start the forward pass directly at the time contour. This has been done in Chart 10.1 simply by entering an EF of 27 for the completed activities at the time contour and an ES of 27 for the activities in process.

The recalculation in Chart 10.1 proceeds as a normal forward pass followed by a backward pass. The forward pass discloses that, as of day 27, the project completion time will be 71 working days. This would indicate that the job is now only one day behind its original schedule. Thus, the equipment logic change discussed in Section 10.20, and designed to ease the job delay caused by the problems encountered by the pile driving on abutment #2 (activity 140), has proven to be effective. The job delay has now been reduced from five working days to one day. The decision is now to accept the one-day delay and make the backward pass in Chart 10.1, starting with a project duration of 71. The critical activities are indicated by bold-lined boxes, and the critical path, by bold lines. Comparing Chart 10.1 with Chart 5.1a reveals that the initial portion of the critical path has moved from its original position. Such changes in the identification of critical activities are common occurrences as the work progresses in the field. If a project duration of 70 is to be maintained, activity 410, "Contingency," would be reduced to five days and the computed activity times would remain as before. The turnaround values used on update calculations vary with personal preference. The matter is unimportant because there is no substantive difference in the management data generated.

Updated networks, such as that in Chart 10.1, also can be used to record the actual start and finish dates of each activity. Noting these dates above each activity box provides a historical record of actual job progress. This action usually is not required where job bar charts are maintained or where scheduling software is being used, because it only duplicates information available elsewhere. A file of the successive update networks should be kept as a record of how the job progressed during the contract period.

10.23 Scheduling Software

Schedule updating is one of the most useful and important computer applications in the entire project management system. Scheduling software can provide timely management information in an easily understood and immediately usable form. Several software applications are available that generate a wide variety of time status reports. The field progress information is entered into the scheduling software, requiring minimal effort. Scheduling software is highly flexible in that it can deliver just about any type of report the project manager wants. Reports can be selectively prepared for any level of management or supervision. From summary statements generated for top management, to highly detailed reports for field supervisors, the software can produce time-status information for any desired level of job management.

At the field supervisory level, scheduling software produces a wide variety of useful data expressed in terms of calendar days and dates. Typical information generated might be:

- Estimated duration and actual duration of each completed activity.
- Scheduled start dates and actual start dates of all activities completed or in progress.
- Scheduled finish date and the actual finish date of each completed activity.
- Time status of each activity in progress, showing the anticipated finish date and the number of days ahead or behind schedule.
- A revised project completion date with an indication of the projected time overrun or underrun.
- Originally scheduled and revised start and finish dates for each activity not yet begun.
- Identification of critical activities.
- Float values of all activities.

This information can be generated using any desired date as the starting point for the backward pass.

10.24 Project Progress Curves

This chapter has discussed the time or progress monitoring of construction projects in the context of comparing the actual progress of job activities with that which was planned. This procedure provides detailed information concerning the current time status of individual job segments. Overall job progress is also used as a time-monitoring device, either as a stand-alone system or in conjunction with the detailed procedures already discussed. Total project progress as of a given date can be expressed in terms of different cumulative measures, such as total labor cost, total money expended, work quantities put into place, total labor hours used, or possibly others. Actual numbers of units, or percentages of the total, can be used. Progress curves are prepared by plotting cumulative job progress, expressed in terms of one of the measures just cited, to a horizontal time scale. Progress monitoring is accomplished by periodically comparing planned project progress with actual values.

A planned progress curve is obtained by calculating cumulative totals of the chosen progress measure at the end of each successive time unit. Days, weeks, or months can be used in the preparation of progress curves, depending on the nature and extent of the work. If the progress values are plotted at short time intervals, the resulting curves are apt to be irregular and jerky in appearance. Plotting values as of the end of each month normally results in reasonably smooth curves. A common and very effective procedure in the plotting of planned progress curves is to produce two such curves. One is determined on the basis of all project activities starting as early as possible. The other is based on all project activities starting at their late-start dates. When these two curves are plotted, they form a closed envelope. Figure 10.10 shows such an envelope plotted in generalized form. These curves are commonly referred to as S-curves because of their typical appearance.

After two extreme condition curves are drawn, an average curve between them is sketched in. This average curve is shown as a dashed line in Figure 10.10 and is used for purposes of general progress monitoring. During the construction period, actual progress is plotted periodically, with these points forming the actual progress line. The relative position of this line to the planned average progress curve is used to evaluate the time status of the project as a whole. Where actual progress plots above the average progress line, the time status of the job is considered to be satisfactory. When it lies below, time progress is considered to be generally unsatisfactory.

A study of the general geometry of a project envelope can provide interesting information concerning the work at hand. Figure 10.10 illustrates the general form of this envelope for a typical construction project. A rule of thumb often used is that 50 percent of the job is accomplished during the middle one-third of the construction period, with the other 50 percent being about equally divided between the initial and final thirds. The location of the ES and EF curves with respect to one another depends on the

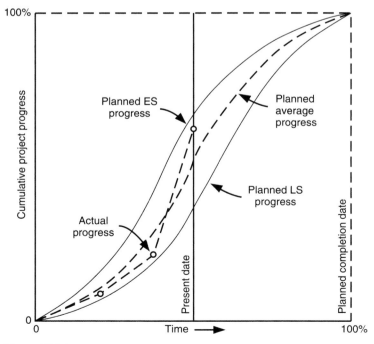

Figure 10.10
Project progress curves

relative amounts of activity float present. If the floats tend to be small, the two curves will be close together, and the shape of the envelope will be long and thin. If relatively large amounts of float are present, the two curves become more widely separated. Consequently, the shape of the envelope gives a quick visual indication of the degree of time control required to keep the project on schedule. A narrow envelope will require a considerably more rigorous time-control program than will a project where the envelope indicates the presence of substantial float. A narrow envelope also will present more difficulties in resource allocation because of the limited float.

Project progress curves, in themselves, are of limited effectiveness insofar as presenting the time status of a construction project is concerned. Although they do indicate the time status of the project as a whole, they are of no assistance in solving time-slippage problems. However, for the owner, architect-engineer, and top project management, such curves can be very useful in affording a quick grasp of the overall time condition of the work. For this reason, progress curves often are superimposed on the project bar chart. In this way, all the progress information is brought together into a single graphic display.

Key Points and Questions

Key Points

- The foreman plays a critical role in planning at the activity level.
- A key part of the implementation of lean construction in the field is found in the Last Planner Process©, which focuses on the reliability of foreman planning.
- Well-thought-out checklists help prevent mistakes and omissions, and they are also a valuable and time-saving aid in quality control.
- The entire process of project time management can be enhanced considerably by scheduling periodic job progress meetings.
- Without frequent updating, schedules quickly lose the ability to properly represent the project and become useless or worse.

Review Questions and Problems

1. What are the three main considerations involved in managing subcontractors?
2. Why is float a valuable resource, and why is free float more valuable than total float?
3. To improve overall production through the application of lean principles, where does the Last Planner Process© focus?
4. What are the four planning quality characteristics that will likely improve the reliability of planning at the activity level?

5. What criteria are considered in deciding on the frequency of progress reporting for a project?
6. Why is an independent review of work accomplishment preferred over having the foreman responsible for the work do the assessment?
7. The entire process of project time management can be enhanced considerably by scheduling periodic job progress meetings. What is the objective of such a meeting, and who should participate?
8. What is the basic objective of a schedule update, and what does it achieve?

11 Project Cost System*

11.1 Introduction

The focus of this chapter shifts from managing time to managing cost. Cost and time, of course, are not unrelated. As we saw earlier, there is a most efficient time in which to complete an activity. Earlier completion can be achieved but, typically, will cost money invested in expediting completion of the activity. Later completion will not necessarily save money, and in most cases will cost money due to extending the project and increasing overhead costs.

In this chapter, we consider the various elements of the project cost cycle, starting with the estimate and moving through the project to collection of actual unit costs to be incorporated into the company cost database for use in starting the cycle again for a future project. We will also return to the relationship between time and money. Although the details of a specific cost-control system vary substantially from one construction firm to another, the ensuing treatment can be regarded as being reasonably typical of current practice within the construction industry.

Learning objectives for this chapter include:

❑ Understand how the cost cycle flows throughout the entire project.
❑ Recognize project cost accounting as the key component of the project cost system.

---

* Portions of this chapter have been adapted from Richard H. Clough, Glenn A. Sears, and S. Keoki Sears, *Construction Contracting*, 7th ed. (Hoboken, NJ: John Wiley & Sons, 2005), chap. 12.

❏ Understand the critical importance of accurately capturing data from the field for cost control and estimating.

❏ Recognize that cost report detail needs to be tailored to the project and the managers receiving the reports.

❏ Learn the value of earned value analysis in considering the combined effects of both time and money.

11.2 The Construction Cost Cycle

The cost cycle for a construction project is shown in Figure 11.1. It begins with an estimate based upon cost and production data specific to the company and developed on previous projects. This historic cost database is highly proprietary and greatly valued by the company. Based on actual production records and costs from previous jobs, it gives the company the best basis for pricing new work so that it can be competitive and yet protected as much as possible from underpricing work.

The estimate, discussed in some detail in Chapter 3, "Project Estimating," is a cost-based model of the project, just as the drawings are a graphical model of the project, the specifications are a verbal model of the project, and the schedule is a time-based model of the project. As the project progresses, this cost model is modified to reflect its purpose at any given point. The initial estimate is developed so that the company will know how much it will cost to complete the project. The initial estimate is then adjusted and used as a basis for negotiating or bidding the project. The objective of this proposal or bid estimate is to successfully enter into a contract to execute the job.

Once the job is under contract, the owner will want a schedule of values from the contractor. The term schedule in this case does not refer to the time-based models that have been discussed previously in the book but, rather, to a listing of costs to be used as a basis for billing throughout the project. Since the schedule of values is the basis for periodic billings, as the project manager develops the schedule of values, the estimate will be tailored in order to design a more advantageous cash flow. Cash flow is critical to the contractor and will be further discussed in Chapter 12, "Project Financial Management."

Another modification to the project estimate results in the contractor's project budget. The project manager, often working with the field supervisor, will refine the initial estimate based on the actual contract amount that was negotiated and on how the project team anticipates investing that amount. It is at this point that the project team has the opportunity to review the initial estimate in detail and reallocate funds from those work items that they feel they can construct more efficiently than estimated to cover shortfalls in work items that they anticipate to be more costly to build than estimated. This budget then becomes the basis for the cost-control system that will be used to track costs throughout the project.

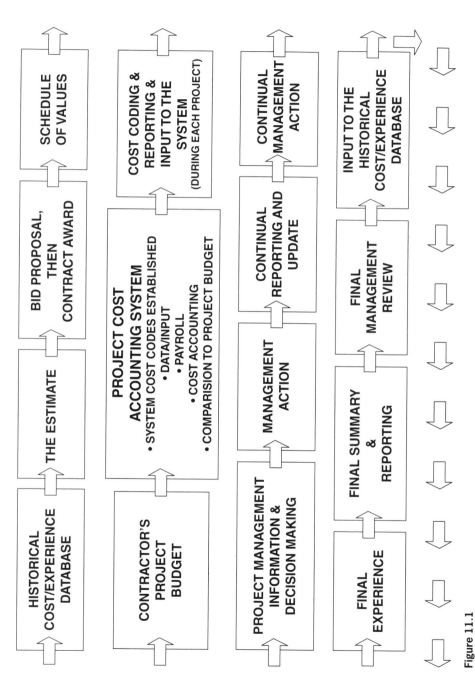

Figure 11.1
Cost cycle for a construction project

As physical work is begun in the field, a repetitive cycle begins in which work is executed, cost and progress are documented and entered into the project cost system, and periodic reports are produced. Based upon the reports, the project manager monitors activity costs, determines where corrective action is required, designs the corrective action, sees that the corrective action is executed, and continues to monitor costs in future reports to verify that the corrective action has achieved what it was designed to do without causing unanticipated consequences. This represents the core of the cost-control process for the project manager.

As the project progresses, key cost and production information is passed on to the estimator with the objective of maintaining the historic cost database using the most current information. When the project is completed, part of the closeout of the project includes a final review of actual costs as compared with estimated costs. Where actual costs were less than estimated, the company can learn what was done to improve the costs with the objective of improving company operations across future jobs. Where actual costs exceeded estimated, the company can determine root causes for estimating failures and improve company operations by eliminating similar failures on future projects. Both building upon successes and eliminating failures contribute to continuous improvement in the company, which is vital, not only for survival, but to excel as a contractor.

11.3 Objectives of a Cost System

The project cost system is designed to accomplish two important objectives. One is to develop labor and equipment production information that is useful to project management and supervision in controlling construction costs and maintaining them within the budget. The second is to develop labor and equipment cost information in a form suitable to support the accurate estimating of the cost of future work. The ability to estimate construction costs accurately and then to execute work within the budget established by that estimate is a key element in the success of any construction firm. Regardless of the type of contract arrangements agreed on with the owner, it is important that the contractor exercise the maximum control possible over its field costs during the construction period. A functioning and reliable cost system plays a vital role in the proper management of a construction project.

How the costs of a construction project are controlled varies with its size and character. A large, complex job requires a detailed reporting and information system to serve project management needs. Simpler and less elaborate cost systems are sufficient for smaller and simpler projects. In any event, the only justification for the expense of a project cost system is the value of the management data that it provides. If the information produced is not used, or if it is not supplied in a usable form or timely fashion, then the cost system has no real value and its cost cannot be justified. Properly designed and implemented, a project cost system is an investment rather than an expense.

11.4 Project Cost Control

As previously indicated, project cost control actually begins with the preparation of the original cost estimate and the subsequent construction budget. Keeping within the cost budget and knowing when and where job costs are deviating are two factors that constitute the key to a profitable operation. As the work proceeds in the field, cost accounting methods are applied to determine the actual costs of production. The costs as they actually occur are continuously compared with the budget. In addition to monitoring current expenses, periodic reports are prepared that forecast final project costs and compare these predicted costs with the established budget. Field costs are obtained in substantial detail because this is the way jobs are originally estimated and also because excessive costs in the field can be corrected only if the exact cause of the overrun can be isolated. Establishing cost overruns is not especially helpful unless the factors leading to the overrun can be distilled and targeted for improvement.

Cost reports are prepared at regular time intervals. These reports are designed to serve as management-by-exception devices, making it possible for the contractor to determine the cost status of the project and to pinpoint those work classifications where expenses are excessive. In this way, management consideration is quickly focused on those job areas that need attention. Timely information is required if effective action against cost overruns is to be taken. Detection of excessive costs once the work is finished leaves the contractor with no possibility of taking corrective action.

11.5 Data for Estimating

As discussed in Chapter 3, "Project Estimating," when the cost of a project is being estimated, many elements of cost must be evaluated. Labor and equipment expenses, in particular, are priced in the light of past experience. In essence, historical production records are the only reliable source of information available for estimating these two job expenses. The company cost system provides a reliable and systematic way of accumulating labor and equipment productivity and costs for use in estimating future jobs.

With regard to feedback information for cost-estimating purposes, there is some variation in the form of production data that are recovered. It can be argued that production rates, discussed in Section 2.11, are fundamental to estimating labor and equipment costs. However, as seen in Chapter 3, "Project Estimating," costs per unit of production, or unit costs, are widely used for estimating labor and equipment because of the convenience of their application. Such unit costs are, of course, determined from production rates and hourly costs of labor and equipment. Such unit costs can be kept up-to-date by being adjusted for changes in hourly rates and production efficiencies.

Information generated for company estimating purposes can, therefore, be in terms of labor and equipment production rates, unit costs, or both. The difference in usage seems to be at least partially a matter of the extent of a contractor's operations. Smaller contractors frequently work exclusively in terms of unit costs, whereas larger companies must, almost of necessity, base at least portions of their estimating on production rates. Both procedures were used in Chapter 3, "Project Estimating," to estimate various segments of the highway bridge. It is obvious that the feedback system to estimating must be designed to produce information in whatever form or forms are compatible with company needs and procedures.

11.6 Project Cost Code

Each individual account of a contractor's financial accounting system is assigned its own code designation as a means of classification and identification. Here, only the cost codes for job expense accounts are involved. Many construction firms use their own customized project cost codes that they have developed and tailored to suit their individual preferences and needs. Many forms of alphabetical, numerical, and mixed cost codes are in use.

Cost codes are essential because they tie the entire cost system together. An item of work will typically have four different elements that contribute to the cost of that item: labor, material, equipment, and subcontract. Labor hour information is collected on the time card (as we will see later in this chapter). Material or installed equipment items come from a supplier such as a lumberyard or a steel fabricator. Construction equipment is either owned or rented, and the cost is assigned to a specific item of work based on the number of hours that equipment is used on that specific work item. Subcontract items are typically billed out based on the percent of the subcontract work completed during the billing period. However, in all four cases, the cost is assigned to the specific work item through the cost code designated for that item.

The project cost code for the highway bridge is typical of systems widely used by heavy civil and industrial contractors. Each work type involved on this job is assigned its unique numerical designation. To illustrate, each item of expense that is incurred on the highway bridge will carry a 16-digit cost code identification. An example might be:

200808 05 03157.20.3

This cost code number is the labor cost of placing and stripping the abutment forms on the highway bridge. The cost code number contains four data groups.

1. *Project number.* The first six digits identify the particular project on which the cost is incurred and to which the cost is to be charged. Different construction companies have their own ways of doing this. Here, the Example Project is designated as Project No. 200808, meaning the eighth project started in the year 2008. If desired, other information can be included, such as the form of contract (unit price) and kind of work (heavy construction).

2. *Area code.* The area or location within the project is indicated by the next two digits in the cost code number. This concept applies only to large projects, where there usually are distinctive features that naturally divide the work into separate geographical areas or major physical parts. The heavy civil example project consists of several major segments, including the earth dam, highway bridge, pipeline relocation, and others. In this case, the highway bridge is assigned an area code designation of 05. One of the values of the area code is that the code makes it possible to associate field costs with the specific supervisors and managers responsible for them.

3. *Work-type code.* This seven-digit number comes after the area code and is taken from the contractor's master list. Costs are divided into 50 divisions corresponding to the way technical specifications are organized by the Construction Specifications Institute (CSI) MasterFormat. To illustrate, if the type of work involved is placing and stripping abutment concrete, the number 03100 is used. The 03 comes from division 3 of the specifications, concrete work. The 100 comes from the specifications section dedicated to formwork. Code numbers for specific forming systems and materials are obtained by changing the last digit in the master account number. For purposes of this chapter, wooden forms are designated by 03157. Project account numbers to indicate different form applications on the highway bridge are obtained by placing a decimal point after the number 03157 and adding digits to the right of the decimal point. Abutment forms are correspondingly identified by 03157.20 (footing forms are 03157.10, and deck forms are 03157.30).

4. *Distribution code.* To serve a variety of accounting purposes in addition to that of project cost accounting, it is necessary to indicate the category of expense involved. This is done by means of a standard company distribution code. For example, "1" is total, "2" is material, "3" is labor, "4" is equipment, and "5" is subcontract; thus, the final numeral 3 in this example cost code indicates that it is a labor charge. With regard to the distribution code, "total" merely indicates an all-inclusive cost that may include any combination of labor, material, equipment, or subcontract cost.

A related method of developing cost codes is derived from the work breakdown structure (WBS). In this case, each level of the outline or WBS constitutes a cost code grouping, as shown earlier. When all levels and all possible attributes within a level are taken together, they form a WBS dictionary covering all project costs and schedule activities. The WBS dictionary is a narrative definition of the work scope represented by each element on the project. Cost codes can be developed at the lowest WBS level and then summarized at any level within the WBS hierarchy. This method of cost coding is often an advantage on larger, complex projects as it coincides with the structure used for developing all of the schedule activities. Specifically, when a WBS is used, a structured relationship exists between schedule activities and cost accounts. This relationship allows costs to be applied and tracked for each individual schedule activity. Correlating both budget and cost to each schedule activity is vital to project earned value calculations, discussed in Section 11.21.

11.7 Usage of Project Cost Code

Labor, materials, supplies, equipment charges, subcontract payments, overhead costs, and other items of expense are charged to the project on which they are incurred. To ensure that each such expenditure is properly charged to the correct cost account, every expenditure is coded in accordance with the project cost code system. The use of "general" or "miscellaneous" cost accounts is poor practice and should be avoided. The job expense accounts provide the basic information for the periodic project cost reports, which will be discussed.

For the project cost code to serve its basic purpose, it must be understood and used consistently by all company personnel. It is mandatory that the project be broken down into the same established work types and that the project cost code be used consistently for the purposes of project estimating, general financial accounting, and field cost accounting. When the estimator prices a new project, the work is subdivided into the standard classifications. Reference to Figure 3.6 shows that the estimator identifies each work quantity by its work-type code when the estimator enters the result of the quantity take-off on the summary sheets. The project number and area code are assigned at a later time. The same cost code numbers and elementary work types are used from initial estimate to project completion. Although the project number and area code are unique to a given job, the work-type code and distribution code are used consistently on all projects. An important point here is that each cost account must consistently contain the same elements of cost. For example, labor costs always must either include, or exclude, applicable indirect costs, such as fringe benefits and payroll taxes and insurance.

In the usage of day-to-day cost forms, and in the preparation of periodic project cost reports, the project number and area code are not usually included as a part of the individual cost account numbers. Rather, they appear only in the heading of the form or report. This is also true when cost records of completed projects are filed. Permanent files of cost records from past projects are a very valuable estimating resource. The fact that historical job costs can be associated with specific projects makes them even more beneficial. They enable the user of the cost files to associate production rates and costs with specific project circumstances and conditions.

11.8 Project Cost Accounting

Project cost accounting is the key ingredient in the project cost system. It provides the basic data required for both cost control and estimating. Cost accounting differs substantially from financial accounting, however. Cost accounting relates solely to determining the detailed makeup of productivity and costs associated with the production of a construction product in the field, including the necessary overhead expense. Project cost accounting is not separate from the contractor's general system of accounts; rather, it is an elaboration of the basic project expense accounts. Cost accounting involves the continuous determination of productivity and cost data, the analysis of this information, and the presentation of the results in summary form.

It is thus seen that project cost accounting differs from the usual accounting routines in that the information gathered, recorded, and analyzed is not entirely in terms of dollars and cents. Construction cost accounting is necessarily concerned with costs, but it is also concerned with man-hours, equipment-hours, and the amounts of work accomplished. The systematic and regular checking of costs is a necessary part of obtaining reliable, time-average production information. A system that evaluates field performance only intermittently in the form of occasional spot checks does not provide trustworthy feedback information, either for cost control or for estimating purposes.

Project cost accounting must strike a workable balance between too little and too much detail. If the system is too general, it will not produce the detailed costs necessary for meaningful management control. Excessive detail will result in the objectives of the cost system being obscured by masses of data and paperwork and will needlessly increase the time lag in making the information available. The detail used in this book with regard to the estimating and cost control of the highway bridge is reasonably typical of actual practice in the industry.

A project cost accounting system supplements field supervision; it does not replace it. In the final analysis, the best cost-control system that a contractor can have is skilled, experienced, and energetic field supervision. It is important for field supervisors to realize that project cost accounting is meant to assist them by the early detection of troublesome areas. Trade and site supervisors are key members of the cost-control team. Without their support and cooperation, the job cost system cannot and will not perform satisfactorily.

11.9 Labor and Equipment Costs

It is a basic accounting principle for construction contractors that project income and expenses be recorded by individual job. The contractor's financial accounting system includes project expense accounts that are used to record every item of expense charged to any given project. Job costs associated with materials, subcontracts, and nonlabor items of project overhead are of a reasonably fixed nature, and cost control of these kinds of expenses is effected mainly by disbursement controls applied to purchase orders and subcontracts. (Section 12.16 provides more on this.) Barring oversight or mistake during the estimating process, these costs are determined with reasonable accuracy when the job is priced, and such costs seldom tend to vary appreciably from their budgeted amounts. For this reason, the cost information available from the monthly cost forecast reports (see Section 11.19) pertaining to materials, subcontracts, and nonlabor field overhead items is normally suitable and timely enough for ordinary cost-control purposes.

Labor and equipment costs are an entirely different matter, however. These two categories of job expense are characterized by considerable uncertainty and can fluctuate substantially during the construction period. These are the only categories of job expense that the contractor can control to any extent, and they merit and need constant management attention. However, the usual monthly cost forecast reports do not normally suffice for labor and equipment cost control. This is because the information concerning these two categories of field expense is not reported in adequate detail or at sufficiently short time intervals. Detailed cost accounting methods must be used in conjunction with labor and equipment expense if effective management control over them is to be realized. Accordingly, the discussion of project cost accounting in this chapter is essentially limited to these two categories of job expense and describes the process of determining at regular intervals how much work is being accomplished in relation to the amounts of labor and equipment being invested.

11.10 Cost Accounting Reports

Summary labor and equipment cost reports must be compiled often enough so that excessive project costs can be detected while there is still time to remedy them. Cost report intervals are very much a function of project size, duration, and nature of the work, and the type of construction contract involved. Obviously, some kind of balance must be struck between the cost of generating the reports and the value of the management information received. Often, daily cost reports are prepared on complex projects involving multiple shifts. Most jobs, however, do not profit from such frequent cost reporting. Some very large projects involving relatively uncomplicated work classifications find monthly or even longer intervals to be satisfactory. For most construction projects, however, cost reports are needed more often than this.

It is generally agreed that weekly labor and equipment cost summaries are about optimum for most construction operations, and this is the basis for discussion here. The cutoff time can be any desired day of the week, although contractors often match their cost system to their usual payroll periods and also to their time-monitoring system. This matter was discussed in Section 10.15. The cutoff times used here for weekly labor and equipment cost reports will be Wednesday afternoons. We assume that this is when the payroll periods end and weekly work quantity measurements are made.

11.11 Labor Time Reporting

The source documents for labor costs and cost code allocations are labor time cards, the same cards used for payroll purposes. They report the hours of labor time for every tradesman and the project cost codes to which the labor is applied. Figure 11.2 shows a typical daily time card and Figure 11.3,

						Daily Labor Time Card				
Project	Highway Bridge			Project No.		200808-05		Weather	Cloudy-windy	
Date	July 20							Prepared by	G.A.S.	
						Cost Code				
Employee Number	Name	Craft	Time Classification	Hourly Rate	03157.10.3	03157.20.3			Total Hours	Gross Amount
132	Winnowich, N.	F	ST	34.00		8			8	$272.00
			OT							
221	Clouten, S.	C	ST	$31.00		8			8	$248.00
			OT							
248	Schluder, L.	C	ST	$31.00		8			8	$248.00
			OT							
319	Mills, R.	O	ST	$33.00		2			2	$66.00
			OT	0						
143	Gibson, E.	L	ST	$22.00		8			8	$176.00
			OT							
417	Sears, K.	L	ST	$22.00		8			8	$176.00
			OT							
176	Sibanda, M.	L	ST	$22.00	2	6			8	$176.00
			OT							
	Total Hours				2	48			50	
	Total Cost				$44.00	$1,318.00				$1,362.00

Figure 11.2
Highway bridge, daily labor time card

11 Project Cost System

	Weekly Labor Time Card							
	Craft C							

Name Clouten, S.
Employee No. 221
Week Ending July 21

Project Highway Bridge
Project No. 200808-05
Prepared by S.K.S.

Cost Code	Time Classification	Hourly Rate	Thursday	Friday	Monday	Tuesday	Wednesday	Total Hours	Total Cost
03157.10.3	ST	$31.00	8					8	$248.00
	OT								
03157.20.3	ST	$31.00		8	8	8		24	$744.00
	OT								
03311.10.3	ST	$31.00					8	8	$248.00
	OT								
	ST								
	OT								
	ST								
	OT								
Total Hours	ST		8	8	8	8	8	40	
	OT								
Gross Amount			$248.00	$248.00	$248.00	$248.00	$248.00		$1,240.00
Weather			Clear-hot	Clear-hot	Cloudy-gusty	Cloudy-windy	Cloudy-windy		

Figure 11.3
Highway bridge, weekly labor time card

a weekly time card. Though these refer specifically to the bridge, similar labor time cards would be used for building construction. Which of these two is used depends on company preference and internal routines. The standard operating procedures of many construction companies require the use of daily time cards because weekly time cards are more susceptible to putting off the recording of time until the end of the week when it becomes very difficult to remember accurately to which activity each worker was assigned and what specifically was the duration of the work.

When daily time cards are used, they are filled in and forwarded each day to the company's payroll office. The top of the card provides for entry of the project name and number, date, weather conditions, and the name of the person preparing the card. The body of the time card provides the employee number, name, and craft for each person covered. In Figures 11.2 and 11.3, the designation "F" indicates foreman, "C" indicates carpenter, "L" means laborer, and "O" denotes equipment operator. Hours are reported as straight time (ST) or overtime (OT), as the case may be. For each person listed, several slots are provided for the distribution of hours to specific cost codes. Absolute precision in allocating labor time is not possible; nevertheless, the need for care and reasonable accuracy in ascribing each person's time to the appropriate cost codes cannot be overemphasized. The accuracy (and hence value) of reports on the specific job is compromised, and even more critical, the accuracy of historic estimating information is negatively impacted if the labor hours collected in the field are inaccurate.

Contractors vary in how they charge their craft supervisors' time. Some prefer to charge it directly to the work classifications on which the supervisor spent time. Others charge all foreman time to a supervisory account in project overhead. Probably the most common procedure is to allocate the time of craft supervisors who regularly work with their tools to the work-type account. Otherwise, the time is charged to job overhead.

If weekly time cards are used, a separate card for each worker is prepared to record the hours and cost codes for that week. Although the diagrammatic arrangement of the weekly time card is different from that of the daily time card, it contains the same payroll and cost accounting information concerning the individual craft worker.

11.12 Time Card Preparation

The distribution of each worker's time among cost codes is normally the responsibility of the field supervisor (foreman), who is in the best position to know how each worker's time was actually spent. Usually, this information is first recorded in the foreman's pocket time book or digital device, with the work performed by the individuals in the foreman's crew often being described in word form rather than by code numbers. In some environments, the foreman also enters the hourly rate for each worker because it is not unusual for a person to be assigned to work that requires different hourly rates of pay during a week, or even in a single day. However, the foreman does not usually extend hours and wage rates into totals. The hourly rates shown on the time cards in Figures 11.2 and 11.3 are base wage rates only and do not include any indirect labor costs such as payroll taxes, insurance, or fringe benefits. Nor do labor unit costs derived from such time card information include any indirect labor costs, a point dealt with in Section 3.16. In this text, all labor unit costs are derived from basic wage rates only.

The importance of accurate and honest time reporting cannot be overemphasized. On the basis of the allocation of labor (or equipment) time to the various account numbers, cost and production data are generated. If this information is inaccurate or distorted, it can be seriously misleading when used for estimating and cost-control purposes. Loss items must be identified as such without any attempt at cover-up by charging time to other cost accounts whose performance has been good. By moving work hours and costs between cost codes to hide performance problems, field managers are guaranteeing that future estimates will contain the same overly optimistic production rates from which they are currently suffering.

Often, the project timekeeper, cost engineer, or project manager fills in the formal time cards rather than the foreman. The timekeeper, for example, can collect the foreman's time books at some convenient time each day and fill out the time cards, adding the necessary information for payroll and cost accounting and making the extensions.

Even if weekly time cards are used, it is preferable that the labor distribution be made each day. Setting down the time and its allocation to different cost codes on a daily basis eliminates the undesirable practice of the foreman ignoring the matter until the end of the week and then trying to enter the information from memory. Completing the labor record at the end of every working day will materially improve the accuracy of the distribution. It is for this and other reasons that many contractors favor the use of daily time cards.

11.13 Measurement of Work Quantities

To determine productivity rates and unit costs, it is necessary to obtain the hours and costs expended as well as the quantity of work put into place. On some types of work, it may be feasible and convenient for the field supervisors to report work quantities accomplished as of the end of each day or each shift. It is a more common practice, however, for project work measurements to be made at longer intervals, such as at the end of each weekly payroll period, which is the basis for the discussion here. Although labor costs are now being discussed, the weekly measurement of work quantities includes all work items performed, whether accomplished by labor, equipment, or a combination of the two. Consequently, the same weekly work quantity determination serves for both labor and equipment cost accounting.

Work quantities can be obtained in a variety of ways, depending on the nature of the work involved and company management methods. Direct field measurement, estimation of percentages completed, computation from the contract drawings, use of the estimating sheets, and determination from network activity progress reports are all used. Direct measurement on the job is common. This procedure is simple and straightforward on projects that involve only a few cost code classifications. Many contracts of the heavy, highway, or utility category are of this nature. Often, there are instances where quantities can be approximated with reasonable accuracy by applying estimated percentages of completion to the total work quantities. Although not as accurate as direct measurement, this procedure can provide usable measures so long as more accurate determinations are made occasionally, such as for monthly pay requests.

The field measurement of work quantities on projects that involve many cost code classifications can become a substantial chore. Most building and industrial projects entail a substantial number of different cost classifications. One convenient procedure in such cases is to mark off and dimension the work advancement in colored pencil on a set of project drawings reserved for that purpose. The extent of work put into place can be indicated as of the end of each day or each week, as desired. By using different colors

	Weekly Work Quantity Report				
Project Highway Bridge			Project No. 200808-05		
Week Ending Wednesday, July 21 (working day 27)			Prepared by S.K.S.		
Cost Code	Work Type	Unit	Total Last Report	Total This Week	Total to Date
31361.10	Piling, steel, driving	lf	1,456	560	2,016
03159.10	Footing forms, strip	sf	0	360	360
03157.20	Abutment forms, place	sf	0	1,810	1,810

Figure 11.4
Highway bridge, weekly work quantity report

and dating successive stages of progress, work quantities can be determined from the drawings or estimating sheets as of any date desired.

The field measurement of work quantities can be, and often is, done by the field supervisors. However, on large projects, and especially those with many work codes, it may be desirable for the cost engineer or project manager to carry out this function because of the time and effort required. Weekly reports of work done are submitted on standard forms such as that in Figure 11.4. Preprinted cards with cost codes, work description, and units of measure already entered can be used for reporting work quantities when cost reports are prepared by computer.

11.14 Work Quantities from Network Activities

Project network activities used for planning, scheduling, and time control can serve as convenient packages for determining work quantities accomplished in the field. The weekly progress report (see Figure 10.5), used for project time management, submits information concerning completed activities and percentages of completion of activities in process. However, the activity completion information contained in the usual weekly progress report must be translated into cost code quantities (Figure 11.4). This can be readily accomplished by obtaining the work-type quantities associated with each activity from the estimating sheets or project drawings.

A note of caution is needed here, however, concerning the use of activities to determine work quantities accomplished. Usually, activity work quantities are taken from the project drawings. For most activities, quantities indicated by the drawings, and those actually required in the field, are essentially the same. However, there may be a difference at times, especially on unit-price contracts. When field quantities can differ to any extent from those indicated by the drawings, actual amounts of work items must be measured in the field.

When using activities as the basis for work quantity determination, the completed amounts of different work classifications can be established readily and accurately for those activities that have been completed. The situation may be considerably different, however, where activities in progress are involved. In the case of such an activity, its reported percentage of completion is based more on its remaining time to completion than on the quantities of work physically in place. The usual assumption is that the rate at which work is accomplished is a linear variable with time. That is, if an activity is reported as being 30 percent complete, this is normally taken to indicate that 30 percent of the work units have been completed. For activities of limited extent and that involve only one cost code classification, this is usually a reasonable assumption.

However, relating work accomplishment linearly with activity time is sometimes deceptive, especially with activities that entail more than one elementary work classification. For instance, on the highway bridge, there are activities that involve both forming for concrete and placing reinforcing steel. Such an activity is accomplished by the carpenters erecting the outside forms, the ironworkers tying the rebar, and the carpenters then putting up the inside forms and bulkheads. At the 30 percent point in time required for the activity—and this is what the usual progress completion report is based on—perhaps 45 percent of the carpentry work is now done, but none of the reinforcing steel has been placed. Written notes that routinely accompany the weekly progress reports normally clarify such matters. Activities that involve more than one elementary work classification must be broken down into their component cost code classifications when the weekly work quantity report is being prepared.

Another difficulty in relating work accomplishment linearly with activity time occurs with longer, more complex activities, where most of the work is in long continuous runs, but some of the work is in small, noncontinuous segments. A canny foreman will assign the quick and easy segments to the crew early on but will leave the small, complex pieces to come back to later. Thus, the bulk of the work is completed with high productivity, indicating that production units are being exceeded, yet as the last bits are completed, production goes way down and the projected cost saving on the activity are never realized because the final units of work could not meet the earlier production rates. Inexperienced project managers are often fooled by high production rates that are projected to activity (and job) completion but cannot be maintained.

11.15 Cost Records and Reports

The procedures being used, whether manual or computerized, dictate the methods and forms in which the field costs and production data are recorded, analyzed, and reported. If the cost system is maintained by hand, as can be the case with some small construction firms, some form of workbook is needed for the recording and analysis of the cost and quantity data. A separate record is kept of each cost account in which an entry is made, of all labor and equipment charges allocated to that account number, and of all quantities of that work type accomplished in the field. These cost accounting worksheets are maintained separately from the usual project expense ledgers that are used for recording all items of job cost.

If cost procedures are computerized, as is true with most of today's construction contractors, labor and equipment cost reports are generated as an integral part of the overall payroll and accounting system. However, regardless of whether manual or computer methods are used, the basic data that go in, and the summarized production and cost information that comes out, are essentially the same.

The detail used in cost reporting is adjusted according to the level of management for which it is intended. The cost information routinely provided to managers or supervisors must be tuned to the scope and nature of their job responsibilities. The site superintendent on the highway bridge would be provided with very detailed cost information concerning all aspects of its job. The project manager of the example project, however, would receive only general information concerning major aspects of the highway bridge. Obviously, the project manager can request and receive additional details where the cost performance on some job item has been unsatisfactory.

11.16 Weekly Labor Reports

Weekly labor reports can be prepared on either a labor-hour or cost basis. That is, labor productivity can be monitored in terms of either labor-hours per unit of work (production rates) or cost per unit of work (unit prices). Which of these methods is used is a function of project size, type of work involved, and project management procedures. Where labor-hour control is used, a budget of labor-hours per unit of work is prepared. In this regard, total labor-hours usually are used with no attempt to subdivide labor time by trade specialty, such as so many hours of carpenter time and so many of ironworker time. Total labor-hour estimates are based on an "average" crew mix for each work type.

During field operations, actual labor-hours and work quantities are obtained. This makes possible a direct comparison of actual to budgeted productivity. Such an approach, of course, reflects productivity but not cost. The approach is simple to implement and avoids many problems associated with labor cost analysis. One such problem occurs on projects that require long periods of time to complete. Wage rates may be increased several times during the life of such a job. When the project cost is first being estimated, educated guesses are made about what the wage raises are likely to be. What this means is that labor costs on long-term jobs often are estimated without exact knowledge of the wage rates that actually will apply during the construction process. As a result, actual labor wage rates during the work period may turn out to be different from those rates used in preparing the original control budget. Thus, labor costs produced by the project cost system are not directly comparable to the budget. To make valid cost comparisons, it is necessary either to adjust these costs to a common wage-rate basis or to work in terms of labor-hours rather than labor costs.

In the forgoing case, labor-hours can serve as a very effective basis for labor cost control. However, for most construction applications, including smaller projects such as the highway bridge, labor cost analysis is more widely applied than is labor-hour analysis. For this reason, the labor cost reports discussed here are all based on costs.

11.17 Weekly Labor Cost Report

Once a week, labor costs obtained from the time cards are matched to the work quantities produced. The results of this analysis are summarized in a weekly labor cost report, two different forms of which are illustrated in Figures 11.5 and 11.6. These labor reports classify and summarize all labor costs incurred on the highway bridge through the effective date of the report (July 21). The labor costs in these two reports are direct labor costs only and do not include indirect labor costs. These reports provide job management with detailed information concerning the current status of labor costs and indicate how these costs compare with those estimated. Both of the labor report forms are designed to identify immediately those work classifications that have excessive labor costs and to give an indication of how serious those overruns are. Labor cost reports vary considerably in format and content from one construction company to another, although all such report forms are designed to convey much the same kind of management information.

Figure 11.5 is a comprehensive weekly cost report for the highway bridge, which summarizes labor costs as budgeted, for the week being reported and to date. Not all cost report forms include costs for the week being reported. These values can be of significance, however, in indicating downward or upward trends in labor costs. The labor report form in Figure 11.5 involves work quantities as well as labor expense and yields unit costs for each work type. Unit costs, obtained by dividing the total labor cost in each work category by the respective total quantity, enable direct comparisons to be made between actual costs and costs as budgeted. In Figure 11.5, the budgeted quantity, budgeted direct labor cost, and budgeted labor unit cost for each work type are taken from the project budget, Figure 3.10a. The other quantities and labor costs are actual values, either for the week reported or to date.

Weekly Labor Cost Report

Project: Highway Bridge
Week ending: July 21 (working day 27)
Project No.: 200808-05
Prepared by: S.K.S.

Cost Code (1)	Work Type (2)	Unit (3)	Quantity Budget (4)	Quantity This Week (5)	Quantity To Date (6)	Direct Labor Cost Budget (7)	Direct Labor Cost Budget to Date (8)	Direct Labor Cost This Week (9)	Direct Labor Cost To Date (10)	Labor Unit Cost Budget (11)	Labor Unit Cost This Week (12)	Labor Unit Cost To Date (13)	To Date Savings (14)	To Date Loss (15)	Projected Savings (16)	Projected Loss (17)
31220.10.3	Excavation, unclassified	cy	1,667	0	1,667	$2,643	$2,643	$0	$2,459	$1.59	-	$1.48	$184		$184	
31222.10.3	Excavation, structural	cy	120	0	120	$2,985	$2,985	$0	$3,353	$24.88	-	$27.94		$368		$368
31350.00.3	Pile-driving rig, mob. & demob.	job	-	-	-	*$4,374	*$4,374	$0	$4,004	$4,374	-	$4,004	$370		$370	
31361.10.3	Piling, steel, driving	lf	2,240	560	2,016	$8,224	$7,402	$6,373	$15,762	$3.67	$11.38	$7.82		$8,360		**$9,293
03150.10.3	Footing forms, fabricate	sf	360	0	360	$936	$936	$0	$963	$2.60	-	$2.68		$27		$27
03150.20.3	Abutment forms, prefabricate	sf	1,810	0	1,810	$3,548	$3,548	$0	$2,780	$1.96	-	$1.54	$768		$768	
03157.10.3	Footing forms, place	sf	720	0	360	$360	$180	$0	$165	$0.50	-	$0.46	$15		$30	
03159.10.3	Footing forms, strip	sf	720	360	360	$158	$79	$48	$48	$0.22	$0.13	$0.13	$31		$62	
03157.20.3	Abutment forms, place	sf	3,620	1,810	1,810	$7,312	$3,656	$3,504	$3,504	$2.02	$1.94	$1.94	$152		$304	
03311.10.3	Concrete, footing, place	cy	120	0	60	$1,200	$600	$0	$554	$10.00	-	$9.23	$46		$92	
												Subtotals	$1,566	$8,755	$1,810	$9,688
												Totals	®	$7,189	®	$7,878

*67% of budget for mobilization
**Based on last week, will probably be $10,087

Figure 11.5
Highway bridge, weekly labor cost report (#1)

Weekly Labor Cost Report

Project	Highway Bridge								Project No.	200808-05	
Week ending	July 21 (working day 27)								Prepared by	S.K.S.	
Cost Code (1)	Work Type (2)	Unit (3)	Total Quantity Budgeted (4)	Total Quantity to Date (5)	Percent Completed (6)	Budgeted Direct Labor (7)	Budgeted Labor to Date (8)	Actual Labor to Date (9)	Cost Difference (10)	Deviation (11)
31220.10.3	Excavation, unclassified	cy	1,667	1,667	100.00%	$2,643	$2,643	$2,459	($184)	0.93
31222.10.3	Excavation, structural	cy	120	120	100%	$2,985	$2,985	$3,353	$368	1.12
31350.00.3	Pile-driving rig, mob. & demob.	job	–	–	100%	*$4,374	$4,374	$4,004	($370)	0.92
31361.10.3	Piling, steel, driving	lf	2,240	2,016	90%	$8,221	$7,399	$15,762	$8,363	2.13
03150.10.3	Footing forms, fabricate	sf	360	360	100%	$936	$936	$963	$27	1.03
03150.20.3	Abutment forms, prefabricate	sf	1,810	1,810	100%	$3,548	$3,548	$2,780	($768)	0.78
03157.10.3	Footing forms, place	sf	720	360	50%	$360	$180	$165	($15)	0.92
03159.10.3	Footing forms, strip	sf	720	360	50%	$158	$79	$48	($31)	0.61
03157.20.3	Abutment forms, place	sf	3,620	1,810	50%	$7,312	$3,656	$3,504	($152)	0.96
03311.10.3	Concrete, footing, place	cy	120	60	50%	$1,200	$600	$554	($46)	0.92
	Totals to Date						$26,400	$33,592	$7,192	1.27

*67% of budget for mobilization

Figure 11.6
Highway bridge, weekly labor cost report (#2)

When the total quantity of a given work item has been completed, its to-date and projected savings or loss figures are obtained merely by subtracting its actual total labor cost from its budgeted total cost. When a work item has been only partially accomplished, the to-date savings or loss of that work item is obtained by multiplying the quantity in place to date by the underrun or overrun of the unit price. The projected savings or loss for each work type can be obtained in different ways. In Figure 11.5, it is determined by assuming that the unit cost to date will continue to the completion of that work type. Multiplying the total estimated quantity by the underrun or overrun of the unit price to date yields the projected savings or loss figure. Another way the projected values can be computed is through the use of unit-price trends. Recent unit costs can be used to forecast final cost variances.

The projected savings and loss figures in columns 16 and 17 of Figure 11.5 afford a quick, informative summary of how the project is doing insofar as labor cost is concerned. Those work types with labor overruns are identified, together with the financial consequences if nothing changes. To illustrate, the effect of the underground boulders on the pile driving for abutment #2 (discussed in Section 10.16) is now obvious and probably will account for a direct labor cost overrun of more than $8,000. Some labor cost reports indicate the trend for each cost code; that is, whether the unit cost involved has been increasing or decreasing. This information can be helpful in assessing whether a given cost overrun is improving or worsening and in evaluating the efficacy of cost reduction efforts.

Again, consider the case of the underground boulders and their effect on pile driving at abutment #2. Although the computer may project future costs based on unit costs to date, the cost engineer should consider overriding this calculation based on field observations. It appears that pile-driving productivity is not going to improve and the unit cost for the most recent week is a better indicator of the cost to finish the work. In such a case, the cost overrun on this account is likely to be $10,087.

Figure 11.6 is an alternative form of weekly labor cost report that shows actual and budgeted labor costs to date for each cost classification. In this figure, the budgeted total quantities and budgeted total labor cost for each cost code are obtained from the project budget, Figure 3.10a. The actual work quantities and labor costs to date are cumulative totals for each work classification obtained from the time cards and weekly quantity reports. Column 10 of Figure 11.6 shows the cost difference as column 8 minus column 9, with a positive difference indicating that the cost, as estimated, exceeds the actual cost to date. Hence, in column 10, a positive number is desirable; a negative number is undesirable. The deviation is the actual cost to date (column 9) divided by the estimated cost to date (column 8). A deviation of less than 1 indicates that labor costs are within the budget, whereas a deviation of more than 1 indicates a cost overrun.

Although column 10 does indicate the magnitude of the labor cost variation for each cost code, it does not indicate the relative seriousness of the cost overruns. The deviation is of value in this regard because it shows the relative magnitude of labor cost variance. For those work types not yet completed, the cost differences listed in column 10 of Figure 11.6 do not always check exactly with the to-date savings and loss values in Figure 11.5. These small variations are caused by the rounding off of numbers and are inconsequential.

11.18 Equipment Cost Accounting

When equipment costs constitute a substantial portion of the cost of construction, the determination and analysis of field equipment expenses are an important part of project cost accounting. Equipment accounts for a substantial proportion of the cost of highway, heavy, and utility construction, and the control of equipment expense on such projects is as important as the control of labor costs. Equipment expense, like labor cost, is inherently uncertain and subject to unpredictable variations. The management and control of equipment expense requires a comprehensive equipment cost accounting system. The objectives of equipment cost accounting are the same as those discussed for labor costs. Management requires timely cost and production information for effective project cost control, and estimators need equipment productivity data to estimate the costs of future work. Only major equipment items require detailed cost study, however. Less expensive items, such as power saws, concrete vibrators, and hand-operated soil compactors, do not merit or receive detailed cost analysis.

Monthly Cost Forecast Report

Period Ending: Highway Bridge, July 31

Cost Code (1)	Description (2)	Materials				Labor			
		Budget (3)	Cost to Date (4)	Est. to Complete (5)	Variance (6)	Budget (7)	Cost to Date (8)	Est. to Complete (9)	Variance (10)
01	General Requirements	$0	$0	$0	$0	$12,360	$7,787	$4,248	$325
31	Sitework	$72,279	$79,176	$0	($6,897)	$23,253	$28,127	$0	$4,874
03	Concrete	$50,748	$26,201	$23,291	$1,256	$38,639	$16,856	$21,193	$590
05	Metals	$82,764	$0	$82,764	$0	$5,467	$0	$5,467	$0
09	Finishes	$0	$0	$0	$0	$0	$0	$0	$0

Figure 11.7
Highway bridge, monthly cost forecast report

11.19 Monthly Cost Forecast

As expressed in Chapter 3, "Project Estimating," because estimating and cost accounting for major equipment used on a job is highly specialized, depending on the industry sector and the way a specific company operates, project managers who would like a detailed analysis of pricing and cost accounting for heavy equipment should reference specific texts that focus on heavy equipment costs.

11.19 Monthly Cost Forecast

The discussions so far in this chapter have described how weekly labor cost reports are generated and the information they contain. Material and subcontract expenses also are incurred on the project, but these two cost categories are not as volatile as labor and do not require constant and detailed monitoring. Nevertheless, all project costs must be summarized and reported at regular intervals, monthly being common. The report used for this purpose also forecasts the final total job cost. Figure 11.7 presents a form of cost forecast report used to inform project management each month concerning the overall financial position of the job.

To prepare a cost forecast report, the information shown in Figure 11.7 pertaining to materials, labor, equipment, and subcontracts is first computed for each elementary work type on the job. These values are then combined and costs are presented only for the major work classifications, as indicated by Figure 11.7. Summary costs are listed for materials, labor, equipment, and subcontracts. Using labor as an example, column 7 lists the total budgeted labor cost for each major cost code listed. Following this is the total labor cost to date and the estimated total labor cost to complete each work account. The variance, column 10, is obtained by subtracting

Project No. 200008-05
Prepared by S.K.S.

Equipment				Subcontracts				Totals					
Budget (11)	Cost to Date (12)	Est. to Complete (13)	Variance (14)	Budget (15)	Cost to Date (16)	Est. to Complete (17)	Variance (18)	Budget (19)	Cost to Date (20)	Est. to Complete (21)	Estimated Final Cost (22)	Estimated Variance (23)	Percent Complete (24)
$4,444	$3,018	$1,240	$186	$0	$0	$0	$0	$16,804	$10,805	$5,488	$16,293	($511)	66%
$13,323	$19,735	$0	$6,412	$0	$0	$0	$0	$108,855	$127,038	$0	$127,038	$18,183	100%
$9,579	$5,422	$4,082	$75	$0	$0	$0	$0	$98,966	$48,479	$48,566	$97,045	($11,921)	50%
$2,520	$0	$2,520	$0	$0	$0	$0	$0	$90,751	$0	$90,751	$90,751	$0	0%
$0	$0	$0	$0	$8,550	$0	$8,550	$0	$8,550	$0	$8,550	$8,550	$0	0%
								$323,926	$186,322	$153,355	$339,677	$15,751	55%

the budgeted cost (column 7) from the sum of costs to date (column 8) and the estimated cost to complete (column 9). Variance is thus the difference between the anticipated actual cost and budget. A negative value of variance is the amount by which the actual cost is expected to exceed the estimated cost. A positive variance is, of course, just the opposite. The costs shown under materials, equipment, and subcontracts are obtained in the same fashion.

Column 19 in Figure 11.7 is obtained as the sum of the budgeted labor, material, equipment, and subcontract costs. Columns 20 and 21 are similarly calculated. The estimated total final cost (column 22) for each cost category is obtained by adding the total cost to date (column 20) to the total estimated cost to complete (column 21). The variance (column 23) is the total budgeted cost (column 19) subtracted from the estimated total final cost (column 22). The percent completed (column 24) for each major cost code is obtained by dividing the total cost to date (column 20) by the estimated total final cost (column 22). The algebraic sum of the total variance values (column 23) is the amount by which, it now appears, the actual cost will be less than, or more than, the budgeted cost for the total project. In Figure 11.7, it appears that, as of July 30, the actual total cost of constructing the highway bridge will be $15,751 more than had been originally estimated.

For completed items, the actual final cost is used for the estimated total final cost. The usual assumption for work under way is that it will be finished at its current cost rate and that unstarted work will be completed as budgeted. Exceptions to this occur when better cost information is available or reduced unit costs reflecting learning curve effects are in order.

This method of cost reporting does not differentiate between or identify whether schedule issues or cost issues or a combination of the two has caused the $15,751 variance. The introduction of new cost analysis tools, such as time-cost envelopes and the Earned Value Management System

11.20 Time-Cost Envelope

provide managers with a means of objectively identifying the root causes of reported variances on the project that go well beyond the capabilities of standard monthly cost forecasts.

In this chapter, methods have been discussed that enable project management to check actual field costs against an established budget. The principal emphasis has been to provide detailed cost information and forecasts for use by field supervisors. There is also a need for a more general kind of cost report that provides a quick and concise picture of the overall cost status of the project. This summary information is for use by top project

11.20 Time-Cost Envelope

management, owners, lending agencies, and others. The monthly cost forecast report of the previous section can serve in this regard. Project time-cost envelopes are graphical cost summaries that are also used for job-cost monitoring.

A cost envelope is obtained by plotting cumulative estimated job costs against construction time. Two such curves are drawn, one on the basis of an early-start schedule and the other on a late-start basis. This concept was discussed in Section 10.24. Figure 11.8 shows a typical project time-cost envelope. Each month, the total actual cost to date is plotted. These values are shown in Figure 11.8 as small circles. Alone, this information is inconclusive, but when the budgeted cost of the work accomplished to date is plotted concurrently, several aspects of the project cost and time status emerge. The budgeted costs to date are shown as crosses in Figure 11.8.

To illustrate how the cost information in Figure 11.8 is interpreted, note the project cost status as of the end of the third month. This is indicated as "present date" in the figure. The actual cost to date plots above the budgeted cost to

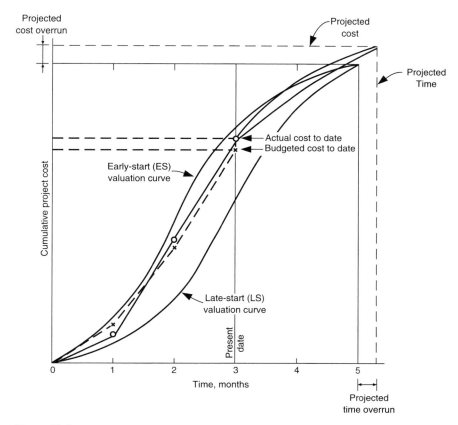

Figure 11.8
Time-cost envelope

date, indicating a cost overrun for the project as a whole. The same situation existed at the end of the previous month, and the cost overrun is increasing. The budgeted cost to date appears near the early-start curve of the cost envelope, indicating that the project is well within the schedule time limits.

Based on the monthly cost forecast report as of the end of the third month, a new cost envelope can be generated from the "present date" through the end of the job. This projection in Figure 11.8 shows that there is a projected project cost overrun and a projected early completion. Time-cost envelopes, such as shown in Figure 11.8, are excellent for graphically displaying time and cost trends for the total project.

11.21 Earned Value Management System

The Earned Value Management System (EVMS), sometimes referred to as Cost/Schedule Control Systems Criteria, is an extension of the time-cost envelope principle. Originally developed by industrial engineers to gauge plant output performance, EVMS compares the physical work accomplished to date with both the actual cost expenditure and the planned work to be accomplished. This three-dimensional comparison determines the earned value on a project and forms the basis for a more accurate projection of final project cost than can be calculated simply by contrasting actual costs with planned funds.

Earned value calculations and project performance projections are based on three fundamental variables:

1. *Budgeted cost of work performed (BCWP).* The cumulative budgeted value of all of the work activities accomplished to date. It is analogous to the budgeted cost of the work accomplished discussed in Section 11.20 and is calculated by adding up the cost budgets assigned to each of the schedule activities accomplished as of the present date. The BCWP is therefore the "earned value" of the project to date.

2. *Budgeted cost of work scheduled (BCWS).* The cumulative planned value of all of the work activities accomplished in the baseline schedule to date. As with time-cost envelopes, this can be any value between the early-start valuation curve and the late-start valuation curve but is most often taken to be the valuation calculated by accumulating planned progress against the early dates in the schedule.

3. *Actual cost of work performed (ACWP).* The cumulative actual cost for work completed to date. In Section 11.20, this was referred to as the actual cost to date and is simply the accrued total cost for all work accomplished through the present date.

As with time-cost envelopes, Figure 11.9 illustrates that these variables can be plotted on a time-cost graph to graphically demonstrate project status and performance. By comparing the BCWP, BCWS, and ACWP, the project manager can calculate to-date variances in schedule, forecasted schedule slippage at project completion, to-date variances in cost, and any forecasted cost over expenditures. The cumulative cost variance (CV) on the project is calculated by subtracting the ACWP from the BCWP. Similarly, cumulative schedule variance (SV) is calculated by subtracting the BCWS from the BCWP. This variance is calculated as a currency value rather than as a division of time. This is because it measures the difference between the planned value scheduled and the earned value achieved, with value measured as a currency.

In order to normalize the measurement of project performance effectively, schedule and cost variances can be converted into unitless indices used to benchmark projects of differing volumes or to forecast completion performance using interim measurements taken at regular updates during the project. The Cost Performance Index (CPI) and the Schedule Performance Index (SPI) are calculated in this way:

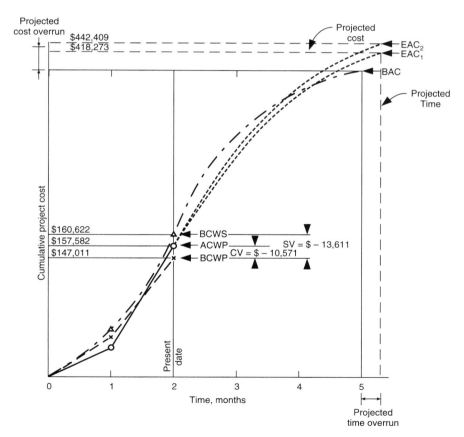

Figure 11.9
Earned value management system

$$CPI = BCWP / ACWP$$

$$SPI = BCWP / BCWS$$

Therefore, an index value of greater than 1 indicates positive performance against plan, whereas a value of less than 1 demands management intervention. A CPI of 0.80 would indicate that for every dollar spent on the project, only 80 cents of planned value was achieved, with a 20 percent pure cost overrun, while an SPI of 1.20 indicates that for every dollar of work planned to be accomplished on the project, $1.20 worth of work was completed during that time period, with work being completed 20 percent faster than planned. Taken together, these indices would suggest that the project has been accelerated to achieve an early completion date at additional project cost.

This result illustrates a key differentiator between standard budget versus actual cost comparisons and EVMS. The ability of EVMS to clearly separate variances into their cost and schedule components provides the project manager with a quantitative and objective means of identifying the root cause of project performance issues.

11.22 Forecasting Final Project Results Using the EVMS

The CPI and SPI can also be used to forecast the project's estimate at completion (EAC). This often is accomplished by using both the CPI and SPI to calculate a statistical range for the most likely project cost at completion. This range occurs between EAC_1 and EAC_2, and is calculated in this way:

$$EAC_1 = ACWP + (BAC - BCWP) / CPI$$

$$EAC_2 = ACWP + (BAC - BCWP) / (CPI \times SPI)$$

where BAC is the budget at completion.

Therefore, the EAC range for the highway bridge using a calculated CPI of 0.93 and an SPI of 0.92 assessed on day 27 would be:

$$EAC_1 = \$157,582 + (\$390,214 - \$147,011) / 0.93 = \$418,273$$
$$EAC_2 = \$157,582 + (\$390,214 - \$147,011) / (0.93 \times 0.91)$$
$$= \$442,409$$

This projection indicates that the estimate at completion will most likely range between $418,273 and $442,409. Stated another way, this calculation assumes that the cumulative project cost inefficiency to date will continue through the end of the project but may be further aggravated by continued delays to the project schedule.

The project manager generally will use a combination of empirical and statistical data to forecast the project estimate at completion. At each project update, once the project manager has forecasted the EAC, EVMS can be used to calculate the required CPI needed to achieve this estimated target. This is referred to as the to-complete CPI (TCPI) and is calculated as:

$$TCPI = (BAC - BCWP) / (EAC - ACWP)$$

TCPI is useful in helping focus project teams on the road ahead. By allocating TCPI to specific activities assigned to individual project supervisors, the project manager's completion strategy can be cascaded in a meaningful way to those in the best position to take appropriate action.

11.23 Special Cost Accounting Problems

Numerous practical considerations give rise to special problems concerning the proper operation of a project cost accounting system. Satisfactory solution of these problems can be fundamental to the success or failure of a project cost system. It follows, therefore, that project management must be aware of such difficulties and take appropriate steps to solve them. The discussion here is meant merely to afford some specific examples and not to be an exhaustive review of all such cases.

Usually, equipment expense is directly chargeable to a single project work account. However, often there are occasions where this is not true, and equipment costs must be accumulated in "suspense" or "clearing" accounts until such time as they can be distributed to the proper cost accounts. Many examples of this could be cited, but one might be a central concrete-mixing plant, consisting of many separate equipment units, that is producing concrete for several different cost accounts. Expenses for equipment of this type can be collected into a suspense account and be periodically distributed equitably to the appropriate cost accounts on the basis of quantities involved. Thus, each cost account is charged in the same manner as if the concrete had been purchased from an outside company. Ultimately, all suspense account costs must be charged to the end-use accounts, leaving a zero balance.

A different case involves an equipment item that serves many different operations. A tower crane on a high-rise building might be such an example. Over a short period of time, this crane could handle structural steel, concrete, masonry, reinforcing steel, door frames, electrical conduit, pipe, and any number of other materials. Allocation of this crane's time to specific cost codes would be an all-but-impossible task. A common approach is to establish a special cost category for the crane in project overhead. All crane costs are charged to this account with no effort made to distribute its time to specific work classifications.

Accounting for bulk materials is another common example of the use of clearing accounts. Stock items, such as lumber, pipe, and unfabricated reinforcing steel, ultimately may be used in conjunction with many different cost accounts. In such cases, a clearing account is created to accumulate all costs associated with the materials until they are put to use. As they are used, the suspense account is credited and the appropriate work accounts are charged.

A different kind of cost accounting problem is encountered when an owner imposes on the contractor a set of cost codes that differ from the contractor's own system. Owners sometimes have a need for particular cost information for purposes of capital depreciation, taxation, and insurance. Some owners, such as public utilities, may require cost values that fit standard account classifications imposed by regulatory agencies. Construction contracts of this kind can require the contractor to maintain project costs in two or even three different coding systems; however, computers usually can automate the conversion from one system to another.

11.24 Production Cost Reduction

The project cost reports generated by the cost accounting system make it possible for job management to assess the cost status of the work and to pinpoint those work areas where expenses are excessive. In this way, management attention is focused quickly on those work classifications that need it. If the project expense information is developed promptly, it may be possible to bring the offending costs back into line. In actual fact, of course, project cost control starts when the job is first priced, because this is when the control budget is established. No amount of management expertise or corrective action can salvage a project that was priced too low in the beginning. In this regard, there likely will always be some work classifications whose actual costs will exceed those estimated. The project manager is primarily responsible for getting the total project built for the estimated cost. If some costs go over, it is anticipated that savings in other areas can counterbalance them.

Having identified where production costs are excessive, project management must then decide just what to do about them, if indeed anything can be done. Certainly, the hourly rates for labor and equipment are not controllable by management. The only real opportunity for cost control resides in the area of improving production rates. This element of work performance can, to a degree, be favorably influenced by skilled field supervision, astute job management, energetic resource expediting, improved makeup of labor crews, and judicious selection of equipment.

Any efforts to improve field production must be based on detailed knowledge of the pertinent facts. If the cause of excessive costs is not identified specifically, a satisfactory solution is not likely to be found. It is impossible to generalize on this particular matter, but certainly the treatment must be gauged to the disease. In the usual case, full cooperation between the field supervisors and project management is needed before any real cost improvement can be realized. Field supervisors play a key role in implementing corrective procedures. There are no precise guidelines for reducing excessive project costs. The effectiveness of corrective procedures depends largely on the ingenuity, resourcefulness, and energy of the people involved.

Production costs often are too high early in the construction process but tend to become lower as the work progresses. For instance, the unit costs of forming abutment #1 on the highway bridge would be expected to be higher than those for the second abutment. One reason for this is that, on repetitive operations, costs have a tendency to decline as experience increases. Also, some work done on the first abutment, such as cutting wales to length, cutting and splicing braces, and drilling holes for form ties, will not have to be repeated for the second abutment. Production rates generally have a tendency to improve as the job proceeds because craft workers become familiar with the job and learn how to work with one another as a team.

11.25 Information for Estimating

Estimating needs production rates and unit costs that are a balanced time average of good days and bad days, and high production and low production. For this reason, information for estimating is normally not recovered from the cost accounting system until after project completion, or at least not until all the work type being reported has been finished. In this way, the best possible time-average rates will be obtained. Permanent files of cost and productivity information are maintained, providing the estimator with immediate access to data accumulated from prior projects.

Both production rates and unit costs are available from the project cost accounting system. To be of maximum value in the future, however, it is important that such productivity data be accompanied by a description of the project work conditions that applied while the work was being done. Knowledge of the work methods, equipment types, weather, problems, and other job circumstances will make the basic cost and productivity information much more useful to an estimator. Such a written narrative becomes part of the total historical record of each cost account.

11.26 Postproject Evaluation

Experience is a valuable commodity in the construction contracting business. In a high-risk industry, experience enables contractors to reduce costs, improve production, and succeed at projects. Contractors must have an effective means of gaining, documenting, and maintaining project experience for utilization on future jobs. The project cost system is an important component of this process. It records the costs and production achieved under actual project conditions. Yet other, less tangible project experiences cannot be easily captured through the cost system alone. These include organizational and operational lessons, management issues, and the development of successful tools and procedures. In order to capture these learned lessons, contractors must have a system for evaluating their projects after completion.

This evaluation is best accomplished by reconvening key project staff upon project completion for a brainstorming session. A technique frequently used by groups to facilitate beneficial feedback is called the *Plus/Delta technique*. Project staff are asked to list on individual sticky notes all of the positive elements of their project experience as well as all of the elements they would change if given a second opportunity. These notes are then collected and organized by the team into similar categories. Broad descriptive statements are drawn up describing each category. The categories are then analyzed by the group, and the results are formally documented.

In addition to brainstorming, the project team should list in flowchart format any specialized procedures or processes developed and utilized on the job. Sometimes staff will manufacture a tool to fulfill a specialized requirement; other times they will assemble a database or spreadsheet to automate repetitive project tasks. This information is often applicable to other projects and therefore should be collected, documented, and distributed to other project managers.

Postproject evaluation should appraise the success of the project planning and scheduling effort. The project team must decide whether the project plan and its associated schedules were useful during construction. If they were not, further analysis must be undertaken to determine why and how to make the planning more meaningful on future work.

An underlying principle of the postproject evaluation is that the goal is to learn from the current project as an element of continuous improvement for the company. The objective is not to identify scapegoats and place blame. If full and enthusiastic participation is to be elicited from each participant, the environment must be nonthreatening, with the focus on what went well, what could be improved upon, and what should be avoided on future projects. If necessary, accountability can be assessed later with appropriate disciplinary action, but for the postproject evaluation, a positive environment must be maintained.

This final step in the evolution of a project is fundamental to the advancement of a construction company. Yet, in most cases, the lessons learned on a project are closed out with the work and are never added to the experience of the company. Even on poorly performing projects—in fact, especially on poorly performing projects—learning from experience assures improved performance on the next job. It is said in the construction industry that good judgment comes from good experience but good experience comes from bad judgment.

11.27 Software Applications

Computers are extensively used by construction contractors in conjunction with their project cost systems. Cost accounting and analysis can become laborious and time consuming, even for relatively small operations. The computer offers the advantages of speed, economy, and accuracy. In addition, the computer provides a cost system with flexibility and depth that manual systems often cannot match. This does not mean that job costs cannot be developed satisfactorily by hand. Many small contractors have perfectly adequate manual cost systems. Experience indicates, however, that few contractors of substantial size are able to manually generate timely field cost reports in the detail required to serve a genuine cost-control purpose. Contractors often find that manual methods serve well as a generator of estimating information but not for cost control. A common experience in this regard is that the project is finished before the contractor knows the profit status of the work. It is realistic to recognize that most contractors find computer support to be a necessary part of their project cost system.

Cost accounting software is now very flexible in the sense that just about any cost report or information that project management may desire can be generated. Cost reports in the same general formats as those presented in this chapter are produced by several current accounting systems. The programs commonly used by contractors actually perform a broad series of cost accounting and financial accounting functions. After input of cost and production information, the computer generates payroll checks, keeps payroll records, maintains the equipment accounts, and performs other functions, as well as producing a variety of productivity and cost reports and project cost forecasts.

11.28 Accuracy of Estimating

A functioning project cost system acts as a detailed check on estimating accuracy. However, many contractors do not consistently maintain a viable project cost system. In addition, the essential thrust of a cost accounting system is not directed toward evaluating the overall "accuracy" of the estimating process itself, with "accuracy" here referring to the magnitude of the deviation between the estimated cost and the actual cost of the project as a whole, as well as each line item in the estimate. Accuracy in the computation of construction costs is crucial to the success of a construction firm, and studies of estimating performance are an important element in making periodic reviews of company performance. An occasional analysis of estimating accuracy will provide a construction company with revealing information of great value in assessing the efficacy of its estimating procedures and practices.

Key Points and Questions

Key Points

- The cost code ties the entire cost-control system together throughout the entire project.
- The labor and equipment elements of cost are the primary focus of the project manager, though material, subcontract, and job overhead costs also need to be borne in mind.
- Accurate and honest labor reporting is of fundamental importance to cost control and estimating.
- The Earned Value Management System enables the project manager to have a broader view of project status than either the scheduling or cost-control systems alone.
- An honest postproject evaluation is critical to continuous improvement for a construction company.

Review Questions and Problems

1. The construction team uses various models to describe the project that is to be built. What are the various models used by the construction team, and what does each depict?
2. What are the two important objectives of the project cost-control system?
3. What essential role does the cost-coding system play?
4. Why does the project manager focus primarily on labor and equipment costs and less on materials, subcontracts, and project overhead?
5. Current technology gives project managers instant access to cost accounting systems virtually anytime. Why is weekly reporting still the most common cycle for construction cost accounting reports?
6. Why is accurate and honest time reporting critical?
7. What advantage does an Earned Value Management System have over more traditional cost-control and schedule-control systems?
8. Why is production cost reduction the primary means to deal with cost overruns?
9. Why does the placement of blame have no place in the postproject evaluation process?

12 Project Financial Management

12.1 Introduction

Previous chapters have discussed the essential elements of construction time, cost, and resource control. One additional major feature of the project management system remains to be treated: financial control. The project manager bears the overall responsibility for financial management of the work. This includes carrying out such fiscal duties as may be imposed by the construction contract and implementing appropriate monetary procedures according to the dictates of good business practice. Project financial management can involve a broad range of responsibilities.

Learning objectives for this chapter include:

- ❏ Understand the broad financial responsibilities of the project manager.
- ❏ Recognize the variety of ways in which progress payments are made, depending upon the specific type of contract.
- ❏ Recognize the importance of cash flow to the contractor and how the project manager can positively (or negatively) impact project cash flow.
- ❏ Understand the importance of properly managing the change processes.

12.2 Financial Control

Construction contracts normally require that contractors perform prescribed duties of a financial nature. For example, they are made responsible for certain aspects of the payment process. This can include project cost breakdowns, the forecasted schedule of progress payments, preparation or approval of periodic pay estimates, and documentation required for final payment. Construction contracts prescribe specific procedures to be followed by the contractor with regard to payment for extra work, extensions of time, processing of change orders, claims, and settlement of disputes.

The project manager is also responsible to the company for implementing and maintaining standard fiscal procedures. One of the most important of these is monitoring project cash requirements during the contract period. Even a highly profitable job will require a considerable amount of cash to meet payrolls, purchase materials, and meet other project obligations. The size and timing of these cash demands is a serious matter for the contractor, and appropriate financial forecasts must be made. In fact, one of the most common causes for construction companies failing is lack of operating cash, which puts the contractor out of business even if the companies have jobs that would be profitable if they could be completed. A system of disbursement control is needed to regulate and control payments to material vendors, subcontractors, and others.

Another aspect of financial control is that of maintaining a complete and detailed daily record of the project. Such a job log can be invaluable in the settlement of claims and disputes that may arise from the work. This job history includes names, dates, places, and documentation of everything that happens as well as everything that fails to happen.

12.3 Progress Payments

Construction contracts typically provide that the owner shall make partial payments of the contract amount to the prime contractor as the work progresses. Payment at monthly intervals is the usual proviso. Depending on the type of work and contract provisions, the monthly pay requests may be prepared by the contractor, the architect-engineer, or the owner. In any event, a pay request is prepared periodically, and the cost of the work accomplished since the owner made the last payment to the contractor must be compiled. Typically, this compilation is done in practice by determining the total value of work actually performed to date and then subtracting the sum of the previous progress payments made by the owner.

The total value of work done to date is obtained in different ways, depending on the type of contract. Under lump-sum contracts, progress customarily is measured in terms of estimated percentages of completion of major job components. The quantities of work done on unit-price contracts are determined by actual field measurement of the bid items put into place. In either type of contract, materials stored onsite usually are taken into account, as well as any prefabrication or preassembly work that the contractor may have done at some location other than the job site.

In accordance with the terms of the contract, the owner usually retains a prescribed percentage of each progress payment. A retainage of 10 percent is common, although other percentages are also used. To an increasing extent, construction contracts provide that retainage shall be withheld only during the first half of the project. After that, if the work is progressing satisfactorily and with the consent of the surety, all subsequent progress payments are made in full, thus effectively reducing retainage to 5 percent by the end of the job. The retainage is held by the owner until the work receives final certification by the architect-engineer, the owner accepts the project, and the contractor submits any required affidavits and releases of lien. Final payment is then made to the contractor, including the accumulated retainage.

Negotiated contracts of the cost-plus variety usually provide for the contractor's submission of payment vouchers to the owner at specified intervals during the life of the contract. A common provision is weekly reimbursement of payrolls and monthly reimbursement of all other costs, including a pro rata share of the contractor's fee. It is not uncommon under this type of contract for the owner to pay all vouchers in full without deducting any percentage as retainage. Some contracts provide for the retention of a stated percentage of the contractor's fee. Others provide that the owner make full reimbursement to the contractor up to some designated percentage (80 percent sometimes is used) of the total estimated project cost. Further payments then are withheld until some specified amount of money has been set aside. The owner retains this reserve until the project has been satisfactorily completed.

12.4 Pay Requests for Unit-Price Contracts

Payment requests under unit-price contracts are based on actual quantities of each bid item completed to date. This is the case with the highway bridge that was bid, as shown in Figure 3.9, as 12 different bid items. The measurement of quantities in the field is done in several different ways, depending on the nature of the particular bid item. When cubic yards of aggregate, tons of asphaltic concrete, or bags of portland cement are set up as bid items, these quantities are usually measured as they are delivered to the work site. Delivery tickets or fabricator's certificates are used to establish tons of reinforcing or structural steel. Other work classifications, such as cubic yards of excavation, lineal feet of pipe, or cubic yards of concrete, are measured in place or computed from field dimensions. Survey crews of the owner and of the contractor often make their measurements independently and adjust any differences.

Many owners use their own standard forms for monthly pay estimates. On unit-price contracts, the owner often prepares the pay request and sends it to the contractor for checking and approval before payment is made. On unit-price contracts that involve a substantial number of bid items, each monthly pay request is a sizable document consisting of many pages. In essence, the total amount of work accomplished to date of each bid item is multiplied by its corresponding contract unit price. All of the bid items are totaled, and the value of materials stored on the site is then added. The prescribed retainage then is subtracted from this total. The resulting figure represents the entire amount due the contractor for its work completed to date. Then the sum of all prior progress payments that have already been paid is subtracted, yielding the net amount of money payable to the contractor for its work that month. Figure 12.1 is the pay request for the highway bridge, covering the work accomplished during the month of July.

Payment Request

Project Highway Bridge
Periodic Payment No. 2 for Period July 1–July 31

Item No.	Description	Quantity	Unit	Unit Price	Amount
1	Excavation, unclassified	1,667	cy	$3.83	$6,384.61
2	Excavation, structural	120	cy	$50.72	$6,086.40
3	Backfill, compacted	0	cy	$17.77	$0.00
4	Piling, steel	2,240	lf	$66.46	$148,870.40
5	Concrete, footings	120	cy	$195.62	$23,474.40
6	Concrete, abutments	140	cy	$380.87	$53,321.80
7	Concrete, deck slab, 10 in.	0	sy	$153.03	$0.00
8	Steel, reinforcing	57,150	lb	$1.045	$59,721.75
9	Steel, structural	0	lb	$1.725	$0.00
10	Bearing plates	0	lb	$2.49	$0.00
11	Guardrail	0	ft	$93.59	$0.00
12	Paint	0	ls	$12,140.39	$0.00
	Construction performed to date				$297,859.36
	Materials stored onsite (schedule attached)				$0.00
	Total work performed and materials stored				$297,859.36
	Less 10% retention				$29,785.94
	Net work performed and materials stored				$268,073.42
	Less amount of previous payments				$1,441.00
	Balance due this payment				$266,632.42

Figure 12.1
Highway bridge, monthly pay request

As explained, monthly pay requests on unit-price contracts are based on the bid-item quantities actually accomplished. One way to obtain progress measurements is from work quantities assigned to the network activities in the project time schedule. However, while activities can serve well as a means of obtaining work quantities put into place for internal cost accounting purposes, owners usually do not consider this an acceptable basis for compiling monthly pay requests. Unit-price contracts often do not have well-defined quantities shown on the construction drawings. Activity quantities taken from the construction drawings before work commences may not be accurate representations of the work amounts actually needed to accomplish the project. Under unit-price contracts, payment to the contractor must be based on field-determined quantities of the individual bid items. Even though schedule activity quantities work well to establish cost control on a weekly basis, occasionally they must be checked and corrected by actual field measurement. The field quantity measurements made monthly for payment purposes also accomplish this goal.

12.5 Project Cost Breakdown

Construction contracts for lump-sum projects normally require that a cost breakdown of the project be submitted by the contractor to the owner or architect-engineer for approval, before submission of the first pay request. This cost breakdown, also called a schedule of values, serves as the basis for subsequent monthly pay requests. The breakdown, which is actually a schedule of costs of the various components of the structure, is prepared to assist the owner or architect-engineer in checking the contractor's pay requests. It must be noted here that the example projects used in this text so far to illustrate costs have been unit-price projects. For illustrative purposes, it is necessary, at this point, to introduce data for a lump-sum job taken from an external source. Figure 12.2, which presents a typical monthly pay request for a lump-sum job, shows the cost breakdown of the project in column 1. Occasionally, the owner will specify the individual work items for which the contractor is to present cost figures. In the absence of such instructions, it is usual to prepare the schedule of values using the same general items as they appear in the specifications and on the final recap sheet of the estimate. This practice minimizes the time and effort required to compute the breakdown values and offers maximal accuracy in the results.

The cost shown in column 1 of Figure 12.2 for each work item consists of the direct expense of the item plus a share of the cost of job overhead, taxes, markup, and bond. The cost of these four items, often called job burden, may be on a strictly pro rata basis, or there may be some unbalancing. It is common practice to include a disproportionate share of the burden in those items of work that are completed early in the construction period.

12 Project Financial Management

THE BLANK CONSTRUCTION COMPANY, INC.
1938 Cranbrook Lane
Portland, Ohio

Periodic Estimate for Partial Payment

Project Municipal Airport Terminal Building   Location Portland, Ohio

Periodic Estimate No. 4   For Period September 1, 20XX to September 30, 20XX

Item No.	Item Description	Total Cost (1)	Completed to Date (2)	Cost to Complete (3)	Percent Complete (4)
1	Clearing and Grubbing	$22,508	$22,508	$0	100
2	Excavation and Fill	$67,559	$57,425	$10,134	85
3	Concrete and Forms				
	Footings	$108,572	$77,086	$31,486	71
	Grade Beams	$108,898	$63,161	$45,737	58
	Beams	$65,960	$6,596	$59,364	10
	Columns	$21,306	$6,392	$14,914	30
	Slabs	$383,093	$45,971	$337,122	12
	Walls	$59,566	$21,444	$38,122	36
	Stairs	$43,380	$0	$43,380	0
	Sidewalks	$35,662	$0	$35,662	0
4	Masonry	$734,584	$22,038	$712,546	3
5	Carpentry	$57,025	$0	$57,025	0
6	Millwork	$129,284	$0	$129,284	0
7	Steel and Misc. Iron				
	Reinforcing Steel	$129,218	$67,193	$62,025	52
	Mesh	$28,382	$4,257	$24,125	15
	Joist	$140,334	$0	$140,334	0
	Structural	$269,908	$45,884	$224,024	17
8	Insulation	$38,782	$0	$38,782	0
9	Caulk and Weatherstrip	$8,667	$0	$8,667	0
10	Lath, Planter, and Stucco	$296,270	$0	$296,270	0
11	Ceramic Tile	$33,651	$0	$33,651	0
12	Roofing and Sheet Metal	$312,079	$12,483	$299,596	4
13	Resilient Flooring	$39,381	$0	$39,381	0
14	Acoustical Tile	$48,771	$0	$48,771	0
15	Painting	$145,071	$0	$145,071	0
16	Glass and Glazing	$108,478	$0	$108,478	0
17	Terrazzo	$138,859	$0	$138,859	0
18	Miscellaneous Metals	$120,238	$0	$120,238	0
19	Finish Hardware	$83,304	$0	$83,304	0
20	Plumbing, Heating, Air Cond.	$939,150	$84,524	$854,627	9
21	Electrical	$592,057	$29,603	$562,454	5
22	Clean Glass	$4,873	$0	$4,873	0
23	Paving, Curb, and Gutter	$123,381	$0	$123,381	0
		$5,438,251	$566,565	$4,871,686	
24	Change Order No. 1	$5,240	$0	$5,240	0
	Total Contract Amount	$5,443,491	$566,565	$4,876,926	10.4%

A	Cost of Work Performed to Date	$566,565
B	Materials Stored onsite (Schedule Attached)	$102,207
C	Total Work Performed and Materials Stored	$668,772
D	Less 10% Retainage	$66,877
E	Net Work Performed and Materials Stored	$601,895
F	Less Amount of Previous Payments	$272,309
G	Balance Due This Payment	$329,586

*Adapted from Richard H. Clough, Glenn A. Sears, and S. Keoki Sears, *Construction Contracting*, 7th ed.(Hoboken, NJ: John Wiley & Sons, 2005), p. 264.

Figure 12.2
Periodic estimate for partial payment, lump-sum contract

This procedure, called front-end loading, serves to help to reimburse the contractor for its initial costs of moving in, setting up, and commencing operations. If the owner will accept a specific pay item for move-in costs, then such unbalancing of the cost breakdown is not necessary.

12.6 Pay Requests for Lump-Sum Contracts

Figure 12.2 illustrates the form of a typical pay request for a lump-sum building contract. Usually prepared by the contractor, it includes all subcontracted work as well as that done by the contractor's own forces. For each work classification done by the contractor, an estimate is made of the percentage completed and in place. From invoices submitted by its subcontractors, suitable percentage figures are entered for all subcontracted work. These percentages are shown in column 4 of the figure. The total value of each work classification is multiplied by its completion percentage, and these figures are shown in column 2. To the total of completed work is added the value of all materials stored onsite but not yet incorporated into the work. The cost of stored materials includes that of the subcontractors and customarily is set forth in a supporting schedule. The retainage is subcontracted from the total of work in place and materials stored onsite. This gives the total amount of money due the contractor up to the date of the pay request. From this is subtracted the amount of the progress payments already made. The resulting figure gives the net amount now payable to the contractor.

Although the pay request procedure for lump-sum contracts has been in general use for many years, it has one serious defect. As shown in Figure 12.2, the project is divided for payment purposes into major work classifications, most of which are extensive and often extend over appreciable portions of the construction period. This situation can make it difficult to estimate accurately the percentages of the various work categories completed. Actual measurement of the work quantities accomplished to date is the key to accurate percentage figures, but this task can become very laborious. Therefore, most of the percentages are established by a visual appraisal and negotiations between the project manager and the architect-engineer or owner. Contractors want these estimates to be fair representations of the actual work achieved, but understandably, they do not want them to be too low. Hence, most of their percentage estimates are apt to be on the generous side.

This circumstance continues to produce vexing problems for both the contractor and the architect-engineer or owner. If it is difficult for the contractor to estimate the completion percentages accurately, it is at least equally difficult for the architect-engineer or owner to check these reported values. This situation presents the architect-engineer with a difficult problem. In the interest of its client, it must make an honest effort to see that the monthly payments made to the contractor are reasonably representative of the actual progress of the job. In addition, architect-engineers have been sued by sureties in cases where defaulting contractors had received excessive progress payments. In such cases, the surety has claimed that the architect-engineer was negligent in approving progress payments that were substantially in excess of the value of work actually accomplished. Architect-engineers are at times casual with the processing of a pay request, feeling that a delay in payment will offset the generous nature of the completion percentages provided by the contractor. Although effective, this procedure is often at odds with contract provisions regarding payment and hardly gets to the basis of the problem.

Although retainage helps to protect against excessive owner payments, it is not inconceivable, nor probably unusual, that contractors' progress payments are occasionally more than they should be. To protect its client and itself, the architect-engineer sometimes will delay payment to the contractor or reduce the amount of payment requested. The unfortunate part about this entire matter is that the architect-engineer is acting more on hunch or intuition than on solid evidence of inflated payment figures.

12.7 Use of Time-Control Activities for Pay Requests

One possible answer to the pay request dilemma on lump-sum jobs is the use of time-control network activities or groups of activities for the project cost breakdown. Although this scheme undoubtedly increases the work necessary to make the initial cost breakdown and increases the length of the individual pay request, the advantages can justify the additional effort required. If direct costs have been assigned to each activity by the contractor for cost-control purposes, the activity format for pay requests is a natural extension of the basic cost system. If activities are not used for cost control, the direct cost (labor, equipment, materials, and subcontracts) of each activity, or for an activity group, can be obtained from the estimating sheets. It is an easy matter to distribute the project burden among the activities. Perfection is not required, and a certain amount of front-end loading probably will be done anyway to help the contractor recover more quickly its substantial mobilization costs.

When the total cost of each activity is known, the routine weekly progress reports can serve as a convenient and accurate basis for the monthly pay requests. For payment purposes, all activities finished by the end of the month would be reported as 100 percent complete. The percentages complete reported for the activities in progress as of the end of the month would be the same as those already available from the latest weekly progress report (Figure 10.5).

The advantages of using activities for pay requests on lump-sum projects are several. From the contractor's point of view, although more pay items are involved, it is an easy matter to compile the total cost of the work accomplished to date. No additional field measurements or inspections are required. The time-control information already available is all that is needed. The compilation of the total value of the work completed to date is more accurate than it would otherwise be, and the cost figures for each pay item, and the percentages completed, are much easier to check. Use of activities can eliminate many of the most troublesome aspects of monthly pay requests. On lump-sum jobs, basing pay requests on network activities seems to have much to commend it, and this procedure is now being used by many public agencies.

As discussed in Sections 11.21 and 11.22, the Earned Value Management System (EVMS) is a convenient way to tie cost and schedule together. EVMS is becoming a popular basis for pay requests on larger projects with more sophisticated owners. Many owners prefer this approach, as it removes much of the uncertainty inherent in the more traditional approaches to developing pay requests.

12.8 Pay Requests for Cost-Plus Contracts

Negotiated contracts of the cost-plus variety provide numerous methods for making payments to contractors. In many cases, contractors furnish their own capital, receiving periodic reimbursement from owners for costs incurred. Other contracts provide that the owner will advance the contractor money to meet payroll and to pay other expenses associated with the work. In one scheme, the contractor prepares estimates of expenses for the coming month and receives the money in advance. Then, at month's end, the contractor prepares an accounting of its actual expenses. Any difference between the estimated expenses and actual expenses is adjusted with the next monthly estimate. Other contracts provide for zero-balance or constant-balance bank accounts. In this case, checks are written by the contractor and funds are furnished by the owner. In any event, the contractor must periodically account to the owner for the cost of the work, either to receive direct payment from the owner or to obtain further advances of funds.

The matter of periodic payments to the contractor under a cost-plus type of contract is not based on quantities of work done but on expenses incurred by the contractor in prosecuting the work. Consequently, such pay requests consist primarily of the submission of original cost records. Invoices, payrolls, vouchers, and receipts are submitted to substantiate the contractor's payment requests. In addition to cost records of direct expense, the periodic pay requests customarily include a pro rata share of the negotiated fee. If the contractor receives advances for construction costs, the owner customarily is credited with all cash discounts.

Because of the sensitive nature of cost reimbursement, it is common practice to maintain a separate set of accounting records for each cost-plus

12.9 Payments to Subcontractors

When the general contractor receives monthly invoices from its subcontractors, it has its own problems of verifying the requested amounts. In many instances, the general contractor does not require or receive cost breakdowns from its subcontractors, and the project manager must expend considerable time and effort in checking subcontractor invoices. Even then, unless it happens to be experienced in each construction specialty, the project manager usually has no real basis for accurately evaluating the progress of a subcontractor's work on a project.

Many prime contractors require their subcontractors to submit appropriate cost breakdowns for payment purposes. The contractors use these cost schedules to evaluate the pay requests submitted by their subcontractors. One form of price breakdown that can be used is to have each subcontractor place a price tag on every network activity with which it is involved. It is not difficult for the project manager to check the reasonableness of the reported amounts for each activity. Once the contractor and the subcontractor have agreed that prescribed sums are due on the completion of designated activities, the analysis and verification of monthly invoices from subcontractors are readily performed.

General contractors are eager to prepare their requests for payment and transmit them to the owners as soon after the first of the month as possible. With a breakdown of subcontractor prices for each activity, the project managers can determine the amount due each subcontractor from their own evaluation of activity progress. This evaluation can be substantiated with the subcontractor and placed in the prime contractor's request for payment. This process can reduce the time necessary to prepare the pay estimate and send it on to the owner. Subcontractors also benefit because earlier payment to the general contractor means earlier payment to them.

12.10 Schedule of Payments by Owner—Unit-Price Contract

A common contract provision requires the contractor to provide the owner, before the start of construction, with an estimated schedule of monthly payments that will become due during the construction period. The owner needs this information so that it can have cash available to make the necessary periodic payments to the contractor. Because the owner sometimes must sell bonds or other forms of securities to obtain funds with which to pay the contractor, it is important that the anticipated payment schedule be as accurate a forecast as possible.

When a unit-price contract is involved, the payment schedule depends entirely on the quantity of each bid item that is planned for completion at the end of each month during the construction period. The highway bridge was bid on the basis of unit prices, and Figure 3.9 is a schedule of the bid items and their respective quantities and bid amounts. Once the job schedule has been established, it is an easy matter to make a compilation of total bid-item quantities that are scheduled to be completed as of the end of each month. Using Chart 5.3a on the companion website, the schedule of progress payments shown in Figure 12.3 is made quickly and directly.

It should be noted in Figure 12.3 that the total value of work accomplished as of the end of each month is compiled on the basis of bid-item quantities completed and their corresponding bid unit prices. Any variation between the actual quantities and the estimated quantities will make the schedule of payments inaccurate and will result in a future modification of the payment schedule. It has been assumed on the highway bridge that a retainage of 10 percent will be withheld by the owner until contract completion. In accordance with this, the total amount due the contractor as of the end of each month is multiplied by a factor of 0.90. Because the highway bridge is to be completed in September, the payment for that month has been shown to be the final payment, including accumulated retainage. In fact, the monthly payment amounts obtained in Figure 12.3 usually will not be paid at the end of the month. Rather, construction contracts frequently provide that the owner shall make payment to the contractor 10 days after a suitable pay request has been submitted. Contractors normally submit such pay requests at the end of each month, so payment by the owner is not made until about the tenth of the following month.

The monthly payment schedule presented in Figure 12.3 was prepared with all activities starting as early as possible, with the exception of activity 40, "Move in," and the next three activities 80, 90, and 100. As discussed in Section 7.21 and Section 8.6, the starts of these four activities are being purposely delayed by nine working days. You may recall that this move will not delay the project in any way and will eliminate a substantial period during which the project would otherwise be at a complete standstill, awaiting delivery of steel pilings and reinforcing steel. It has been amply demonstrated in this text that, although early-start schedules serve as optimistic goals, there are many practical reasons why such schedules often must be modified. Deviations from the early-start schedule are almost always in the nature of delays and later starts. As a result, the monthly payment schedule obtained in Figure 12.3 may well indicate that the early monthly payments are too high and the later payments are too low. If monthly payment schedules based on an early-start schedule are submitted to the owner, the first

Item No.	Bid Item	Estimated Quantity	Bid Unit Price	June 30 Estimated Quantity Complete	June 30 Total Value to Date	July 31 Estimated Quantity Complete	July 31 Total Value to Date	August 31 Estimated Quantity Complete	August 31 Total Value to Date	September 30 Estimated Quantity Complete	September 30 Total Value to Date
1	Excavation, unclassified	1,667	$3.83	418	$1,601	1,667	$6,385	1,667	$6,385	1,667	$6,385
2	Excavation, structural	120	$50.72			120	$6,086	120	$6,086	120	$6,086
3	Backfill, compacted	340	$17.77					340	$6,042	340	$6,042
4	Piling, steel	2,240	$66.46			2,240	$148,870	2,240	$148,870	2,240	$148,870
5	Concrete, footings	120	$195.62			120	$23,474	120	$23,474	120	$23,474
6	Concrete, abutments	280	$380.87			140	$53,322	280	$106,644	280	$106,644
7	Concrete, deck slab, 10 in.	200	$153.03					200	$30,606	200	$30,606
8	Steel, reinforcing	90,000	$1.045			57,150	$59,722	90,000	$94,050	90,000	$94,050
9	Steel, structural	65,500	$1.725					65,500	$112,988	65,500	$112,988
10	Bearing plates	3,200	$2.49					3,200	$7,968	3,200	$7,968
11	Guardrail	120	$93.59					40	$3,744	120	$11,231
12	Paint	job	$12,140.39					20%	$2,428	100%	$12,140
			Totals		$1,601		$297,859		$549,284		$566,484

	Monthly Progress Payment	Total Payment on Contract
June 30	1,601 × 90% = $1,441	$1,441
July 31	$297,859 × 90% − $1,441 = $266,633	$268,073
August 31	$549,284 × 90% − $268,073 = $226,283	$494,356
September 30	$566,484 × 90% − $494,356 = $15,480	$509,836
Final Payment	$56,648	$566,484

Figure 12.3
Highway bridge, schedule of progress payments

few pay requests submitted by the contractor may be smaller than those forecasted because of inconsequential schedule slippages. This is apt to lead the owner and architect-engineer to conclude that the work is falling behind schedule even when things are progressing very well.

If schedule adjustments must be made to effect project shortening or the leveling of resources, the monthly payment schedule can be made to reflect these. Otherwise, making allowances for the inevitable schedule slippages is very much a matter of intuition and judgment. One approach to this matter would be to compile a payment schedule on the basis of a late-start schedule. That is to say, assume that every activity will start as late as possible. This would, of course, produce a payment schedule with low initial payments and high later payments. Presumably, the actual payment schedule will be somewhere between the early-start and the late-start payment schedules.

12.11 Schedule of Payments by Owner—Lump-Sum Contract

The payment schedule for a lump-sum project often is computed using the traditional bar chart—one that is not network based. An approximate cost is established for each bar on the chart, and this cost is distributed uniformly over the length of the bar. From this, a cumulative total project cost curve is computed for the contract period. The problem with this procedure is the inaccurate way the costs of the project segments are distributed over the contract period. Although the total cost of each major job segment can be established with reasonable accuracy, the time rates at which these expenses are incurred can involve such variation that the payment schedule derived therefrom may be seriously in error. To illustrate, bar charts often have a category called "Electrical" or "Mechanical," whose bar extends from project start to finish. The total value of this work, which is usually subcontracted, is distributed uniformly over its duration. This is seldom a realistic time allocation of expense.

A much more dependable procedure is to compute cumulative job cost values using the network-based project schedule and cost of each activity as a basis. Closely associated activities can be combined for simplicity. The total cost (direct cost, plus a pro rata share of job overhead, tax, markup, and bond) of each activity is required. This is the same activity cost that would be needed for monthly payment purposes in Section 12.7. Dividing the total cost of each activity by its estimated duration gives the cost of the activity per working day. Because the duration of each activity is short compared with a monthly pay period, the error induced by the uniform distribution of cost is negligible. Using the activity cost figures and the project schedule, the cumulative project expense can be compiled. One way this can be accomplished is to use a worksheet with a horizontal time scale in terms of working days. The activities can be listed vertically, with the daily cost of each activity entered for the appropriate workdays. The costs of activities involving the procurement of job materials are charged as a lump sum on the last day of their duration. Adding the costs cumulatively from left to right produces the daily total project cost to date.

Although the process just described produces a cumulative total project cost up through each working day, usually only those costs as of the end of each month are of real concern. Therefore, an alternative procedure can be used that is more direct. As of the end of each month, simply add up the values of all activities scheduled to be completed on or before the end of that month. For activities in process as of that time, include an appropriate percentage of their costs. This is a rapid, accurate way to determine the total value of the work that will have been accomplished as of the end of any specified month.

12.12 Final Payment

The steps leading up to acceptance of the project and final payment by the owner vary with the type of contract and the nature of the work involved. In a typical procedure, the contractor, having achieved substantial completion, requests a preliminary inspection. The owner or its authorized representative, in company with general contractor and subcontractor personnel, inspects the work. A punch list is made up describing all the observed deficiencies. After the work has been finalized and all deficiencies remedied, a final inspection is held. Following this, the owner makes formal written acceptance of the project and the contractor presents its application for final payment. Under a lump-sum form of contract, the final payment is the final contract price less the total of all previous payment installments made. With a unit-price contract, the final total quantities of all payment items are measured and the final contract amount is determined. Final payment is again equal to the contract price less the sum of all progress payments previously made. In all cases, final payment by the owner includes all retainage that has been withheld.

Construction contracts often require that the contractor's request for final payment be accompanied by a number of different documents. For example, releases or waivers of lien executed by the general contractor, all subcontractors, and material suppliers are common requirements on privately financed jobs. Other contracts call for an affidavit certifying that all payrolls, bills for materials, payment to subcontractors, and other indebtedness connected with the work have been paid or otherwise satisfied. Construction contracts frequently require the contractor to provide the owner with as-built drawings, various forms of written warranties, maintenance bonds, and literature pertaining to the operation and maintenance of job machinery. Consent of surety to final payment is an almost universal prerequisite. The project manager is responsible for obtaining the owner's acceptance and for conforming with contractual provisions pertaining to final payment.

12.13 Cash Flow

Construction projects can make substantial demands on a contractor's cash. Initially incurred are the usual start-up costs of moving in workers and equipment. These include erecting the field office, storage sheds, fences, and barricades. They also include job layout and installation of temporary electrical, water, telephone, sanitary, and other services. The premiums for performance and payment bonds, as well as for certain types of permits and project insurance, must be paid at the inception of field operations. There is seldom a pay item pertaining specifically to these start-up costs, and the contractor recovers these expenses only as the work progresses.

The contractor's investment in the job is increased even further after the work gets under way. It must meet its payroll costs on a weekly basis and will want to take advantage of cash discounts when paying material bills if it possibly can. The contractor does receive progress payments at monthly intervals from the owner. However, these payments are not normally due until sometime during the month following their submittal, and their amounts are reduced by the amount of the retainage.

As a consequence, the contractor's expense on a project typically exceeds its monthly progress payment income over an appreciable part of the construction period. The cash deficit on the project must be made up from the contractor's working capital, or money must be borrowed to provide the necessary operating funds. *Cash flow* refers to a contractor's income and outlay of cash. The net cash flow is the difference between disbursements and income at any point in time. A negative net cash flow means disbursements are exceeding income, a normal situation on even a highly profitable project during the greater part of its duration. A determination of the future rates of cash disbursements and cash income, together with their combined effect on the project cash balance, is called a *cash flow forecast*.

The comptroller or financial vice-president of a construction company is concerned with the combined effects of the cash flow forecasts of all of the company's projects. Since the cash flow forecast for the company is simply the sum of the forecasts for the individual projects, it is the responsibility of the project manager to determine the cash flow of the project and to make regular revisions to it as the job progresses.

Cash flow forecasting is equally important for both large and small construction companies. Large firms with many concurrent projects try to use the positive cash flows of one project to handle the negative cash flows of another. Where cash demands exceed the normal working capital, arrangements for short-term loans are made and repayment schedules established. Where cash income exceeds the demand, short-term investments are made, with a liquidation schedule to fit future demands. Cash forecasting can be done with no more than approximate accuracy, but it is a useful device for controlling a company's cash position. The corporate profit of a large company can be greatly enhanced by proper cash management.

Small construction companies as well as growing firms have a special cash flow situation. Working capital in these companies is almost always in short supply. There are continual requirements for additional equipment and tools to handle the growing size and number of projects. Lending institutions are hesitant to make large loans to small companies; this is especially true when the funds are needed for working capital. As a result, the growth of small firms often is limited by their cash flow position. The ability to forecast cash flow needs accurately and to manage cash as a resource can increase a firm's growth rate and its annual turnover of projects.

As workers work and materials are purchased on a construction project, their costs accumulate on the contractor's books. These costs are referred to as accrued costs. When payment is made for the labor and materials, these are cash disbursements. Accrued costs and cash disbursements are exactly equal in value, but their timing is different. Generally, disbursements follow accrued costs.

As workers work and materials are put into place, the value of a construction project increases. This increase in the value of the construction represents an accrued income to the contractor. Periodically, the contractor and the architect-engineer evaluate the value of the construction, and the contractor prepares a progress payment request. When the owner sends a check to the contractor, the contractor's accrued income becomes a cash receipt. Accrued income and cash receipts are equal in value, but as in the case of accrued costs and cash disbursements, their timing is different. In this case, the cash receipts usually follow the accrued income.

It is the intent of the contractor to accrue income faster than it accrues costs. In the end, this difference represents the contractor's profit. The problem is that the cost of the work put into a project is not directly related to the resulting income. Additionally, the time delays between the accrued costs and the cash disbursement are not the same as the time delays between the accrued income and the cash receipts. To forecast the amount of money the contractor must invest in the project, it is necessary to estimate the amount and the timing of the cash disbursements and the cash receipts.

12.14 Cash Disbursement Forecasts

Cash disbursements by contractors on construction projects can be divided into three classifications. First, there are the up-front costs, or initial expenses necessary to start the project. These include various payments that must be made before construction starts, such as the costs of bonds, permits, insurance, and expenses of a similar kind. The second group of disbursements involves the payment of direct job expenses. These include costs associated with payrolls, materials, construction equipment, and subcontractor payments. The third classification relates to payments for field overhead expense and tax.

The original cost estimate and project schedule provide the basic cost and time information needed to make a cash disbursement forecast. As described in Section 11.20, anticipated time-cost envelopes for the project can be developed for early-start (ES) and late-start (LS) schedules. As the project progresses, actual costs can be plotted to the current date, and then projections can be developed from that point forward. The actuals can be plotted on a weekly basis to correspond with payroll but more commonly are plotted on a monthly basis unless there is a compelling reason for more accuracy. Once these cumulative project costs are plotted, they are connected by a smooth S-curve, as shown in in Figure 12.4. Because the contractor pays its bills throughout each month, the smooth expense curve shown is considered to be an acceptable representation of project cash disbursements.

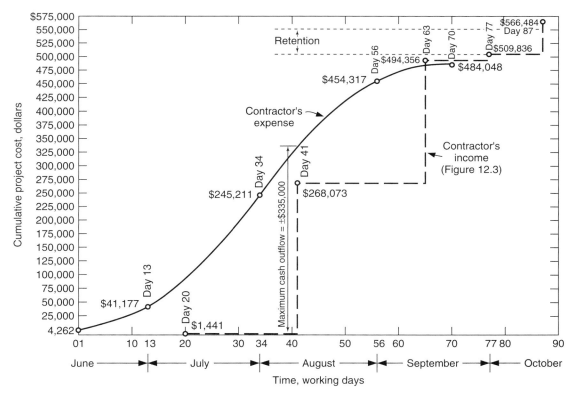

Figure 12.4
Highway bridge, contractor's expense and income

This procedure for calculating the time rate of disbursements on a construction job involves many approximations. Greater accuracy could be obtained by separating the direct job expenses into labor, materials, equipment, and subcontractor categories. For each of these cost components, different delays apply between the time the expense is incurred and the time it actually is paid. Such a procedure can be quite complicated, however. The assumption of a zero time delay for all direct costs is conservative and suitably accurate for most purposes.

12.15 Cash Income Forecasts

Cash income to the contractor consists of progress payments from the owner. Estimated amounts of the monthly payments will initially be consistent with billings for all work anticipated to be completed each month minus the retainage. However, as has been pointed out, these payments are not received by the contractor until about the tenth of the following month, and the delay could be much longer, depending on how the contract is written. Consequently, when the values are plotted, they must be displaced to the right by the amount of anticipated delay. Notice that the cash income curve is stepped, with income occurring only once each month.

The curves in Figure 12.4 are typical of most construction projects. At any point in time, the vertical distance between the two curves represents the amount by which the contractor's project expense exceeds its receipts. Figure 12.4 reveals that the contractor typically has a considerable amount of money invested in the owner's job for the entire construction period. It is not until receipt of the last progress payment that the cash flow becomes positive. The fact that cash flow does not become positive until the project is closed out makes it of paramount importance to the contractor to close out the project as soon as possible.

Since the final payment largely consists of retainage, it may be possible to improve the cash flow by negotiating an early return of some of the retainage prior to final closeout of the project. As an example, the electrical contractor on a building project may be asked to energize the power and lighting systems well before final completion of the project to eliminate the need for temporary power and lights. At this point, the systems are being turned over to the owner for beneficial use in completing the rest of the project. The case can be made that if the systems are energized and turned over by the electrical contractor, it is no longer required to maintain retainage on that part of the electrical contract, which will enable the withheld retainage to be diminished.

It can be clearly seen in Figure 12.4 that at some points during the construction period, the contractor's cash deficit is considerable. This is precisely the reason that some unbalancing, either in unit-price bid-items or in the schedule of values on lump-sum jobs, is such a common practice.

By front-loading the first pay items accomplished on a given project, the contractor can at least reduce the amount of its negative cash flow.

There are other ways of minimizing the amount of negative cash flow. Some of these are within the control of the project manager and some are not. Probably the best way to improve the job cash position is through the attainment of maximum production in the field. Monthly progress billings are based on the value of work accomplished, not on its cost (except on cost-plus jobs). Therefore, if the project can be built for less than the estimate, the progress billings remain as planned and the cost of labor and equipment is reduced. Subcontracting work has a positive effect on cash demands. Subcontractors add to the value of work and therefore add to the progress billing, yet typically they are paid only after receipt of the progress payment by the general contractor. The timing of delivery of large material orders to coincide with the submittal of the contractor's monthly pay estimate can reduce cash requirements. Favorable terms with material suppliers involving 60 and 90 days for payment will decrease cash requirements but may have an adverse effect on profit. Prompt submittal of monthly pay requests and follow-up action to ensure timely payment by the owner is an excellent way to reduce the negative cash flow.

Owners and architect-engineers are often unaware of the amount of money necessary to meet payrolls and other job expenses. As a result, often the contractor's monthly progress payments are delayed, and it must wait for its money, sometimes for substantial periods of time. This is a major problem for contractors that may be faced with the prospect of borrowing large amounts of short-term money at two or three percentage points above prime rate to cover deficits.

12.16 Disbursement Controls

To coordinate the actions of the company accounting office with the project, it is necessary to implement a system of disbursement controls. These controls are directed toward regulating payments made to vendors and subcontractors and require that no such payment be made without proper approval from the field. The basic purpose of disbursement control is twofold: Ensure that payment is made only up to the value of the goods and services received, and see that total payment does not exceed the amount established by the purchase order or subcontract.

Payments made for materials are based on the terms and conditions of covering purchase orders. Copies of all job purchase orders are provided to the project manager for use and information. Purchase order disbursement by the accounting office is conditional on the receipt of a signed delivery ticket or receiving report from the job site. Suitable internal controls are established to ensure that total payments do not exceed the purchase order amount. Any change in purchase order amount, terms, or conditions is in the form of a formal written modification, with copies sent to the job site.

Disbursements to subcontractors follow a similar pattern. Because there are no delivery tickets or receiving reports for subcontractors, all subcontractor invoices are routed for approval through the project manager, who has copies of all the subcontracts. The project manager compares the invoice with the progress determination and approves the invoice or makes appropriate changes. The subcontractor is not actually paid until the owner has made payment to the general contractor. General contractors normally withhold the same percentage from their subcontractors as is retained by owners. If the subcontractor bills for materials stored onsite, a common requirement is that copies of invoices be submitted to substantiate the amounts billed. Any change to a subcontract is accomplished by a formal change order.

12.17 Project Changes

Changes in the work or deviations from the anticipated job-site conditions can stem from a variety of causes. The owner or architect-engineer may decide to add additional work or change certain contract requirements. The contractor may suggest construction changes in accordance with the contract's value-engineering clause. The owner, architect-engineer, or another prime contractor may cause delay in the prosecution of the fieldwork. Job-site conditions may be encountered that are appreciably different from those originally contemplated. Changes of this sort can result in work extra to the contract, extensions of contract time, and claims for additional costs.

The usual construction contract is explicit about how changes are to be handled and the extent of the owner's liability. The project manager has an important responsibility in evaluating the effects of project changes and taking all the steps necessary to protect the contractor's interests. Doing this involves the negotiation of contract change orders, keeping detailed job records, analyzing the time and cost effects of project changes, documentation of extra costs, and timely notification to the owner of all job delays, extra costs, and claims under the contract. In particular, the job manager must be especially careful to proceed in full conformance with the applicable contract provisions.

The time and cost consequences of project changes are often difficult to document, especially in a form meaningful to the owner or architect-engineer, or perhaps that can serve as evidence in arbitration or litigation. The project network is a powerful tool for analyzing the time effects of a project change. This matter was discussed in Section 9.7. In a situation where a substantial change in the work occurs, it is advisable to update the network so that it is completely current but without reflecting the proposed change. Network calculations and a current cost forecast report are used to establish the time and cost status of the project before the change occurs. The project network is then modified to reflect the change in the work. A new set of network calculations is made, and the cost implications are figured into a revised cost forecast report. Any differences in project time and cost are the result of the proposed change.

Care must be taken in interpreting the results of this procedure. A change in the work can have far-reaching implications. As an example, a change in the foundation of a building project can seriously delay all of the work that follows. This may mean that the project cannot be closed out before winter weather, further delaying other trades. Major items of material and project equipment may already be ordered and must be received and stored prior to their need. Job delays can cause work to be rescheduled from periods of favorable to periods of unfavorable weather. On the highway bridge, a delay in the start of the project or trouble with an abutment foundation could push the work into the rainy season, when the normally dry riverbed becomes a swift stream. In some cases, delays can improve a bad-timing situation. Generally, however, the project plan has been designed to take all possible advantage of weather and other conditions. Therefore, delays usually tend to disrupt a project plan, making replanning more difficult and resulting in higher costs to the contractor.

It is always advisable to analyze the time and cost implications of project change as it occurs. One reason is that any revision made to the construction contract that incorporates a change must be based on an accurate assessment of the total effect of that change. Another reason for making a prompt study is that job changes can involve substantial time and cost and occasionally lead to disputes between owner and contractor that are not settled until long after the project is finished. At best, an analysis made long after the fact to demonstrate the effect of a job change produces data of questionable value and authenticity.

12.18 Contract Change Orders

Alterations to the contract involving modifications to the time or price of the project are consummated by formal change orders. These changes may alter the contract by additions, deletions, or modifications to the work, and can be initiated by the owner, architect-engineer, or contractor. The dollar amount of a change is negotiated and, depending on the contract terms, can be expressed as a lump sum, unit prices, or cost plus a fee. Sometimes a number of changes are incorporated into one change order. However, each change needs to be documented at its inception and estimated for cost and time consequences before work commences, and each change must receive written approval to proceed from the owner or architect-engineer. Strict adherence to this change policy greatly reduces misunderstandings.

An important part of a change order is any extension of contract time that may be required as a consequence of the change. In the absence of a project network, the contractor has no real basis for determining the additional contract time required. Usually, the contractor either will request no extra time or will request an extension equal to the full time required to accomplish the extra work itself. As was discussed in Section 9.7, the influence of the change on total project duration can be clearly demonstrated by performing a forward pass on the current network and a forward pass with the change incorporated into the network. Often, the change affects only a noncritical path, and no additional time is justified. Even so, the contractor's scheduling leeway has been decreased, with possible cost implications. If a longer critical path results, the net additional time actually required can be easily determined. Such a project network analysis can be very effective in substantiating a claim for additional contract time.

Part of the change order dollar amount is an allowance for job overhead. Many contracts provide that the cost of extra work shall include an amount for overhead, expressed as a fixed percentage (10 to 15 percent is common) of the direct cost involved. Basically, this is an attempt to reimburse the contractor for the office expense of estimating the cost of the change and processing the paperwork rather than an allowance for additional field overhead expense. This is a satisfactory arrangement when no increase in project time is involved but can be most unsatisfactory when the duration is extended. The job overhead on the highway bridge is $869 per day, and on larger projects, overhead can be a major expense. Network analysis can provide the number of days of additional project overhead that should be charged when such a change is being negotiated. Where a deductive change order is concerned, only direct costs will be involved unless the project duration actually is decreased.

12.19 Claims

During the construction period, disputes sometimes arise between the owner and the contractor concerning claims by the contractor for extensions of time or payment of extra costs. If such claims cannot be settled amicably during the construction period, they must either be dropped by the contractor or be settled by mediation, arbitration, appeal boards, or the courts.

Claims of this type can stem from a variety of conditions and often involve substantial sums of money. Job delays caused by the owner or another of its prime contractors result in many claims against the owner for impact costs or consequential damages. Impact costs are additional expenses incurred by the contractor as a consequence of a delay to the project. Claims for extensions of time and extra costs often result from failure of the owner to furnish sites, make timely decisions, or provide owner-furnished materials.

Errors or oversights on the part of the architect-engineer account for numerous contractor claims.

Another category of claims can result from the contractor's dealings with its subcontractors. The coordination and timing of the work of subcontractors is critical to the overall job schedule. A delay in the work of one subcontractor can have a domino effect on the work of everyone who follows. It was mentioned earlier that subcontractors should have a definite voice in the preparation of the original schedule. This tends to make the schedule realistic from the standpoint of all parties responsible for its execution. At the same time, it will show individual subcontractors how their work fits into the overall construction program. The project network, when kept updated, shows when all subcontractors start and complete their work and what effect any delay on their part will have on the work yet to be done. The updated diagram provides an excellent means of allocating responsibility among delaying parties and assigning the resulting financial responsibility.

Disputes and claims are commonplace in the construction industry. To assist the contractor in pursuing its own claims and in defending itself against claims made against it, the byword is *documentation*. If the contractor's position is to prevail in such matters, full and complete documentation of all pertinent facts and information is an absolute necessity. The as-built diagram discussed in Section 9.3D is one important element of a contractor's project documentation process. The standard dictum of "put everything in writing" is very important. Letters, memoranda, e-mails, drawings, notes, diaries, photographs, and clippings can be useful. Basic to job documentation is the daily job log.

12.20 Daily Job Log

A job log is a historical record of the daily events that take place on the job site. The information to be included is a matter of personal judgment but should include everything relevant to the work and its performance. The date, weather conditions, numbers of workers, and amounts of equipment should always be noted. It is advisable to indicate the numbers of workers by craft and to list the equipment items by type. A general discussion of daily progress, including a description of the activities completed and started, and an assessment of the work accomplished on activities in progress are important. Where possible and appropriate, quantities of work put into place can be included.

The diary should list the subcontractors who worked on the site, together with the workers and equipment provided. Note should be made of the performance of subcontractors and how well they are conforming to the project time schedule. Material deliveries received must be noted together with any shortages or damage incurred. It is especially important to note when material delivery dates are not met and to record the effect of such delays on job progress and costs.

The diary should include the names of visitors to the site and facts pertinent thereto. Visits by owner representatives, the architect-engineer, safety inspectors, union representatives, and people from utilities and governmental agencies should be documented and described. Meetings of various groups at the job site should be recorded, including the names of people in attendance, problems discussed, and conclusions reached.

Complete diary information is occasionally necessary for extra work and is always needed for any work that might involve a claim. The daily diary always should include a description of job problems and a list of the steps being taken to correct them. The job log is an especially important document when disputes result in arbitration or litigation. To be accepted by the courts as evidence, the job diary must meet several criteria. The entries in the log must be original entries made on the dates shown. The entries must have been made in the regular course of business and must constitute a regular company record. The entries must be made contemporaneously with the events being recorded and must be based on the personal knowledge of the person making them. It is also preferable that the log not be kept as loose-leaf pages, because sheets can be added or removed, but should be maintained in a bound booklet or journal. With the evolution of technology, many contractors are moving to an electronic format for the job log. Steps comparable to those previously discussed for the written log must be taken to ensure the integrity of the log in electronic format.

Where these criteria have been met, the courts generally have ruled that the diary itself can be used as evidence, even if the author is not available to testify. Also of importance is the tone and style used in daily diaries. Every attempt should be made to state facts clearly and objectively. Opinions, editorials, and emotion should always be avoided, as they detract from the credibility of the record.

Key Points and Questions

Key Points

- ❏ Timely, accurate pay applications are a key part of cash flow management.
- ❏ Managing cash flow is a critical responsibility of the project manager.
- ❏ The project manager has an important responsibility in evaluating the effects of project changes and taking all the steps necessary to protect the contractor's interests.
- ❏ The job log provides important documentation of the construction processes used on a job.

Review Questions and Problems

1. Why is it important for the project manager to be accurate in the submission of pay applications? What is lost if the amount requested is low? What is the risk if the amount requested is more than earned?
2. What is the importance of the schedule of values in project financial management?
3. Describe ways in which the project manager can positively or negatively impact the project cash flow.
4. Describe steps the project manager can take to protect the contractor's interests in a change order situation.
5. Why is it important for the project manager to expedite closeout of the project?
6. Why is the job log important?

Index

Acceleration, project, see Time acceleration, project
Accidents, planning for, 152
Accounting, cost, see Project cost accounting
Activities, 74–75
   arrow notation for, see Arrow notation
   burst, 106, 110, 112
   and calendar dates, 123
   cost slope of, 166–167
   crash cost of, 166–167
   critical, 113
   definition of, 23
   direct costs of, and time, 165–167
   duration of, 99–112
   early activity times, 106–108
   hammock, 133–134
   lags between, 124–127
   late activity times, 109–112
   logic of, 74–75
   merge, 106, 108
   planning of, 245–246
   precedence notation for, see Precedence notation
   presentation of activity times, 114
   restraints on, 75–76
   sorting, 217–220
Activity number sort, 217
Activity on node notation, see Precedence notation

Actual cost of work performed (ACWP), 300–301
ACWP, see Actual cost of work performed
AOA notation, see Arrow notation
AON notation, see Precedence notation
Architect-engineer, 5–6
Area code, 281
Arrow notation, 80–81
As-built schedules, 216
As-planned schedules, 215

BAC, see Budget at completion
Bar charts, 137–139,
Baseline schedules, 214, 215, 227, 228, 233, 300
BCWP, see Budgeted cost of work performed
BCWS, see Budgeted cost of work scheduled
Beginning-to-end planning, 74, 77
Bidding:
   balanced, 63
   combined, 7–8
   competitive, 6–7
   by subcontractors, 56–57
   unbalanced, 63
BIM, see Building Information Modeling
Bonds, contract, 60–61

BOT contracts, see Build-Operate-Transfer contracts
Brainstorming, 146–147, 171, 306
Budget, project, 26–27, 36, 41, 64–66
Budget at completion (BAC), 302
Budgeted cost of work performed (BCWP), 301
Budgeted cost of work scheduled (BCWS), 301
Build-Operate-Transfer (BOT) contracts, 13
Building Information Modeling (BIM), 4, 13–14, 159
Bulk materials, accounting for, 304
Burst activities, see Activities, burst
Business management, 20
But-for schedules, 227–228

Calendar dates, associating activities with, 123
Calendars, weather, 123–124, 233–237
Camp facilities, 152
Cash flow, 156–157, 323–324
   and disbursement controls, 327–328
   disbursement forecasts, 324–326
   income forecasts, 327–327

335

Change orders, contract, 329–330
Changes, project, 328–329
Checklists, production, 247–249
Claims, 330–331
Clearing accounts, 303–304
Codes, cost, 280–282
Combined bidding, 7–8
Commercial Building example project, 31–33
  cost estimating in, 41–45
  early activity times in, 105–108
  late activity times in, 109–112
  planning in, 77–80
  project outlines, 79
Commissioning, 5
Competitive bidding, 6–7
Complex labor scheduling, 205–206
Components in example projects, 78–79
Computer applications:
  production planning, 143
  project cost system, 307
  project planning, 94–95
  schedule updating, 270–271
  scheduling, 139–140
Construction contract services, 9
Construction industry, 2
Construction management, 10–11
Construction methods, cost estimating and choice of, 46–47
Construction projects:
  complexity of, 3
  stages of, 4–5
  team for, 3
  uniqueness of, 3
Contract bonds, 60–61
Contract change orders, 329–330
Contractors, 2
"Contractor's estimate," 36
Contracts:
  build-operate-transfer, 13
  for construction services, 9
  cost-plus-fee, 12
  design-construct, 10

  fixed-sum, 11–12
  lump-sum, 11
  negotiated, 7
  separate vs. single, 6
  turn-key, 10, 13
  unit-price, 11–12
Cost accounting, project, *see* Project cost accounting
Cost estimating, project, 26, 35–68
  and choice of construction methods, 46
  and contract bonds, 60–61
  equipment costs, 48, 55
  for bridge example project, 41, 44–45
  and field supervisory team, 46
  final cost estimate, 38–39
  function, cost per, 37
  index number estimate, 37
  labor costs, 52–53
  and management input, 45–46
  and markup, 59–60
  material costs, 52
  and general time schedule, 47–48
  overhead (indirect) expenses, 57–59
  panel unit cost, 37–38
  parameter cost, 38
  partial takeoff estimate, 38
  preliminary cost estimates, 36–38
  progress estimate, 67
  and project budget, 64–67
  quantity survey for, 41–45
  recap sheet for, 61–63
  subcontractor bids, compilation/analysis of, 56–57
  "summary sheets" for, 49–51
  unit area cost, 37
  unit volume cost, 37
Cost models, 63–64
Cost performance index (CPI), 301–302
Cost per function estimate, 37

Cost-plus contracts, 12
  pay requests for, 317
  progress measurements under, 310–311
Cost/schedule control systems criteria, 300
Cost slope, *see* activities, cost slope of
Cost system, project, 26, 275–307
  codes, cost, 280–282
  computer application in, *see* computer applications, project cost system
  equipment costs, 284, 296–297
  and estimating, 305, 307
  estimating, cost, 279–280
  labor costs, 284–288, 291–296
  monthly cost forecasts, 297–298
  and postproject evaluation, 305–306
  project cost accounting in, 283
  and project cost control, 279
  records, 290–291
  reduction, cost, 304–305
  reports, cost accounting, 284–285, 290–291
  special problems in, 303–304
  and time-cost envelope, 298–300
  weekly reports, 291–292
  work quantities, measurement of, 298–299
Cost variance (CV), 301
CPI, *see* Cost performance index
CPM, *see* Critical Path Method
Craftspeople, 151, 154, 158
Crash cost, 166–167
Critical activities, 112
  parallel performance of, 171
  subdivision of, 171–172
Critical Path Method (CPM), 24–25, 71–75
  bar charts in, 137–139
  and job logic, 75
  PERT vs., 98–99
  and scheduling, 98–99, 103, 113
CV, *see* Cost variance

Index

Definition stage (of construction projects), 4
Delays:
  schedule analysis to determine project, 261–265
  weather, 104, 123–124, 233–237
Dependency lines (in precedence notation), 83
Design-bid-construct procedure, 8
Design-construct contracts, 10, 20
Design phase, 4, 14–15
Design professional, 5–6
Detailed schedules, 242–243
Direct costs, 165–166
Disbursement controls, 327–328
Distribution code, 281
Documentation, 250–251, 305–306. *See also* Reports
Drug testing policies/programs, 153–154

EAC, *see* Estimate at completion
Early activity times, *see* Activity, early activity times
Early finish (EF), 118–127
Early start (ES), 118–127
Early start schedules, 118–119
Early start sort, 217
Earned value (EV), 300
EF, *see* Early finish
Earned value management system (EVMS), 300–302
Engineer's estimate, 36
Equipment management, 206
Equipment planning, 158–159
Equipment estimating, 55–56
Equipment restraints, 76, 87
ES, *see* Early start
Estimate at completion (EAC), 302
Estimation:
  of activity durations, 99–112
  cost, *see* Cost estimating, project
Events, interface and milestone, 182

EV, *see* Earned value
EVMS, *see* Earned value management system
Example project, *see* Highway Bridge example project
Expediting actions, 174–180, 182, 210–211
Extensions, project, 184

Fast tracking, 9
Field construction, management of, 15, 21–22
Field maintenance, 206
Field progress narratives, 258
Field supervisors, 242–243, 253, 265–266, 270–271
Field supervisory team, 46
Final cost estimate, 38–39
Final payment, 310–311, 322–323
Financial management, project, 28–29, 309–332. *See also* Cost system, project
  cash flow, 323–327
  changes, project, 328–329
  and claims, 330–331
  cost–plus contracts, 317–318
  disbursements, cash, 324–326
  final payment, 322–323
  income, cash, 326–327
  lump–sum contracts, 315–317, 321–322
  pay requests, 311–318
  and progress payments, 310–311
  subcontractors, payments to, 318
  unit-price contracts, 311–313, 318–321
Fire emergencies, 152–153
Fixed-sum contracts, 11–12, 20
Float, 112–117, 135–137
Force-account system, 12–13
Free float, 114, 135–137

Gantt charts, 137, 220
General contractors, *see* Prime contractors

Hammock activities, 133–134
Highway Bridge example project, 31–33
  bid-item summary sheets, 49–52
  cash flow in, 324–327
  cost estimating in, 41–45
  early activity times in, 105–108
  late activity times in, 109–112
  planning in, 77–80
  project outlines, 79
  time acceleration in, 174–178
  time management in, 254–266
Home offices, overhead for, 58–59

Impacted baseline schedules, 224–227
Index number estimate, 37
Indirect labor costs, 53–54
Indirect (overhead) expenses:
  estimating, 57–59
  and project acceleration, 166–169
Interface events, *see* Events, interface
Integrated Project Delivery (IPD), 13–14, 159

Job activities, *see* Activities
Job logic, 75
Job restraints, 75–76

Key-date schedules, 187–188

Labor and material payment bonds, 60–61
Labor costs:
  accounting for, 283–288
  estimating, 52–55
  indirect, 53–54
  and resource management, 195–197
  unit costs, 54–55
Labor requirements, 195–206
  complex labor scheduling, 205–206

Labor requirements (*continued*)
  daily labor needs, projected, 197
  manpower leveling, 198–204
  and restricted labor supply, 204–205
  tabulation of, 195–197
  and variation in labor demand, 198
Lag relationships (in precedence notation), 84
Lags (between activities), 124–127
Last Planner Process©, 216, 246–247
Late activity times, 109–112
Late finish (LF), 105, 109–112, 129–131, 260–261
Late finish sort, 217
Late start (LS), 105, 105–108, 129–131, 260–261
Late start sort, 217
Lean construction, 147–148, 154, 155, 216, 246,
LF, *see* Late finish
Line-of-balance schedules, 221–224
Logic diagrams, 74, 221
Longest time path, shortening of, 164–165
Look ahead schedules, 216, 249–250
LS, *see* Late start
Lump-sum contracts/projects, 11–12
  financial management with, 315–316
  owner, schedule of payments by, 321–322
  pay requests for, 315–316
  progress measurements under, 311
  summary sheets for, 49–51

Management:
  construction, 10–11
  during design phase, 14–15
  by exception, 22, 26, 279
  of field construction, 15
  input of, in cost estimating, 45–46
Manpower leveling, 198–201
Manual methods, 29–30
Markup, 59–60
Material management, 195, 208–209
Material restraints, 76
Materials:
  bulk, 304
  estimating cost of, 52
  handling of, 157
  ordering/expediting of, 155–156
  planning for acquisition of, 148–150
  storage/protection of, 157
Material Safety Data Sheets (MSDS), 153
Material scheduling, 208–210
Merge activities, 106–108
Milestone events, 134–137, 182–183
Mobilization costs, 317
Move-in date, adjustment of, 188–189
MSDS (Material Safety Data Sheets), 153

Need for project management, 19–20
Negotiated contracts, 7

On-the-job training, 152
Operational schedules, 215–216
OT (overtime), 52–53, 205, 262, 287
Overhead, *see* Indirect expenses
Overtime (OT), 52–53, 205, 262, 287
Owner(s), 5
  approval of schedule by, 231–232
  schedule of payments by, 318–322

Panel unit cost estimate, 37–38
Paperwork, planning for, 250–251
Parallel performance (of critical activities), 171
Parameter cost estimate, 38
Partial takeoff estimate, 38
Payment(s):
  final, 322
  progress, 310–311
  scheduling of, 318–322
  to subcontractors, 318
Pay requests:
  for cost-plus contracts, 317–318
  for lump-sum contracts, 315–316
  for unit-price contracts, 311–313
Performance bonds, 61
Personnel planning, 151–152
Phased construction, 9
Planning, 23–24, 71–95. *See also* Production planning
  and activities, 74–75
  beginning-to-end planning, 77
  computer applications to, 94–95
  for example projects, 84–87
  and job logic, 75
  and lag relationships, 84
  methodologies of, 74
  network format for, 83–84
  phase, 72–74
  precedence notation/diagrams for, 83–91
  and repetitive operations, 88–91
  and restraints, 75–76
  steps in, 73–74
  and time acceleration, 170–171
  top-down planning, 77–80
Planning stage (of construction projects), 4
Postproject evaluation, 305–306

PPP, see Public Private Partnerships
Precedence notation, 83–91, 220–221
  arrow notation vs., 80–81
  dependency lines in, 83
  diagram, precedence, 81–82
  for example projects, 84–87
  lag relationships in, 84
  network format in, 83–84
  repetitive operations in, 89–91
  subnetworks in, 93–94
  value of, 88
Preliminary cost estimates, 36–38
Prefabrication, 147, 151
Prime contractor, 6
Private owners, 5
Procurement, 4–5
Production planning, 143–159
  and activity planning, 245–246
  and assembly processes, 158
  checklists, use of, 247–248
  and documentation, 250–251
  equipment and materials, 151, 155–157
  look ahead schedules, use of, 216, 249–250
  and paperwork, 250–251
  personnel, morale/training of, 151–152
  and quality control system, 155–156
  and re-engineering, 146–147
  safety program, implementation of, 152–153
  scope of, 147–148
  team for, 144–146
  technical problems, prevention/anticipation of, 150–151
Production rate, 53
Program Evaluation Review and Technique (PERT)
  CPM vs., 98
  and scheduling, 98, 103
Progress Estimate, 67
Progress measurement, 25, 251–252

Progress payments, 310–311
Progress reporting, 253–258, 271–273
Project budget, 64–66
Project changes, 328–329
Project cost accounting, 27–28, 283
Project cost codes, 280–283
Project cost control, 279
Project management:
  need for, 19–20
  procedures for, 21–22
  scope of, 21
Project manager, 15–16
Project number, 281–283
Project outline, 278–279
Project progress curves, 271–273
Project time acceleration, see Time acceleration, project
Public owners, 5
Public Private Partnerships (PPP), 5

Quality control system, 154–155
Quantity survey, 41–44

Recap sheet (project cost estimating), 61–62
Reengineering, 146–147
Repairs, 206
Repetitive operations:
  bar charts for, 221–222
  in precedence notation, 88–90
  summary diagrams for, 129–130
Reports:
  cost, 290–291
  labor time, 291–296
  progress, 253–261, 271–273
  project cost accounting, 27–28, 283
  weekly labor, 291–292
  work quantity, 288–290
Resource-loaded schedules, 232
Resource management, 28, 191–211
  aspects of, 194–195

Index **339**

  basic objective of, 192–193
  equipment requirements, 206
  expediting of resources, 210–211
  labor requirements, 195–206
  materials, 208–210
  of subcontractors, 244–245
Resource restraints, 75–76
Resource leveling, 198–200
Restraints, planning, 75–76
Risk analysis, 145–146

Safety planning, 152–154
Safety restraints, 75–76
Schedule(s), 213–239
  as-built, 216, 227–228
  baseline, 224–227
  but-for, 227–228
  and determination of project delays, 261–265
  impacted baseline, 224–227
  legal aspects of, 229–233
  line-of-balance, 221–224
  operational, 215–216
  owner approval of, 231–232
  presentation of, 220–221, 238–239
  resource-loaded, 232
  role of, 214
  short-term, 216
  unknowns, handling of, 233–237
  updated, 215–216
  and weather effects, 233–237
Schedule performance variance (SPI), 301–302
Schedule variance (SV), 301
Scheduling, project, 23–24, 97–140. See also Time acceleration, project
  with bar charts, 137–139
  calendar dates, 118–123
  computer application to, 139–140
  corrective actions, 265–267
  and cost estimating, 47
  and critical path, 113–114

Scheduling, project (*continued*)
  detailed schedules, 242–244
  early activity times, 105–108
  early start schedules, 118–119
  equipment scheduling, 206
  and estimation of activity durations, 99–103
  and estimation of project duration, 109–112
  field supervisors, role of, 73
  and float paths, 117–118
  and free float, 114, 136
  of hammock activities, 133
  interfaces, computations of, 131–132
  key-date schedules, 187–189
  and lags between activities, 124–127
  late activity times, 109–112
  look ahead schedules, 249–250
  material scheduling, 201–210
  of milestone events, 134
  network time, computations of, 104–105
  payment scheduling, 318–322
  and presentation of activity times/float values, 117–123
  procedure for, 98–99
  for repetitive operation projects, 127–130
  sorting of activities, 217–220
  subcontractors, 244–245
  tabular time schedules, 119–123
  and time contingency, 103–104
  with time-scaled networks, 134–135
  and total float, 112, 135–137
  updating, 267–269
  and weather delays, 104, 123–124
Short-term schedules, 249–250, 216
Single-contract system, 6, 13
Sorts (various types), 217–219
Specialty contractors, 2
Speculative construction, 14

SPI, *see* Schedule performance index
Straight time (ST), 285–286
Subcontractors, 6, 8, 12, 195
  estimating costs of, 56–57
  payments to, 318
  project time acceleration through use of, 173
  scheduling of, 244–245
Subdivision (of critical activities), 171–173
Subnetworks (in precedence notation), 93–94
Summary sheets, 49–51
Supply chain, 147–148, 207–208
Suspense accounts, 303
SV, *see* Schedule variance

Tabular time schedules, 119–123
TCPI, *see* To-complete cost performance index
Technical problems, prevention and anticipation of, 150–151
TF, *see* Total float
Time acceleration, project, 161–189
  and direct costs, 165–166
  by expediting, 174–178, 182–183
  in Highway Bridge example, 169–183
  and indirect costs, 168
  limitations on, 178–180
  longest time path, shortening of, 164–165
  manual approach to, 168–169
  milestone/interface events, dates of, 182–183
  need for, 162–163
  parallel, performance of critical activities in, 171
  practical aspects of, 168–169
  procedure for, 163–164
  project extension vs., 184
  and restudy of project plan, 170–171
  and subcontracting, 173

  and subdivision of critical activities, 171–172
  and variation of total project cost with time, 180–182
Time cards, 285–288
Time-cost envelope, 298–300
Time-cost trade-off, 180–182
Time management, project, 22–23, 25, 161–189. *See also* Time acceleration, project
  aspects of, 185–187
  bar charts for, 254–256, 258–260
  computer applications in, 270–271
  and corrective actions, 265–267
  detailed schedules for, 242–244
  field progress narrative for, 258
  key-date schedules for, 187–188
  move-in date, adjustment of, 188–189
  and network updating, 267–270
  progress measurements/reporting, 251–252, 271–273
  project progress curves for, 271–273
  system for, 184–185
  weekly progress reports for, 256–258
Time reduction, *see* Time acceleration, project
Time-scaled logic diagrams, 221
Time-scaled networks, 134–135
To-complete cost performance index (TCPI), 303
Top-down planning, 74, 77–80
Total float sort, 217–218
Total float (TF), 112, 135–137, 261
Training, on-the-job, 151–152
Turn-key contracts, 10, 13

Unbalanced bidding, 63
Unit area cost estimate, 37

Unit-price contracts/projects, 11–12
  bidding for, 61–63
  owner, schedule of payments by, 318–319
  pay requests for, 311–313
  summary sheets for, 49–51
Unit volume cost estimate, 37
Updated schedules, 215–216
Updating, network, 267–269

WBS, *see* Work breakdown structure
Weather calendars, 123–124, 233–237
Weather delays, 233–237
  allowances for, 104
  calendars for, 123–124
Weekly labor cost reports, 292–296
Weekly labor reports, 291–292

Weekly progress reports, 256–258
Work breakdown structure (WBS), 282
Work quantities:
  measurement of, 288–289
  network activities, determination from, 289–290
Work type code, 281